AF358797

Feature Papers for Land Innovations—Data and Machine Learning

Feature Papers for Land Innovations—Data and Machine Learning

Editor

Chuanrong Zhang

Basel • Beijing • Wuhan • Barcelona • Belgrade • Novi Sad • Cluj • Manchester

Editor
Chuanrong Zhang
Department of Geography &
Center for Environmental
Sciences and Engineering
University of Connecticut
Storrs, CT
USA

Editorial Office
MDPI
St. Alban-Anlage 66
4052 Basel, Switzerland

This is a reprint of articles from the Special Issue published online in the open access journal *Land* (ISSN 2073-445X) (available at: https://www.mdpi.com/journal/land/special_issues/FP_land_data).

For citation purposes, cite each article independently as indicated on the article page online and as indicated below:

Lastname, A.A.; Lastname, B.B. Article Title. *Journal Name* **Year**, *Volume Number*, Page Range.

ISBN 978-3-7258-0083-4 (Hbk)
ISBN 978-3-7258-0084-1 (PDF)
doi.org/10.3390/books978-3-7258-0084-1

Contents

About the Editor

Chuanrong Zhang

Chuanrong Zhang is currently a professor at the Department of Geography and Center of Environmental Sciences and Engineering, University of Connecticut, Storrs, USA. Zhang is a broadly trained geographer with substantive interests in geospatial technologies. Her research is interdisciplinary. In particular, her research concentrates on geographical information science (GIS), remote sensing, geospatial statistics, geo-computation, and their applications in land use/cover studies, climate change, managing disasters and natural resources, landscape ecology, and environmental planning, as well as transportation studies. She has published articles in leading journals in GIS, geography, remote sensing, soil, and environmental studies. Thus far, she has published almost 200 peer-reviewed journal articles, book chapters, and conference proceedings. Her work has been supported by different prestigious agencies such as the National Science Foundation and the Department of Energy. She has served as a committee chair or member of several international or national organizations in the geospatial community.

Review

Land Use and Land Cover Mapping in the Era of Big Data

Chuanrong Zhang [1,*] and Xinba Li [2]

[1] Department of Geography & Center for Environmental Sciences and Engineering, University of Connecticut, Storrs, CT 06269, USA

[2] Department of Economics, Vanderbilt University, Nashville, TN 37235, USA

* Correspondence: chuanrong.zhang@uconn.edu

Abstract: We are currently living in the era of big data. The volume of collected or archived geospatial data for land use and land cover (LULC) mapping including remotely sensed satellite imagery and auxiliary geospatial datasets is increasing. Innovative machine learning, deep learning algorithms, and cutting-edge cloud computing have also recently been developed. While new opportunities are provided by these geospatial big data and advanced computer technologies for LULC mapping, challenges also emerge for LULC mapping from using these geospatial big data. This article summarizes the review studies and research progress in remote sensing, machine learning, deep learning, and geospatial big data for LULC mapping since 2015. We identified the opportunities, challenges, and future directions of using geospatial big data for LULC mapping. More research needs to be performed for improved LULC mapping at large scales.

Keywords: land use and land cover mapping; remote sensing; machine learning; deep learning; geospatial big data

1. Introduction

Accurate and timely land use and land cover (LULC) maps are important for a variety of applications such as urban and regional planning, disasters and hazards monitoring, natural resources and environmental management, and food security [1–3]. LULC mapping may help tackle many significant large-scale challenges, such as global warming, the accelerating loss of species habitat, unprecedented population migration, increasing urbanization, and growing inequalities within and between nations [4,5]. Therefore, it is important to produce accurate LULC maps.

The *land use* concept and the *land cover* concept, though related, are distinctly different [6]. Land cover mainly refers to direct observations of terrestrial ecosystems, natural resources, and habitats on the Earth's surface, while land use generally describes a certain land type produced, changed or maintained by the arrangements, activities, and inputs of people. Land use relates to the purpose for which land is utilized by people, but land cover specifies landscape patterns and characteristics. Examples of land use may include multi-family residential homes, state parks, reservoirs, and shopping centers. In contrast, examples of land cover may include forests, wetlands, built areas, water, and grasslands. However, land use and land cover are often used as interchangeable terms in existing research literature. In this article, we discuss LULC mapping without making a specific differentiation between land use and land cover, and the mapping includes both.

Remotely sensed satellite imagery is a valuable source for LULC mapping [7–9]. Many studies have attempted to extract LULC information from remotely sensed imagery [2,10]. Advances in remote sensing technologies have resulted in improvements in spectral, spatial, and temporal resolutions of satellite imagery, all of which benefit LULC mapping. LULC mapping is currently experiencing a transformation from the coarse and moderate scales to much finer scales in order to provide more precise land knowledge. Although remotely sensed imagery has been used in LULC mapping since the launch of Landsat 1 in 1972 [11],

Citation: Zhang, C.; Li, X. Land Use and Land Cover Mapping in the Era of Big Data. *Land* **2022**, *11*, 1692. https://doi.org/10.3390/land11101692

Academic Editor: Linda See

Received: 26 August 2022
Accepted: 27 September 2022
Published: 30 September 2022

it is still difficult to capture complex and diverse LULC information and patterns by using remotely sensed imagery alone [12]. Ancillary data are typically needed as a supplement to remotely sensed imagery in order to accurately identify LULC information, especially the land use information related to socioeconomic aspects [13].

With the development of GPS and data acquisition techniques, the merging of big data with spatial location information—such as social media data, mobile phone tracking data, public transport smart card data, Wi-Fi access point data, wireless sensor networks, and other sensing information generated by Internet of Things devices—may provide useful ancillary data for LULC mapping [14]. Compared to traditional geospatial data acquisition, these geospatial big data are normally obtained at a lower cost and have different coverages and better spatio-temporal resolutions. They contain abundant human activity information and may thus be used to compensate for the lack of socioeconomic attributes of the remotely sensed imagery data for accurate LULC mapping [15]. In fact, the aforementioned geospatial big data were integrated with remotely sensed imagery and other source data for accurate LULC mapping in many studies [16,17].

We are currently living in the era of big data. The volume of collected or archived geospatial data, including remotely sensed data, is increasing from terabytes to petabytes and even to exabytes [18]. For example, the European Space Agency (ESA), the National Aeronautics and Space Administration (NASA), the United States Geological Survey (USGS), and the National Oceanic and Atmospheric Administration (NOAA) provide a huge amount of freely available remotely sensed data and other Geographic Information System (GIS) data for LULC mapping. Social media sites, such as Facebook, Twitter, and Instagram, are generating an enormous volume of data with geospatial location information that can be used for LULC mapping nowadays [19]. Progress in data access and algorithm development in the era of big data provides opportunities for developing improved LULC maps [20]. Figure 1 illustrates the major opportunities of LULC mapping in the era of big data. Databases that offer free access to LULC maps at the global scale have emerged. For example, as a free search engine, "Collect Earth" developed by the Food and Agriculture Organization (FAO) can help derive past and present LULC change information [21].

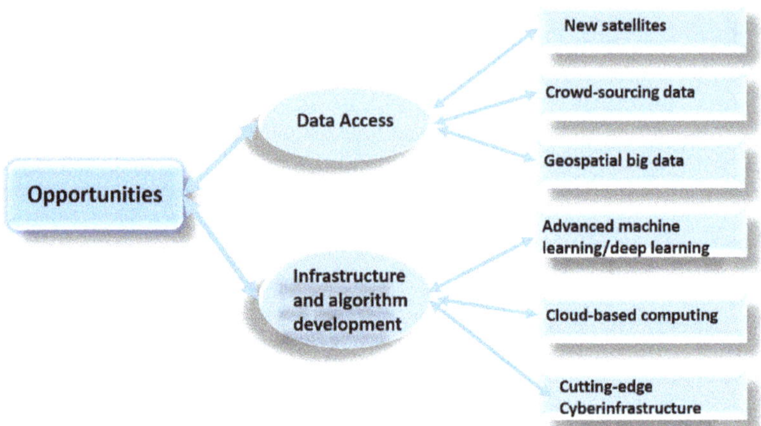

Figure 1. Major opportunities of LULC mapping in the era of big data.

While these geospatial big data provide new opportunities, challenges remain in storing, managing, analyzing, and visualizing these data for LULC mapping [22]. Geospatial big data not only have various forms but are also often associated with unstructured data that are difficult to manage [23]. It is extremely difficult to integrate, analyze, and transform these heterogeneous geospatial big data from different sources into useful values for LULC mapping. Traditional LULC classification or mapping solutions and software face excessive challenges in dealing with these large and complex geospatial big data. New approaches are needed to efficiently process and analyze these data to reveal patterns, trends, and associations related to LULC mapping [24].

Lately, advanced machine learning techniques, especially deep learning (DL), have been developed for large-scale LULC mapping based on multispectral and hyperspectral satellite images or the integration of satellite imagery with other geospatial big data [25]. Deep learning has demonstrated better performance compared to traditional methods, such as random forest (RF) and support vector machine (SVM), e.g., [26–28]. Nevertheless, there are still many issues in applying advanced machine learning or deep learning for accurate LULC mapping using geospatial big data.

The Special Issue "Feature Papers for Land Innovations—Data and machine learning" targets contributions to spatial data science for obtaining, processing, analyzing, harnessing, and visualizing social, economic, environmental, and other land-related data. Particularly, the Special Issue focuses on research in geospatial artificial intelligence and machine learning techniques for dealing with spatial big data. This includes remotely sensed data and social media data. A number of literature review articles related to LULC mapping have been published in the fields of remote sensing, machine learning, deep learning, and geospatial big data since 2015, e.g., [12,13,15,16,18,29–47]. This article summarizes these recent review studies and recent research progress in remote sensing, machine learning, deep learning, and geospatial big data for the Special Issue. As a review article for this Special Issue, the purpose of this paper is to briefly review LULC mapping in the big data era. The method and materials are briefly introduced in Section 2. LULC mapping using remotely sensed imagery is reviewed in Section 3. LULC mapping by integrating geospatial big data and remotely sensed imagery is examined in Section 4. Advanced machine learning, deep learning, and cloud computing for large-scale LULC mapping are summarized in Section 5. Challenges and future directions for the use of geospatial big data for LULC mapping are identified in Section 6. Finally, a brief conclusion is provided in Section 7 at the end of the paper.

2. Method and Materials

We conducted a search of "review articles" from 2015 to now using Google Scholar with combinations of the following keywords: "land use", "land cover", "mapping", "classification", "remote sensing", "geospatial big data", "deep learning", "machine learning", "cloud computing", and "cyberinfrastructure". In case we missed some review articles, we also carried out another search of "any type articles" from 2015 to now using Google Scholar based on combinations of the above keywords. We sorted the search results by relevance. We went through the searched results and selected papers based on their scopes, objectives, and characteristics. Table S1 lists the cited publications since 2015 based on five grouped themes associated with LULC mapping: meta-analysis, remote sensing, big data, machine learning (especially deep learning), and advanced cyberinfrastructure (especially cloud computing). Please note: * indicates a review paper.

3. LULC Mapping from Remotely Sensed Imagery Data

As mentioned previously, remote sensing has become one of the most important methods for LULC mapping [48,49]. Many existing LULC maps were made by the classification of remotely sensed satellite imagery data [50]. Remotely sensed data have multi-source, multi-scale, high-dimension, and non-linear characteristics [51]. Since the advent of remote sensing technology, many satellites have been launched. Every day, a large set of spaceborne and airborne sensors provide a massive amount of remotely sensed data. At present, there are more than 200 on-orbit satellite sensors capturing a large amount of multi-temporal and multi-scale remotely sensed data. For example, NASA's Earth Observing System Data and Information System (EOSDIS) managed more than 7.5 petabytes of archived remotely sensed data and archived a daily data increase of four TB in 2013 [52]. Many satellite imagery data providers release timely remotely sensed data to the public without any cost. USGS, NASA, NOAA, IPMUS Terra, NEO, and Copernicus open access hubs are among the most popular open access remotely sensed data providers.

In the past, many LULC maps were made from coarse spatial resolution satellite imagery data such as advanced very-high-resolution radiometer (AVHRR) and moderate-resolution imaging spectroradiometer (MODIS) [53]. Advances in remote sensing technology and the launch of sensors with moderate spatial resolutions, such as Landsat, Satellite Pour l'Observation de la Terre (SPOT), and Advanced Spaceborne Thermal Emission and Reflection Radiometer (ASTER), have contributed to enhanced LULC mapping, e.g., [54,55]. Lately, detailed LULC maps have been produced from high-resolution imagery data such as QuickBird, IKONOS, and WorldView, which can provide more detailed spatial and spectral information for LULC mapping, e.g., [56–58]. With these high-resolution remotely sensed data, it is possible to identify the detailed geometries, textures, sizes, locations, and adjacent information of ground objects at a much finer scale for LULC mapping [59].

In addition to the different spatial resolutions, the remotely sensed data for LULC mapping also have different spectral and temporal resolutions. Many satellite sensors produce imagery data with very-high spectral resolutions [32]. For example, the Hyperion sensor consists of 220 spectral bands, the AVIRIS system provides 224 spectral bands, the WIS instrument has 812 bands, and the hyperspectral imager equipped in HJ-1A has 128 bands. Furthermore, remotely sensed data may come from different types of satellites. Some satellites use optical sensors such as SPOT, Landsat, and IKONOS; some use microwave synthetic aperture radar (SAR) sensors such as TerraSAR, Envisat, and RADARSAT; while others use multi-mode sensors such as MODIS. While the optical satellite imagery data face challenges in producing LULC maps under cloudy weather conditions, microwave SAR data allow LULC mapping under all weather conditions, including the constantly cloudy weather situation [60–62]. From a temporal resolution perspective, these satellites also have different capabilities to revisit an observation area. Some satellites have a short revisit period of one day (e.g., MODIS and WorldView), while other satellites have a long revisit period of 16 days (e.g., Landsat). Figure 2 shows different types of remotely sensed satellite imagery for LULC mapping. Teeuw et al. [63], Navin and Agilandeeswari [64], and Pandey et al. [40] provided detailed tables for the characteristics of different types of remotely sensed data.

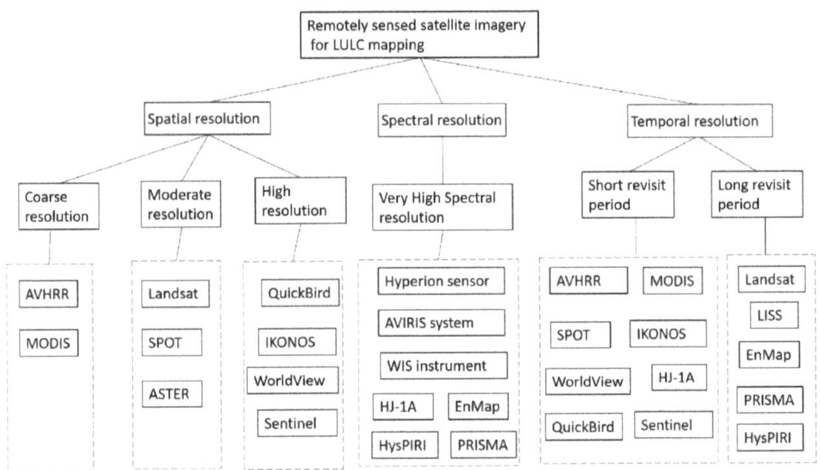

Figure 2. Different types of remotely sensed satellite imagery for LULC mapping.

As illustrated in Figure 3, more types of remotely sensed data have emerged to provide additional observations for LULC mapping [51]. These remotely sensed data provide observations to differentiate LULC types with complex structures, which are difficult to differentiate in the past. For example, as a unique measure of human activities and socio-economic attributes, remote sensing-based nighttime lights (NTL) imagery is especially useful for urban LULC mapping at different spatial and temporal scales, e.g., [65,66]. Light detection and ranging (Lidar) is another type of remotely sensed data for detailed LULC mapping, e.g., [67]. Unlike optical data, airborne Lidar data can capture highly accurate structural information to differentiate LULC types with different structures, components, and compositions [68]. In addition, street-view imagery from Google, Baidu, and Tencent also functions as an additional type of remotely sensed data for LULC mapping, e.g., [69,70]. In contrast to the overhead view captured by most other remote sensing methods, street-view imagery data provide street-level or eye-level observations along the road networks. By providing information about what people typically see at street level on ground, street-view imagery data provide crucial information on the functions of objects conventionally hidden from the view above, e.g., [71]. For example, street-view imagery data have been used for level II or III land use classification (e.g., differentiation of commercial buildings and residential buildings by using text information on buildings from street-view imagery data [72]. Street-view imagery can also be used for ground truth purposes. Recently, unoccupied aerial system (UAS) platforms with small-sized and high-detection-precision sensors have also started producing massive high-resolution images as well, and have been extensively used for high-resolution LULC mapping, e.g., [73,74]. Currently, the amount of data collected by UAS is about to explode.

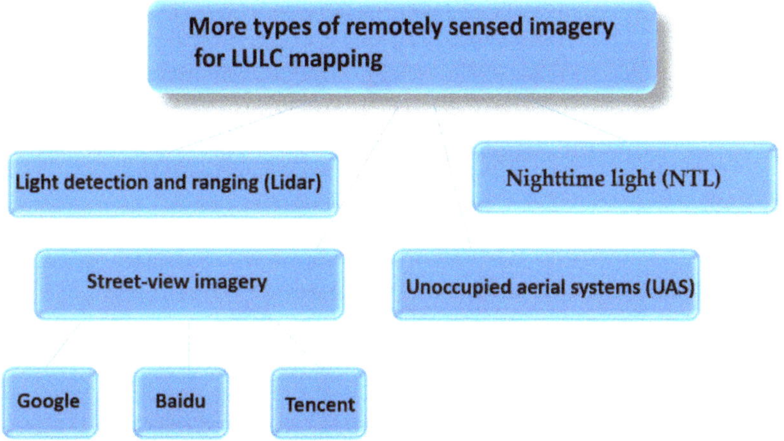

Figure 3. More types of remotely sensed imagery for LULC mapping.

Because of the diversity and high dimensionality of remotely sensed data, LULC mapping from remotely sensed big data becomes complex. It is challenging to identify the right datasets and combine them to make detailed LULC maps at large scales. Although the multi-source optical and microwave remotely sensed data allow us to obtain LULC information from multiple viewpoints, they sometimes cause confusion in deciding which type is the most appropriate for particular LULC mapping. In addition, because of the data representation challenge, it is difficult to integrate the various remotely sensed data with different features (e.g., spectral signatures in optical imagery and electromagnetic radiations in microwave imagery) from various sources. Traditional pixel-level, feature-level, and decision-level fusion cannot be used to integrate remotely sensed imagery with different scales and/or formats [18]. New approaches need to be developed to fuse remotely sensed imagery with other geospatial big data, such as photos from a social network and crowdsourcing spatial data, for LULC mapping.

4. LULC Mapping from Integration of Geospatial Big Data and Remotely Sensed Imagery Data

Although remotely sensed data have become one of the most important data sources for LULC mapping, these have limitations [42,75]. Remotely sensed data are valuable to extract natural and physical land cover information based on spectrum, texture, geometry, context, and temporal information, but they have limitations in capturing the patterns of human activities and socioeconomic environments and describing indirect anthropogenic differences among different land use classes [13]. For example, while the spectral information of remotely sensed imagery data is effective to extract land cover information such as water area, forest land, and built-up area, it is almost impossible to distinguish some land use classes such as some industrial land, residential land, and commercial land using the spectral information of remotely sensed data alone [72].

With the development of mobile positioning techniques, wireless communication, and the Internet of Things, new emerging types of social sensing big data are providing complementary information to differentiate some land use classes caused by human activities and socioeconomic environments [9]. Examples of these emerging social sensing big data include mobile phone data, geo-tagged photos, social media data, traffic trajectories, and volunteered geographic information (VGI) data [76]. These emerging social sensing big data are able to more effectively capture human activities and dynamic socioeconomic environments, and are regarded as complements of remotely sensed imagery data for effectively LULC mapping [77,78]. For example, Geo-Wiki is a crowdsourcing platform for LULC mapping and other tasks, which was used to derive the global LULC reference data via four campaigns [79]. Flickr offers online services for the sharing of digital photos with geographic locations based on social networks, which was used to identify socioeconomic and human activities in LULC mapping [19]. OpenStreetMap (OSM) (as a VGI database) allows the adding, editing, and updating of basic geographic map information with users' experience and knowledge, which was also used to uncover some land use types and patterns, e.g., [80,81]. The points of interest (POIs), as one of the most common categories of crowdsourced data, were explored for land use classification by many scholars, e.g., [82–84]. In addition, a large amount of GPS traffic trajectory data also further enriched the remotely sensed data in excavating human activities at a fine scale for accurate LULC mapping [85].

These emerging social sensing big data improved the existing LULC maps by providing more detailed socioeconomic information and finer spatio-temporal resolutions [86]. Many studies have been conducted to integrate the social sensing big data with remotely sensed data for LULC mapping at different scales and locations, e.g., [87–89]. For example, Hu et al. [90] developed a protocol to identify urban land use functions over large areas using satellite images and open social data. Yin et al. [91] employed both the decision-level integration and feature-level integration of remotely sensed data with social sensing big data for urban land use mapping. Integrating data from these social sensing big data with remotely sensed data may provide a more comprehensive picture of LULC patterns, as shown in Figure 4.

In addition to the integration of remotely sensed data and social sensing big data for LULC mapping, other auxiliary datasets may also be used for LULC mapping [92]. For example, census data including demography, employment, education, housing, and income information may provide valuable information to reveal spatial differences in socioeconomic statures across different land use types, e.g., [24]. Municipal data such as water consumption data may offer important information to identify the socioeconomic functions of land uses and help classify mixed patterns of land uses [93]. In addition, topographic information such as elevation, slope, and aspect information extracted from digital elevation or digital terrain models (DEMs/DTMs) may also be combined with remotely sensed data to increase the accuracy of urban land use classification, e.g., [94].

Ubiquitous sensor networks can constantly obtain spatio-temporal data in days, hours, minutes, seconds, or even milliseconds. These spatio-temporal data allow people to acquire multi-dimensional dynamic information about various land entities and human activities, which may be used for making or updating LULC maps. LULC mapping is expanding from professional aspects to public aspects with the development of The Internet of Things (IoT) and Volunteered Geographic Information (VGI), as evidenced by Geo-Wiki and My Maps feature in Google Maps. However, the non-professional characteristics of IoT and VGI often make the data obtained from them contain data uncertainty such as data loss, noise, inconsistency, and ambiguity [95]. Therefore, it is important to develop quality assurance procedures such as data cleaning and quality inspection for high-quality LULC mapping.

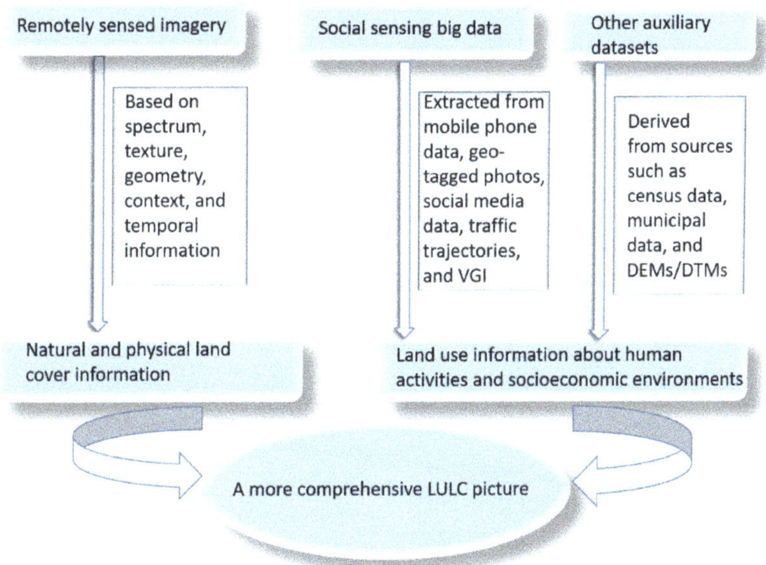

Figure 4. LULC mapping from the integration of geospatial big data and remotely sensed data.

It is still challenging to integrate multi-source remotely sensed data, social sensing big data, and other auxiliary datasets for LULC mapping because of intensive computing and the heterogeneity in spatial data structures, formats, resolutions, scales, and data quality. Novel machine learning including deep learning and cloud computing approaches are urgently needed for LULC mapping.

5. Machine Learning and Cloud Computing for LULC Mapping

Machine learning is a data analysis method and a subset of artificial intelligence based on the idea that computer systems can learn from data to identify patterns and make decisions with minimal human intervention. There are many different machine learning approaches for LULC mapping [96,97], such as support vector machine (SVM), random forest (RF), and K-nearest neighborhood (KNN). The strengths of machine learning include the capacity to handle data of high dimensionality and to map LULC classes with very complex characteristics. With growing volumes and varieties of the available aforementioned remotely sensed imagery and geospatial big data, cheaper and more powerful computational processing tools, and affordable data storage, machine learning has become more popular than ever for analyzing bigger and more complex data and delivering more accurate LULC mapping results at larger scales [10]. Machine learning provides the foundation for autonomously solving data-based LULC mapping problems [98].

Supervised learning, unsupervised learning, and semi-supervised learning are the three main types of machine learning methods for LULC mapping, as shown in Figure 5. Supervised learning algorithms are trained using labeled LULC examples and apply what has been learned in the labeled LULC example data to predict the labels of new LULC data. By inferring methods such as regression and gradient boosting, supervised learning methods use patterns to predict the values of the labels on unlabeled LULC data [99]. Popular supervised learning methods include support vector machine (SVM), random forest (RF), classification and regression tree (CART), radial basis function (RBF), decision tree (DT), multilayer perception (MLP), naive Bayes (NB), maximum likelihood classifier (MLC), and fuzzy logic. Unsupervised learning algorithms are used with data that have no historical LULC labels and computers infer a function to describe a hidden structure from unlabeled LULC data. Unsupervised learning methods are used when it is unclear what the LULC mapping results will look like and computers need to dig through hidden layers of LULC data and cluster data together based on the similarities or differences of LULC classes. Popular unsupervised learning methods include self-organizing maps, k-means clustering, nearest-neighbor mapping, affinity propagation (AP) cluster algorithm, ISODATA (iterative self-organizing data), and fuzzy c-means algorithms. Semi-supervised learning is similar to supervised learning. However, it uses both labeled and unlabeled data for training—usually a small amount of labeled data with a large amount of unlabeled data.

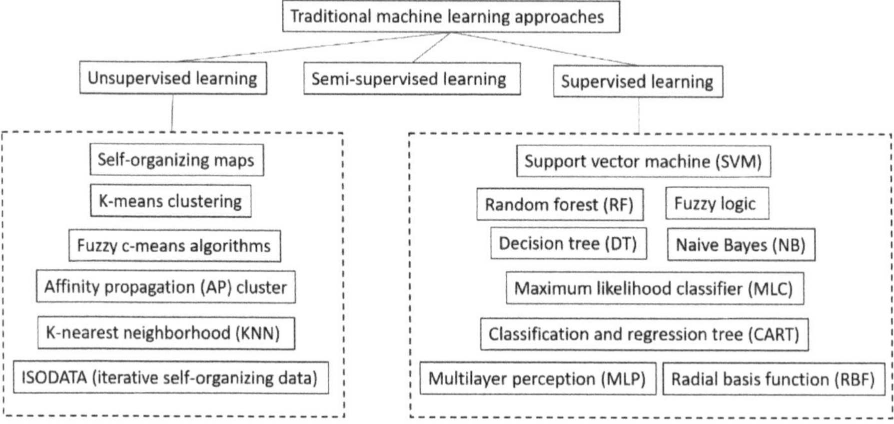

Figure 5. Types of traditional machine learning approaches for LULC mapping.

Recent advances in machine learning for LULC mapping have been accomplished via deep learning approaches [100,101]. As illustrated in Figure 6, deep learning is a subfield of machine learning. All deep learning is machine learning, but not all machine learning is deep learning. Deep learning emerged because shallow machine learning cannot successfully analyze big data for LULC mapping. While basic machine learning models do become progressively better at performing their specific functions as they take in new emergent data, they still need some human intervention. Deep learning algorithms in layers can build an "artificial neural network" (Figure 7) that is able to learn and make intelligent classification decisions on its own [102]. Figure 8 illustrates the differences between traditional machine learning and deep learning. For traditional machine learning, feature extraction and classification are separate processes and humans are needed to perform feature extraction. With a deep learning model, feature extraction is integrated with classification and a classification algorithm can determine whether a class prediction is accurate through its own neural network—beyond the training data and without requiring human help. Deep learning algorithms can be considered as a both sophisticated and mathematically complex evolution of machine learning algorithms [103]. Deep learning algorithms analyze data with a logic structure similar to how a human would draw LULC mapping conclusions. When fed training data, deep learning algorithms would eventually learn from their own errors whether a LULC class prediction is good or whether it needs to adjust.

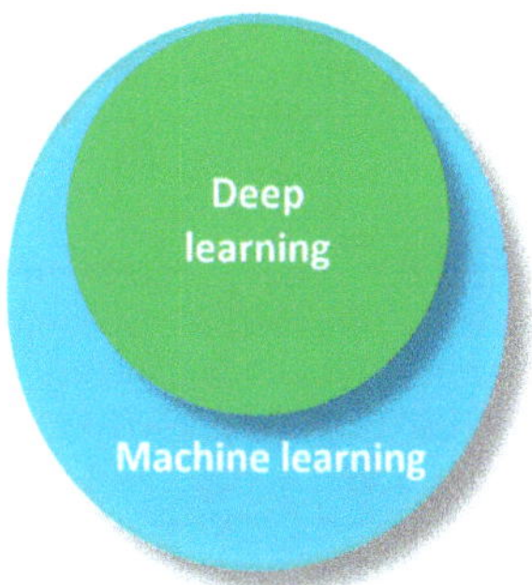

Figure 6. Deep learning is a subfield of machine learning.

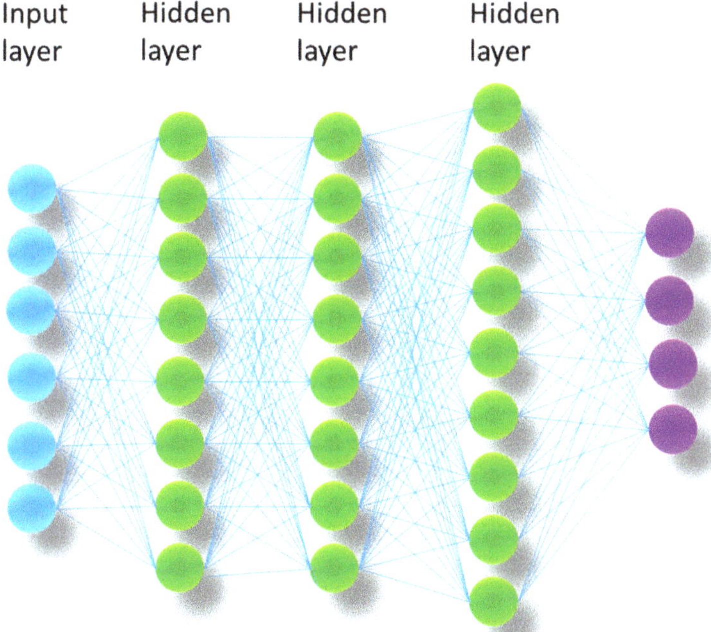

Figure 7. A simple artificial neural network.

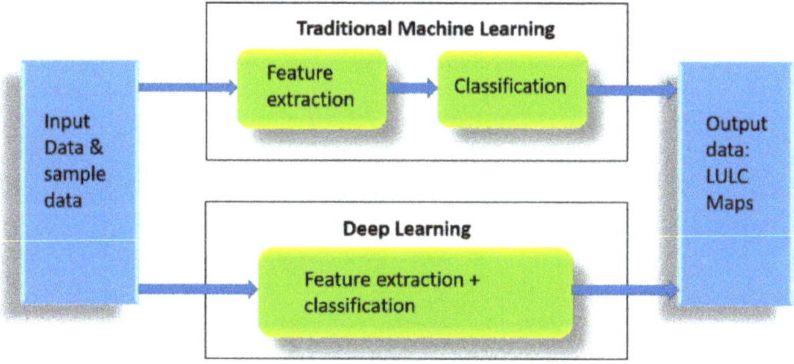

Figure 8. Differences between traditional machine learning and deep learning.

In recent studies, deep learning outperformed other machine learning algorithms in some LULC mapping problems, particularly in detecting fine-scale types such as small artificial objects [29,104,105]. Deep learning algorithms have been used to automatically extract spatial features from very-high-resolution satellite images such as IKONOS, WorldView-3, and SPOT-5, e.g., [101].

There are several different types of deep learning algorithms for LULC mapping, among which the most popular algorithms include convolutional neural networks (CNNs) and recurrent neural networks (RNNs). CNNs are some of the most popular neural network architectures because they can extract low-level features with a high-frequency spectrum, such as the edges, angles, and outlines of LULC objects, whatever the shape, size, or color of the objects are. Therefore, CNNs are well suited for LULC mapping [106]. Some popular CNN architectures used in the literature are LeNet5, AlexNet, VGGNet, CaffeNet, GoogLeNet, and ResNet models. RNNs have built-in feedback loops that allow the algorithms to "remember" past data points. RNNs can use this memory of past events to inform their understanding of current events or even predict the future. RNNs are mainly designed to process time series data and are suitable to detect LULC changes [29].

By performing complex abstractions over data through a hierarchical learning process, deep leaning algorithms have shown great potential for analyzing big datasets for LULC mapping [16]. The hidden layers in deep leaning approaches can discover class structures and patterns in big data and extract valuable class knowledge. Deep learning is also able to handle nonlinear and highly complex big data more effectively than conventional machine learning methods [43,100]. However, compared to traditional machine learning approaches, deep learning requires a vast amount of training data and substantial computing power [27]. A deep learning algorithm requires much more data than a traditional machine learning algorithm to properly conduct LULC mapping. Due to the complex multi-layer structure, a deep learning system needs a large training dataset to eliminate fluctuations and make high-quality class interpretations [43]. Without a large set of training data, deep learning may show a similar or worse performance than classical machine learning techniques such as SVM [107].

The emergence of cloud computing infrastructure and high-performance GPUs (graphic processing units, used for faster calculations) helped to solve the expensive computational problem faced by deep learning [108]. The storage and processing requirements of big data for LULC mapping are greater than that available in traditional computer systems and technologies [109]. The existing cluster-based high-performance computing (HPC) with plenty of computational capacities can be used for storing large remotely sensed data and other big data for LULC mapping. However, it is still challenging to process these big remotely sensed data and other big data for large-scale LULC mapping because system architectures and the tools of the existing cluster-based HPC have not been optimized to process such data. The cluster systems or peta-scale supercomputers are not good at loading, transferring, and processing extremely big remotely sensed data and other data for LULC mapping. A potential solution to this problem is cloud computing. Cloud computing satisfies the two main requirements of LULC mapping using big data analytics solutions: (1) scalable storage that can accommodate growing data; and (2) a high processing capability that can run complex LULC mapping tasks in a timely manner. Cloud computing makes deep learning more accessible, making it easier to manage large datasets and train algorithms for distributed hardware, and deploy them efficiently [110]. It provides access to special hardware configures, including GPUs, field-programmable gate arrays (FPGAs), TensorFlow processing units (TPUs), and massively parallel high-performance computing (HPC) systems.

Cloud computing has been used for storing big remotely sensed data and other data for LULC mapping with good scalability [111]. Three main types of cloud computing services have been used [112]: (1) infrastructure-as-a-service (IaaS), which allows renting IT infrastructures. Servers, virtual machines with storages, networks, and operating systems are completely provided and managed by a cloud provider. Users can pay for what they use; (2) platforms-as-a-service (PaaS): this service is an on-demand style of service where users can obtain a complete development environment required for software applications; (3) software-as-a-service (SaaS): using SaaS, it is possible to deliver software applications over the Internet, such as the 'on demand' or 'subscription' services.

Google AppEngine, Microsoft Azure, and Amazon EC2 are the most popular cloud providers and offer pay-as-you-go clouding computing for storing, processing, and visualizing big remotely sensed data and other data for LULC mapping. GoogleTM developed a geospatial data analysis platform—Google Earth Engine (GEE)—capable of storing and analyzing vast amounts of remotely sensed data for rapid LULC mapping at large scales [113,114]. GEE provides users with free access to numerous remotely sensed datasets including Landsat, Sentinel, and MODIS images. GEE has already proven its capacities for LULC classification and change detection, e.g., [115–124]. Microsoft Azure Cloud Services and Amazon Web Services (AWS) have also been used to improve LULC mapping and monitoring [125]. Microsoft Azure Cloud Services have established artificial intelligence (AI) for an Earth initiative to address environmental challenges. However, Azure only offers Landsat and Sentinel-2 products for North America and MODIS imagery. Amazon Web Services offer open data from more satellites such as Sentinel-1, Sentinel-2, Landsat-8, and China–Brazil Earth Resources Satellite program (CBERS-4), NOAA image datasets, as well as global model outputs.

In addition to machine learning and cloud computing approaches, other advanced cyberinfrastructure techniques, such as novel scalable parallel file systems capable of storing and managing massive data, and NoSQL (Not Only SQL) databases for managing big unstructured or non-relational data have also been developed for LULC mapping with complex characteristics [126].

6. Challenges and Future Research Directions

Despite recent progress, LULC mapping continues to face challenges. There are still many issues remaining to be further explored for LULC mapping, as shown in Figure 9.

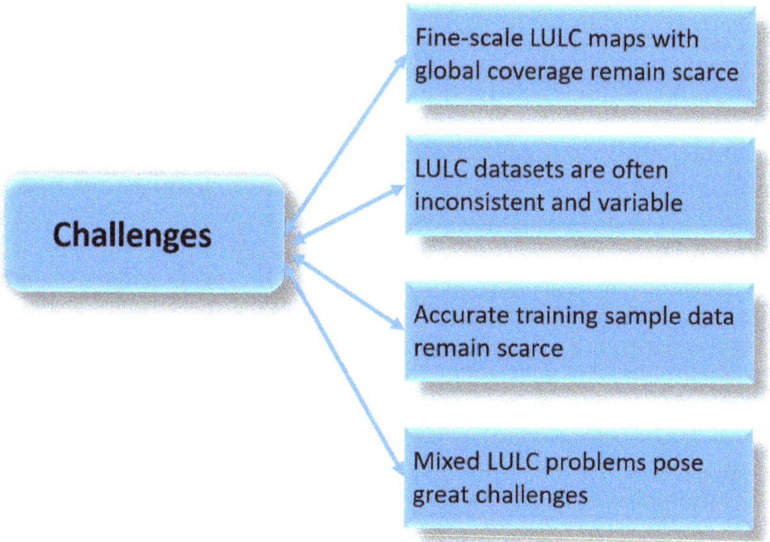

Figure 9. Major challenges of LULC mapping.

First, fine-scale LULC maps with global coverage remain scarce, particularly for developing countries, many of which are experiencing rapid LULC changes. Although many global LULC maps have been developed by different agencies, most of these maps have coarse spatial resolutions [127,128]. For example, the NASA MCD12Q1 dataset has a 500 m resolution, the LULC maps from the European Space Agency (ESA) Climate Change Initiative (CCI) dataset have a 300 m resolution, and the Copernicus Global Land Service (CGLS) Land Cover dataset has a 100 m resolution. There is a lack of detailed LULC maps at the global scale. To the best of our knowledge, the Esri global LULC Maps and the European Space Agency (ESA) WorldCover LULC Maps are the only available global LULC maps with a high (10 m) resolution. The Esri global LULC Maps were derived from ESA optical earth observation data Sentinel-2 imagery for 2018~2022. The ESA WorldCover LULC Maps were produced based on the use of both SAR data Sentinel-1 and optical data Sentinel-2 for 2020. However, there were no historical global LULC maps with a high resolution. Spatial resolution influences various aspects of landscape classification and may significantly affect landscape metrics and landscape pattern analysis [129,130]. Although very-high-resolution images are available for developing detailed LULC maps, they have proven challenging for creating global maps because of the high cost associated with these very-high-resolution images and their incomplete data coverage and small spatial extent (one image only covers a very small study area). In addition, variations of radiometric properties among different sensors, the influence of different acquisition conditions, and different classic atmospheric perturbations also cause challenges in using these very-high-resolution images for detailed LULC mapping at the global scale. Obtaining cloud-free images is challenging and there are often unavailable data in certain seasons, times, or locations. Therefore, there is a lack of well-annotated fine-scale LULC maps at the global scale and even at the country level for some developing countries. Because high-resolution images have only been available recently, there is a lack of well-annotated fine-scale historical LULC maps for change detection.

Second, existing LULC datasets are often inconsistent and variable in time, space, formats, formal validation, or map legends [23,131]. There are various or inconsistent definitions of LULC classes. Different methods and incompatible classification systems are used by different source agencies [132]. It is difficult to compare different legend information from various classification schemes. The land surface is heterogeneous and the mapping standards to acquire, represent, and generalize land characteristics are about as diverse as the land surface itself [133,134]. For example, according to Cruz et al. [135] in Wisconsin, counties and municipalities may maintain different land use codes for their land parcels. This case is particularly interesting because both the city of Madison and the Fitchburg Township are in the same county—Dane county. LULC data have variations in the semantic contents among different research projects and research teams [133]. In addition, different spectral and spatial sensor characteristics, acquisition geometries or illumination conditions, or atmospheric settings also lead to inconsistencies in developing LULC products derived from multi-sensor approaches. It is, therefore, still challenging to combine different LULC products for practices or other applications.

Third, despite more observations from very-high-resolution satellite images, crowd sources and other geospatial big data, accurate training sample data for advanced machine learning or deep learning still remain comparably scarce [43,136]. Currently, field surveys, the visual interpretation of high-resolution images, crowdsourcing technology, and existing labeled land use datasets are common methods to obtain reference or training sample data for LULC mapping. However, all of these methods have some limitations. For example, many sample data generated through visual interpolation contain errors. Field surveys are the most accurate method of generating training sample data; however, it is a labor-intensive and time-consuming task. It is expensive to obtain ground truth data via field survey. In addition, the field survey cannot be conducted in all locations because of inaccessibility issues.

Crowdsourcing technology is one of the latest techniques to obtain sample data, but the spatial and temporal sample data created by this technology may be ad hoc, the quality may be highly variable, and some sample data may contain many uncertainties and errors. The existing labeled LULC datasets can help obtain sample data but they normally have low accuracies and coarse resolutions. Therefore, all of the common methods of obtaining sample data for LULC mapping have some issues. There is a lack of reliable training sample data for LULC mapping using advanced machine learning or deep learning approaches for many locations of the world, especially for some developing countries. In addition, some existing sample data may have imbalance issues [137]. For example, the sample data produced by crowdsourcing technology or geospatial big data such as social media data or Google Street View data only represent human activities well in urban areas [138]. These would perform poorly in measuring land use classes that represent the human activity characteristics in rural areas, where there are low populations and activity densities. Nevertheless, advanced machine learning or deep learning classification methods need a large set of reliable and balanced training sample data, which should cover different classes and areas well, to produce accurate maps, because these methods need to train, test, and classify LULC classes based on the training sample data. A large number of parameters used by these advanced classification methods need to be fine-tuned using a great amount of training sample data. However, obtaining sufficient high-quality training sample data remains a critical issue.

Fourth, mixed LULC pixels also pose great challenges for LULC mapping [16]. Because of the insufficient spatial resolution of remotely sensed imagery, it is quite difficult to differentiate the mixed LULC pixels in the past. With advances in remote sensing techniques, the very-high-resolution images with distinct spatial, temporal, spectral, radiometric, and angular characteristics are emerging and they are available for detailed LULC mapping. However, the mixed LULC problem still exists because LULC classification algorithms compatible with these super-high-resolution multispectral images are still underdeveloped. Various subpixel analysis approaches, such as variations in spectral mixture analysis (SMA), support vector machine (SVM), import vector machine (IVM), convolutional neural networks (CNNs), and deep learning-based subpixel mapping network (DLSMNet) have been developed to solve the mixed LULC problem, e.g., [139–142]. However, these subpixel analysis approaches have still experienced some difficulties in handling the spectral heterogeneity of diverse landscape features. Recent studies have been working on addressing the mixed LULC problem by combining remotely sensed imagery with other ancillary data such as road network data, social sensing data, as well as other environmental and socioeconomic data [143]. However, the data availability limited the large-scale implementation of these kinds of approaches. Although advanced deep learning approaches can transform multispectral image pixels into high-level abstract features and thus may reduce some mixed pixel problems, the model interpretability of these deep learning approaches is a big challenge due to the "black box" nature of the training procedure. In addition, the model transferability is also an important issue faced by these advanced machine learning or deep learning approaches. Because LULC classes are different across different regions, the model developed for one study area using local training data may not be appropriate for classification in other study areas.

Based on the review, although a large set of LULC maps exist, these maps are often connected to considerable uncertainty due to the positional inaccuracy, unreliable input data, and processing algorithms limitations. More research is needed in the future to develop improved LULC maps. A few general recommendations for future research directions of LULC mapping, as illustrated in Figure 10, are suggested in the paragraphs below.

First, more research on fusing multiple big datasets from different sources and across scales for detailed LULC mapping at large scales needs to be done. Specifically, the joint use of multipl150e remotely sensed datasets (lidar, radar, and optical data), big geospatial data such as social media data, other physical and socioeconomic big data such as census data, and sampling survey data may help improve LULC mapping at multiple scales. Big data fusion is necessary for obtaining the full picture of LULC situation. The fused big datasets will typically have better spatial, temporal, spectral, as well as radiometric resolutions and coverages. However, these multiple big datasets usually have the aforementioned heterogeneous problems, such as different spatial and temporal resolutions, formats, and semantics. Therefore, it is usually difficult to ensure that the satisfactory mapping results in a heterogeneous environment. Because of the variances in shape, color, size, and other properties of objects for each class, it is a challenge to use a universal scale parameter for classification. A cross-scale mapping strategy that can is compatible across pixel-, object-, and parcel-scale may be highly necessary. A cross-scale mapping strategy may solve the aforementioned mixed LULC problem via diverse outputs of classification maps that can include the compositions and proportions of mixed LULC pixels. The cross-scale approach, however, may require a great deal of time and effort to determine the appropriate parameters. More research needs to be performed using the cross-scale mapping strategy to fuse multiple big datasets from different sources.

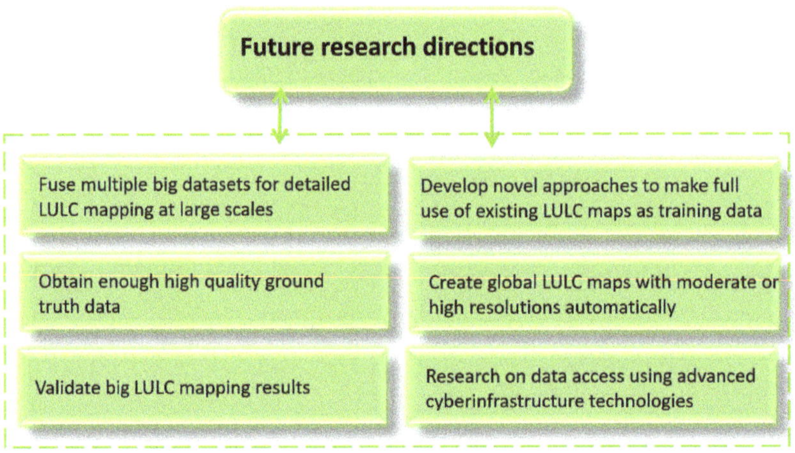

Figure 10. Future research directions of LULC mapping.

Second, novel approaches need to be developed to make full use of existing labeled maps for detailed LULC mapping. Numerous global LULC maps often can be accessed for free. However, most LULC maps typically have a much lower resolution than the current very-high-resolution satellite imagery and also contain many noises; thus, they cannot be directly considered as ground truth or training sample data for detailed LULC mapping using advanced machine learning or deep learning approaches. However, these coarse resolution maps do contain some valuable class information, and thus they may be utilized as indirect auxiliary training data for the same purpose. However, current studies to utilize these existing LULC products are still limited and more strategies or methods need to be designed and developed for this purpose.

Third, as mentioned before, it is still a challenge to obtain enough high-quality ground truth data or training sample data for using advanced machine learning or deep learning for LULC mapping. Existing sample data are not only limited in number but are also limited in terms of variety. The sample data are often not sufficient to train a generalized machine learning or deep learning model, because they are specific to time and location. The classifier trained using one dataset normally does not perform well over other datasets. To improve the performance of a deep learning model, image transformations such as flip, translation, and rotation may be adopted to generate additional and more diversified training data from original data. Transfer learning is another way proposed to deal with the challenge of limited training data [144]. The transfer learning method employs a pre-trained LULC classifier to extract an initial set of representations for a new LULC dataset. Unsupervised learning can also be used to tackle the problem of lacking labelled LULC training data [43]. The use of crowdsourcing sample collections or an open science framework that supports the integration of citizen science and IoT may as well generate more labeled samples. However, all of these methods have limitations. More research needs to be performed to explore these various approaches to obtain enough sample data for LULC mapping. In addition, more research is necessary to determine how to keep the good performances of deep learning methods using fewer training samples. Additionally, more attention from the scientific community is also needed to address the LULC class imbalance issue. LULC classes are normally imbalanced with some majority classes dominating a study area while some minority classes only occasionally occur at some locations. Therefore, some of the classes may have fewer samples than the others, and it is easy to obtain a sample dataset that misses some minority classes. The class imbalance issue may affect the classification accuracy of advanced machine learning or deep learning methods. Future research should pay more attention to the class imbalance issue.

Fourth, more research needs to be performed for automatically creating and updating spatially explicit LULC maps with moderate or high resolutions at the global scale. Currently, most of the available moderate-resolution LULC maps are only available for limited spatial and temporal coverage [3], such as USGS NLCD and LCMAP, BaseVue, GlobelLand30, and GlobeLand10. Although the Esri global LULC Maps and the ESA WorldCover LULC Maps have high 10 m resolution and the recent iMap series of products are available globally at a seasonal cadence with a 30 m resolution, these maps have limited temporal coverage: the Esri Maps cover time periods of 2018~2022 and the ESA Maps only cover the year 2020. In addition, these maps only generated broad classes such as water, trees, grass, crops, scrub/shrub, built area, bare ground, snow/ice, and clouds (unclassified). They lack detailed LULC information, such as Level II LULC classes, to differentiate residential, commercial, and industrial land uses. More research is necessary to generate historical global LULC maps or detailed global LULC maps with Level II or higher level LULC information and fill the data gaps in some locations, particularly those regions or countries where data are extremely deficient.

Fifth, more research is necessary to validate big LULC mapping results. At present, it is still difficult to compare the different LULC mapping results. There is no specifically acceptable accuracy assessment metrics or standards for the evaluation of LULC mapping results. The overall accuracy, user accuracy, and the Kappa coefficient of the confusion matrix are popular methods to validate the mapping results with the ground truth data. However, these evaluation methods are imperfect and only provide evaluations from some perspectives while ignoring other perspectives. For example, many existing LULC mapping studies are only concerned with overall accuracy while ignoring the poor accuracies of rare classes [145]. The confusion matrix is entirely devoid of spatial context [146]. Kappa indices may be misleading and/or flawed for some practical applications in remote sensing [147]. Accuracy validation efforts and standard assessment systems are needed to accurately assess LULC mapping using heterogeneous big data sources.

Finally, research on data access using advanced cyberinfrastructure technologies is also needed. Efforts to develop protocols and platforms to compile, share, visualize, and distribute large LULC datasets including their associated biases and errors over the Internet are urgently needed. Traditional cyberinfrastructure technologies have met a few limits due to the constant growth of data. For example, when LULC data become extremely large, human eyes have difficulty in extracting meaningful information for visualization. It is challenging to present too many data on a limited screen. It is even more challenging to present huge data on mobile devices due to smaller screens and resolutions. The traditional methods of building tile and pyramids are not efficient for visualization at a satisfying speed [148]. We need to explore novel ways to display and visualize large data using various abstraction techniques.

The field of computer vision has seen rapid progress in the last decade, which is in a large part growing to the growth of deep learning. Computer vision tasks such as image classification, object detection, and image segmentation also saw prominent achievements brought by deep learning techniques. Although computer vision tools and methods have been adopted for a wide variety of applications, these tools and methods have yet to be unified and integrated with traditional methods of spatial analysis to deal with geographic/spatial data [149]. Because several factors such as the lack of large-scale annotated sample data and disparate object sizes of remote sensing imagery, the applications of advanced computer vision techniques to the remote sensing domain has lagged behind greatly [150]. The extraction of meaningful LULC information from remotely sensed imagery can be aided by techniques in computer vision [151]. Computer vision methods with neural networks as the underlying framework can be used to identify LULC spatial features and patterns. Computer vision algorithms may also be used to remove noise and enhance satellite and aerial imagery data for LULC classification, change detection, and data fusion. Novel tools and approaches that combine cutting-edge computer vision technology and remotely sensed imagery need to be developed in the near future for LULC mapping.

7. Conclusions

Accurately mapping LULC information is important for many applications such as natural resource and environmental management, urban planning, biodiversity conservation, and health promotion. With the advent of remote sensing and computer technologies, massive data have been generated. New and improved remote sensing earth observations and emerging social sensing big data and auxiliary crowdsourcing datasets together offer great data sources for LULC mapping [152,153]. The recent innovative machine learning and deep learning algorithms as well as cutting-edge cloud computing have proven their powerful capabilities to process big remotely sensed data and other geospatial big data of high dimensionality for LULC mapping [154,155].

Despite considerable recent progress, LULC mapping still has challenges to deal with, including data gaps, inconsistent and heterogeneous data, imbalanced and scarce sample data, mixed LULC, and the model interpretability and transferability issues of advanced machine learning or deep learning approaches. More research is necessary in the future in the following fields for improved LULC mapping: fusing multiple big datasets from different sources using cross-scale approaches, developing novel approaches to make full use of the existing LULC maps, obtaining high-quality sample data, automatically creating and updating LULC maps with moderate or high resolution at the global scale, improving accuracy assessment methods, and developing advanced cyberinfrastructure technologies for sharing, visualizing, and distributing large LULC datasets over the Internet for various applications.

Supplementary Materials: The following supporting information can be downloaded at: https://www.mdpi.com/article/10.3390/land11101692/s1, Table S1: The cited publications since 2015 based on five grouped themes (* review paper).

Author Contributions: Conceptualization, C.Z.; methodology, X.L.; validation, C.Z. and X.L.; formal analysis, C.Z. and X.L.; investigation, C.Z. and X.L.; resources, C.Z.; data curation, X.L.; writing—original draft preparation, C.Z.; writing—review and editing, C.Z. and X.L.; visualization, X.L.; supervision, C.Z.; project administration, C.Z.; funding acquisition, C.Z. All authors have read and agreed to the published version of the manuscript.

Funding: This research is partially supported by USA NSF grants No. 2022036 and No. 2118102. Authors have the sole responsibility for all of the viewpoints presented in the paper.

Data Availability Statement: Not applicable.

Conflicts of Interest: The authors declare no conflict of interest.

References

1. Treitz, P.; Rogan, J. Remote sensing for mapping and monitoring land-cover and land-use change-an introduction. *Prog. Plan.* **2004**, *61*, 269–279. [CrossRef]
2. Rogan, J.; Chen, D. Remote sensing technology for mapping and monitoring land-cover and land-use change. *Prog. Plan.* **2004**, *61*, 301–325. [CrossRef]

3. Mora, B.; Tsendbazar, N.E.; Herold, M.; Arino, O. Global land cover mapping: Current status and future trends. In *Land Use and Land Cover Mapping in Europe*; Manakos, I., Braun, M., Eds.; Springer: Dordrecht, The Netherland, 2014; Volume 18, pp. 11–30. [CrossRef]

4. Meyer, W.B.; Turner, B.L. Human population growth and global land-use/cover change. *Annu. Rev. Ecol. Syst.* **1992**, *23*, 39–61. [CrossRef]

5. Hasan, S.S.; Zhen, L.; Miah, M.G.; Ahamed, T.; Samie, A. Impact of land use change on ecosystem services: A review. *Environ. Dev.* **2020**, *34*, 100527. [CrossRef]

6. Nedd, R.; Light, K.; Owens, M.; James, N.; Johnson, E.; Anandhi, A. A synthesis of land use/land cover studies: Definitions, classification systems, meta-studies, challenges and knowledge gaps on a global landscape. *Land* **2021**, *10*, 994. [CrossRef]

7. Parece, T.E.; Campbell, J.B. Land use/land cover monitoring and geospatial technologies: An overview. In *Advances in Watershed Science and Assessment*; Younos, T., Parece, T., Eds.; Springer: Cham, Switzerland, 2015; Volume 33, pp. 1–32. [CrossRef]

8. Joshi, N.; Baumann, M.; Ehammer, A.; Fensholt, R.; Grogan, K.; Hostert, P.; Jepsen, M.R.; Kuemmerle, T.; Meyfroidt, P.; Mitchard, E.T.; et al. A review of the application of optical and radar remote sensing data fusion to land use mapping and monitoring. *Remote Sens.* **2016**, *8*, 70. [CrossRef]

9. Chen, B.; Tu, Y.; Song, Y.; Theobald, D.M.; Zhang, T.; Ren, Z.; Li, X.; Yang, J.; Wang, J.; Wang, X.; et al. Mapping essential urban land use categories with open big data: Results for five metropolitan areas in the United States of America. *ISPRS J. Photogramm. Remote Sens.* **2021**, *178*, 203–218. [CrossRef]

10. Alqurashi, A.; Kumar, L. Investigating the use of remote sensing and GIS techniques to detect land use and land cover change: A review. *Adv. Remote Sens.* **2013**, *2*, 193–204. [CrossRef]

11. Phiri, D.; Morgenroth, J. Developments in Landsat land cover classification methods: A review. *Remote Sens.* **2017**, *9*, 967. [CrossRef]

12. MohanRajan, S.N.; Loganathan, A.; Manoharan, P. Survey on Land Use/Land Cover (LU/LC) change analysis in remote sensing and GIS environment: Techniques and Challenges. *Environ. Sci. Pollut. Res.* **2020**, *27*, 29900–29926. [CrossRef]

13. Yin, J.; Dong, J.; Hamm, N.A.; Li, Z.; Wang, J.; Xing, H.; Fu, P. Integrating remote sensing and geospatial big data for urban land use mapping: A review. *Int. J. Appl. Earth Obs. Geoinf.* **2021**, *103*, 102514. [CrossRef]

14. Lee, J.G.; Kang, M. Geospatial big data: Challenges and opportunities. *Big Data Res.* **2015**, *2*, 74–81. [CrossRef]

15. Deng, X.; Liu, P.; Liu, X.; Wang, R.; Zhang, Y.; He, J.; Yao, Y. Geospatial big data: New paradigm of remote sensing applications. *IEEE J. Sel. Top. Appl. Earth Obs. Remote Sens.* **2019**, *12*, 3841–3851. [CrossRef]

16. He, J.; Li, X.; Liu, P.; Wu, X.; Zhang, J.; Zhang, D.; Liu, X.; Yao, Y. Accurate estimation of the proportion of mixed land use at the street-block level by integrating high spatial resolution images and geospatial big data. *IEEE Trans. Geosci. Remote Sens.* **2020**, *59*, 6357–6370. [CrossRef]

17. Yi, J.; Du, Y.; Liang, F.; Tu, W.; Qi, W.; Ge, Y. Mapping human's digital footprints on the Tibetan Plateau from multi-source geospatial big data. *Sci. Total Environ.* **2020**, *711*, 134540. [CrossRef]

18. Chi, M.; Plaza, A.; Benediktsson, J.A.; Sun, Z.; Shen, J.; Zhu, Y. Big data for remote sensing: Challenges and opportunities. *Proc. IEEE* **2016**, *104*, 2207–2219. [CrossRef]

19. Shi, Y.; Qi, Z.; Liu, X.; Niu, N.; Zhang, H. Urban land use and land cover classification using multisource remote sensing images and social media data. *Remote Sens.* **2019**, *11*, 2719. [CrossRef]

20. Wulder, M.A.; Coops, N.C.; Roy, D.P.; White, J.C.; Hermosilla, T. Land cover 2.0. *Int. J. Remote Sens.* **2018**, *39*, 4254–4284. [CrossRef]

21. Saah, D.; Johnson, G.; Ashmall, B.; Tondapu, G.; Tenneson, K.; Patterson, M.; Poortinga, A.; Markert, K.; Quyen, N.H.; San Aung, K.; et al. Collect Earth: An online tool for systematic reference data collection in land cover and use applications. *Environ. Model. Softw.* **2019**, *118*, 166–171. [CrossRef]

22. Sitthi, A.; Nagai, M.; Dailey, M.; Ninsawat, S. Exploring land use and land cover of geotagged social-sensing images using naive bayes classifier. *Sustainability* **2016**, *8*, 921. [CrossRef]

23. Zhang, C.; Zhao, T.; Li, W. *Geospatial Semantic Web*; Springer: Berlin/Heidelberg, Germany, 2015.

24. Chen, B.; Xu, B.; Gong, P. Mapping essential urban land use categories (EULUC) using geospatial big data: Progress, challenges, and opportunities. *Big Earth Data* **2021**, *5*, 410–441. [CrossRef]

25. Helber, P.; Bischke, B.; Dengel, A.; Borth, D. Eurosat: A novel dataset and deep learning benchmark for land use and land cover classification. *IEEE J. Sel. Top. Appl. Earth Obs. Remote Sens.* **2019**, *12*, 2217–2226. [CrossRef]

26. Nijhawan, R.; Joshi, D.; Narang, N.; Mittal, A.; Mittal, A. A futuristic deep learning framework approach for land use-land cover classification using remote sensing imagery. In *Advanced Computing and Communication Technologies*; Springer: Singapore, 2019; pp. 87–96.

27. Jozdani, S.E.; Johnson, B.A.; Chen, D. Comparing deep neural networks, ensemble classifiers, and support vector machine algorithms for object-based urban land use/land cover classification. *Remote Sens.* **2019**, *11*, 1713. [CrossRef]

28. Li, W.; Fu, H.; Yu, L.; Gong, P.; Feng, D.; Li, C.; Clinton, N. Stacked Autoencoder-based deep learning for remote-sensing image classification: A case study of African land-cover mapping. *Int. J. Remote Sens.* **2016**, *37*, 5632–5646. [CrossRef]

29. Alem, A.; Kumar, S. Deep learning methods for land cover and land use classification in remote sensing: A review. In Proceedings of the 2020 8th International Conference on Reliability 2020, Infocom Technologies and Optimization (Trends and Future Directions) (ICRITO), Noida, India, 4–5 June 2020; pp. 903–908.

30. Chaves, M.E.D.; Picoli, M.C.A.; Sanches, I.D. Recent applications of Landsat 8/OLI and Sentinel-2/MSI for land use and land cover mapping: A systematic review. *Remote Sens.* **2020**, *12*, 3062. [CrossRef]

31. Ghamisi, P.; Plaza, J.; Chen, Y.; Li, J.; Plaza, A.J. Advanced spectral classifiers for hyperspectral images: A review. *IEEE Geosci. Remote Sens. Mag.* **2017**, *5*, 8–32. [CrossRef]

32. Ghamisi, P.; Yokoya, N.; Li, J.; Liao, W.; Liu, S.; Plaza, J.; Rasti, B.; Plaza, A. Advances in hyperspectral image and signal processing: A comprehensive overview of the state of the art. *IEEE Geosci. Remote Sens. Mag.* **2017**, *5*, 37–78. [CrossRef]

33. Gómez, C.; White, J.C.; Wulder, M.A. Optical remotely sensed time series data for land cover classification: A review. *ISPRS J. Photogramm. Remote Sens.* **2016**, *116*, 55–72. [CrossRef]

34. Grekousis, G. Artificial neural networks and deep learning in urban geography: A systematic review and meta-analysis. *Comput. Environ. Urban Syst.* **2019**, *74*, 244–256. [CrossRef]

35. He, L.; Li, J.; Liu, C.; Li, S. Recent advances on spectral–spatial hyperspectral image classification: An overview and new guidelines. *IEEE Trans. Geosci. Remote Sens.* **2017**, *56*, 1579–1597. [CrossRef]

36. Li, S.; Dragicevic, S.; Castro, F.A.; Sester, M.; Winter, S.; Coltekin, A.; Pettit, C.; Jiang, B.; Haworth, J.; Stein, A.; et al. Geospatial big data handling theory and methods: A review and research challenges. *ISPRS J. Photogramm. Remote Sens.* **2016**, *115*, 119–133. [CrossRef]

37. Li, S.; Song, W.; Fang, L.; Chen, Y.; Ghamisi, P.; Benediktsson, J.A. Deep learning for hyperspectral image classification: An overview. *IEEE Trans. Geosci. Remote Sens.* **2019**, *57*, 6690–6709. [CrossRef]

38. Ma, L.; Liu, Y.; Zhang, X.; Ye, Y.; Yin, G.; Johnson, B.A. Deep learning in remote sensing applications: A meta-analysis and review. *ISPRS J. Photogramm. Remote Sens.* **2019**, *152*, 166–177. [CrossRef]

39. Maxwell, A.E.; Warner, T.A.; Fang, F. Implementation of machine-learning classification in remote sensing: An applied review. *Int. J. Remote Sens.* **2018**, *39*, 2784–2817. [CrossRef]

40. Pandey, P.C.; Koutsias, N.; Petropoulos, G.P.; Srivastava, P.K.; Ben Dor, E. Land use/land cover in view of earth observation: Data sources, input dimensions, and classifiers—A review of the state of the art. *Geocarto Int.* **2021**, *36*, 957–988. [CrossRef]

41. Paoletti, M.E.; Haut, J.M.; Plaza, J.; Plaza, A. Deep learning classifiers for hyperspectral imaging: A review. *ISPRS J. Photogramm. Remote Sens.* **2019**, *158*, 279–317. [CrossRef]

42. Talukdar, S.; Singha, P.; Mahato, S.; Pal, S.; Liou, Y.A.; Rahman, A. Land-use land-cover classification by machine learning classifiers for satellite observations—A review. *Remote Sens.* **2020**, *12*, 1135. [CrossRef]

43. Vali, A.; Comai, S.; Matteucci, M. Deep learning for land use and land cover classification based on hyperspectral and multispectral earth observation data: A review. *Remote Sens.* **2020**, *12*, 2495. [CrossRef]

44. Wang, J.; Bretz, M.; Dewan, M.A.A.; Delavar, M.A. Machine learning in modelling land-use and land cover-change (LULCC): Current status, challenges and prospects. *Sci. Total Environ.* **2022**, *822*, 153559. [CrossRef]

45. Zhang, L.; Zhang, L.; Du, B. Deep learning for remote sensing data: A technical tutorial on the state of the art. *IEEE Geosci. Remote Sens. Mag.* **2016**, *4*, 22–40. [CrossRef]

46. Zhang, X.; Zhou, Y.N.; Luo, J. Deep learning for processing and analysis of remote sensing big data: A technical review. *Big Earth Data* **2021**, *5*, 1–34. [CrossRef]

47. Zhu, X.X.; Tuia, D.; Mou, L.; Xia, G.S.; Zhang, L.; Xu, F.; Fraundorfer, F. Deep learning in remote sensing: A comprehensive review and list of resources. *IEEE Geosci. Remote Sens. Mag.* **2017**, *5*, 8–36. [CrossRef]

48. Bartsch, A.; Höfler, A.; Kroisleitner, C.; Trofaier, A.M. Land cover mapping in northern high latitude permafrost regions with satellite data: Achievements and remaining challenges. *Remote Sens.* **2016**, *8*, 979. [CrossRef]

49. Thyagharajan, K.K.; Vignesh, T. Soft computing techniques for land use and land cover monitoring with multispectral remote sensing images: A review. *Arch. Comput. Methods Eng.* **2019**, *26*, 275–301. [CrossRef]

50. Comber, A.; Wulder, M. Considering spatiotemporal processes in big data analysis: Insights from remote sensing of land cover and land use. *Trans. GIS* **2019**, *23*, 879–891. [CrossRef]

51. Liu, P.; Di, L.; Du, Q.; Wang, L. Remote sensing big data: Theory, methods and applications. *Remote Sens.* **2018**, *10*, 711. [CrossRef]

52. Sugumaran, R.; Hegeman, J.W.; Sardeshmukh, V.B.; Armstrong, M.P. Processing remote-sensing data in cloud computing environments. In *Remotely Sensed Data Characterization, Classification, and Accuracies*; CRC Press: Boca Raton, FL, USA, 2015; pp. 587–596.

53. Feng, M.; Li, X. Land cover mapping toward finer scales. *Sci. Bull.* **2020**, *65*, 1604–1606. [CrossRef]

54. Franklin, S.E.; He, Y.; Pape, A.; Guo, X.; McDermid, G.J. Landsat-comparable land cover maps using ASTER and SPOT images: A case study for large-area mapping programmes. *Int. J. Remote Sens.* **2011**, *32*, 2185–2205. [CrossRef]

55. Alphan, H.; Doygun, H.; Unlukaplan, Y.I. Post-classification comparison of land cover using multitemporal Landsat and ASTER imagery: The case of Kahramanmaraş, Turkey. *Environ. Monit. Assess.* **2009**, *151*, 327–336. [CrossRef]

56. de Pinho, C.M.D.; Fonseca, L.M.G.; Korting, T.S.; De Almeida, C.M.; Kux, H.J.H. Land-cover classification of an intra-urban environment using high-resolution images and object-based image analysis. *Int. J. Remote Sens.* **2012**, *33*, 5973–5995. [CrossRef]

57. Novack, T.; Esch, T.; Kux, H.; Stilla, U. Machine learning comparison between WorldView-2 and QuickBird-2-simulated imagery regarding object-based urban land cover classification. *Remote Sens.* **2011**, *3*, 2263–2282. [CrossRef]

58. Toure, S.I.; Stow, D.A.; Shih, H.C.; Weeks, J.; Lopez-Carr, D. Land cover and land use change analysis using multi-spatial resolution data and object-based image analysis. *Remote Sens. Environ.* **2018**, *210*, 259–268. [CrossRef]

59. Malinverni, E.S.; Tassetti, A.N.; Mancini, A.; Zingaretti, P.; Frontoni, E.; Bernardini, A. Hybrid object-based approach for land use/land cover mapping using high spatial resolution imagery. *Int. J. Geogr. Inf. Sci.* **2011**, *25*, 1025–1043. [CrossRef]
60. Qi, Z.; Yeh, A.G.O.; Li, X.; Lin, Z. A novel algorithm for land use and land cover classification using RADARSAT-2 polarimetric SAR data. *Remote Sens. Environ.* **2012**, *118*, 21–39. [CrossRef]
61. Lv, Q.; Dou, Y.; Niu, X.; Xu, J.; Xu, J.; Xia, F. Urban land use and land cover classification using remotely sensed SAR data through deep belief networks. *J. Sens.* **2015**, *2015*, 538063. [CrossRef]
62. Zhang, R.; Tang, X.; You, S.; Duan, K.; Xiang, H.; Luo, H. A novel feature-level fusion framework using optical and SAR remote sensing images for land use/land cover (LULC) classification in cloudy mountainous area. *Appl. Sci.* **2020**, *10*, 2928. [CrossRef]
63. Teeuw, R.M.; Leidig, M.; Saunders, C.; Morris, N. Free or low-cost geoinformatics for disaster management: Uses and availability issues. *Environ. Hazards* **2013**, *12*, 112–131. [CrossRef]
64. Navin, M.S.; Agilandeeswari, L. Comprehensive review on land use/land cover change classification in remote sensing. *J. Spectr. Imaging* **2020**, *9*, a8. [CrossRef]
65. Hasan, S.; Shi, W.; Zhu, X.; Abbas, S. Monitoring of land use/land cover and socioeconomic changes in south china over the last three decades using landsat and nighttime light data. *Remote Sens.* **2019**, *11*, 1658. [CrossRef]
66. Zhang, Q.; Seto, K.C. Mapping urbanization dynamics at regional and global scales using multi-temporal DMSP/OLS nighttime light data. *Remote Sens. Environ.* **2011**, *115*, 2320–2329. [CrossRef]
67. Yan, W.Y.; Shaker, A.; El-Ashmawy, N. Urban land cover classification using airborne LiDAR data: A review. *Remote Sens. Environ.* **2015**, *158*, 295–310. [CrossRef]
68. Antonarakis, A.S.; Richards, K.S.; Brasington, J. Object-based land cover classification using airborne LiDAR. *Remote Sens. Environ.* **2008**, *112*, 2988–2998. [CrossRef]
69. Li, X.; Zhang, C.; Li, W. Building block level urban land-use information retrieval based on Google Street View images. *GIScience Remote Sens.* **2017**, *54*, 819–835. [CrossRef]
70. Zhang, W.; Li, W.; Zhang, C.; Hanink, D.M.; Li, X.; Wang, W. Parcel feature data derived from Google Street View images for urban land use classification in Brooklyn, New York City. *Data Brief* **2017**, *12*, 175–179. [CrossRef]
71. Li, X.; Zhang, C.; Li, W.; Ricard, R.; Meng, Q.; Zhang, W. Assessing street-level urban greenery using Google Street View and a modified green view index. *Urban For. Urban Green.* **2015**, *14*, 675–685. [CrossRef]
72. Zhang, W.; Li, W.; Zhang, C.; Hanink, D.M.; Li, X.; Wang, W. Parcel-based urban land use classification in megacity using airborne LiDAR, high resolution orthoimagery, and Google Street View. *Comput. Environ. Urban Syst.* **2017**, *64*, 215–228. [CrossRef]
73. Ma, L.; Fu, T.; Blaschke, T.; Li, M.; Tiede, D.; Zhou, Z.; Ma, X.; Chen, D. Evaluation of feature selection methods for object-based land cover mapping of unmanned aerial vehicle imagery using random forest and support vector machine classifiers. *ISPRS Int. J. Geo-Inf.* **2017**, *6*, 51. [CrossRef]
74. Ahmed, O.S.; Shemrock, A.; Chabot, D.; Dillon, C.; Williams, G.; Wasson, R.; Franklin, S.E. Hierarchical land cover and vegetation classification using multispectral data acquired from an unmanned aerial vehicle. *Int. J. Remote Sens.* **2017**, *38*, 2037–2052. [CrossRef]
75. Green, K.; Kempka, D.; Lackey, L. Using remote sensing to detect and monitor land-cover and land-use change. *Photogramm. Eng. Remote Sens.* **1994**, *60*, 331–337.
76. Wu, H.; Gui, Z.; Yang, Z. Geospatial big data for urban planning and urban management. *Geo-Spat. Inf. Sci.* **2020**, *23*, 273–274. [CrossRef]
77. Huang, B.; Wang, J. Big spatial data for urban and environmental sustainability. *Geo-Spat. Inf. Sci.* **2020**, *23*, 125–140. [CrossRef]
78. See, L.; Schepaschenko, D.; Lesiv, M.; McCallum, I.; Fritz, S.; Comber, A.; Perger, C.; Schill, C.; Zhao, Y.; Maus, V.; et al. Building a hybrid land cover map with crowdsourcing and geographically weighted regression. *ISPRS J. Photogramm. Remote Sens.* **2015**, *103*, 48–56. [CrossRef]
79. Fritz, S.; See, L.; Perger, C.; McCallum, I.; Schill, C.; Schepaschenko, D.; Duerauer, M.; Karner, M.; Dresel, C.; Laso-Bayas, J.C.; et al. A global dataset of crowdsourced land cover and land use reference data. *Sci. Data* **2017**, *4*, 170075. [CrossRef] [PubMed]
80. Johnson, B.A.; Iizuka, K. Integrating OpenStreetMap crowdsourced data and Landsat time-series imagery for rapid land use/land cover (LULC) mapping: Case study of the Laguna de Bay area of the Philippines. *Appl. Geogr.* **2016**, *67*, 140–149. [CrossRef]
81. Fonte, C.C.; Minghini, M.; Patriarca, J.; Antoniou, V.; See, L.; Skopeliti, A. Generating up-to-date and detailed land use and land cover maps using OpenStreetMap and GlobeLand30. *ISPRS Int. J. Geo-Inf.* **2017**, *6*, 125. [CrossRef]
82. Andrade, R.; Alves, A.; Bento, C. POI mining for land use classification: A case study. *ISPRS Int. J. Geo-Inf.* **2020**, *9*, 493. [CrossRef]
83. Vargas-Munoz, J.E.; Srivastava, S.; Tuia, D.; Falcao, A.X. OpenStreetMap: Challenges and opportunities in machine learning and remote sensing. *IEEE Geosci. Remote Sens. Mag.* **2020**, *9*, 184–199. [CrossRef]
84. Wu, H.; Lin, A.; Xing, X.; Song, D.; Li, Y. Identifying core driving factors of urban land use change from global land cover products and POI data using the random forest method. *Int. J. Appl. Earth Obs. Geoinf.* **2021**, *103*, 102475. [CrossRef]
85. Ai, T.; Yang, W. The detection of transport land-use data using crowdsourcing taxi trajectory. In Proceedings of the International Archives of the Photogrammetry, Remote Sensing and Spatial Information Sciences, Prague, Czech Republic, 12–19 July 2016; pp. 785–788. [CrossRef]
86. Liu, W.; Wu, W.; Thakuriah, P.; Wang, J. The geography of human activity and land use: A big data approach. *Cities* **2020**, *97*, 102523. [CrossRef]

87. Schultz, M.; Voss, J.; Auer, M.; Carter, S.; Zipf, A. Open land cover from OpenStreetMap and remote sensing. *Int. J. Appl. Earth Obs. Geoinf.* **2017**, *63*, 206–213. [CrossRef]

88. Xing, J.; Sieber, R.E. A land use/land cover change geospatial cyberinfrastructure to integrate big data and temporal topology. *Int. J. Geogr. Inf. Sci.* **2016**, *30*, 573–593. [CrossRef]

89. Yang, D.; Fu, C.S.; Smith, A.C.; Yu, Q. Open land-use map: A regional land-use mapping strategy for incorporating OpenStreetMap with earth observations. *Geo-Spat. Inf. Sci.* **2017**, *20*, 269–281. [CrossRef]

90. Hu, T.; Yang, J.; Li, X.; Gong, P. Mapping urban land use by using landsat images and open social data. *Remote Sens.* **2016**, *8*, 151. [CrossRef]

91. Yin, J.; Fu, P.; Hamm, N.A.; Li, Z.; You, N.; He, Y.; Cheshmehzangi, A.; Dong, J. Decision-level and feature-level integration of remote sensing and geospatial big data for urban land use mapping. *Remote Sens.* **2021**, *13*, 1579. [CrossRef]

92. Copenhaver, K.L. Combining Tabular and Satellite-Based Datasets to Better Understand Cropland Change. *Land* **2022**, *11*, 714. [CrossRef]

93. Guan, Q.; Cheng, S.; Pan, Y.; Yao, Y.; Zeng, W. Sensing mixed urban land-use patterns using municipal water consumption time series. *Ann. Am. Assoc. Geogr.* **2021**, *111*, 68–86. [CrossRef]

94. Chen, B.; Huang, B.; Xu, B. Multi-source remotely sensed data fusion for improving land cover classification. *ISPRS J. Photogramm. Remote Sens.* **2017**, *124*, 27–39. [CrossRef]

95. Haklay, M. Citizen science and volunteered geographic information: Overview and typology of participation. In *Crowdsourcing Geographic Knowledge*; Sui, D., Elwood, S., Goodchild, M., Eds.; Springer: Dordrecht, The Netherland, 2013; pp. 105–122.

96. Laban, N.; Abdellatif, B.; Ebeid, H.M.; Shedeed, H.A.; Tolba, M.F. Machine Learning for Enhancement Land Cover and Crop Types Classification. In *Machine Learning Paradigms: Theory and Application*; Springer: Cham, Switzerland, 2019; pp. 71–87.

97. Kuras, A.; Brell, M.; Rizzi, J.; Burud, I. Hyperspectral and lidar data applied to the urban land cover machine learning and neural-network-based classification: A review. *Remote Sens.* **2021**, *13*, 3393. [CrossRef]

98. Jamali, A. Land use land cover mapping using advanced machine learning classifiers: A case study of Shiraz city, Iran. *Earth Sci. Inform.* **2020**, *13*, 1015–1030. [CrossRef]

99. Schmitt, M.; Prexl, J.; Ebel, P.; Liebel, L.; Zhu, X.X. Weakly supervised semantic segmentation of satellite images for land cover mapping—Challenges and opportunities. *arXiv* **2020**, arXiv:2002.08254. [CrossRef]

100. Kussul, N.; Shelestov, A.; Lavreniuk, M.; Butko, I.; Skakun, S. Deep learning approach for large scale land cover mapping based on remote sensing data fusion. In Proceedings of the 2016 IEEE International Geoscience and Remote Sensing Symposium (IGARSS), Beijing, China, 10–15 July 2016; pp. 198–201.

101. Zhang, P.; Ke, Y.; Zhang, Z.; Wang, M.; Li, P.; Zhang, S. Urban land use and land cover classification using novel deep learning models based on high spatial resolution satellite imagery. *Sensors* **2018**, *18*, 3717. [CrossRef] [PubMed]

102. Storie, C.D.; Henry, C.J. Deep learning neural networks for land use land cover mapping. In Proceedings of the IGARSS 2018-2018 IEEE International Geoscience and Remote Sensing Symposium, Valencia, Spain, 22–27 July 2018; pp. 3445–3448.

103. Srivastava, S.; Vargas-Munoz, J.E.; Tuia, D. Understanding urban landuse from the above and ground perspectives: A deep learning, multimodal solution. *Remote Sens. Environ.* **2019**, *228*, 129–143. [CrossRef]

104. Feizizadeh, B.; Mohammadzade Alajujeh, K.; Lakes, T.; Blaschke, T.; Omarzadeh, D. A comparison of the integrated fuzzy object-based deep learning approach and three machine learning techniques for land use/cover change monitoring and environmental impacts assessment. *GIScience Remote Sens.* **2021**, *58*, 1543–1570. [CrossRef]

105. Digra, M.; Dhir, R.; Sharma, N. Land use land cover classification of remote sensing images based on the deep learning approaches: A statistical analysis and review. *Arab. J. Geosci.* **2022**, *15*, 1003. [CrossRef]

106. Bhosle, K.; Musande, V. Evaluation of deep learning CNN model for land use land cover classification and crop identification using hyperspectral remote sensing images. *J. Indian Soc. Remote Sens.* **2019**, *47*, 1949–1958. [CrossRef]

107. Abdi, A.M. Land cover and land use classification performance of machine learning algorithms in a boreal landscape using Sentinel-2 data. *GIScience Remote Sens.* **2020**, *57*, 1–20. [CrossRef]

108. Amani, M.; Ghorbanian, A.; Ahmadi, S.A.; Kakooei, M.; Moghimi, A.; Mirmazloumi, S.M.; Moghaddam, S.H.; Mahdavi, S.; Ghahremanloo, M.; Parsian, S.; et al. Google earth engine cloud computing platform for remote sensing big data applications: A comprehensive review. *IEEE J. Sel. Top. Appl. Earth Obs. Remote Sens.* **2020**, *13*, 5326–5350. [CrossRef]

109. Kang, X.; Liu, J.; Dong, C.; Xu, S. Using high-performance computing to address the challenge of land use/land cover change analysis on spatial big data. *ISPRS Int. J. Geo-Inf.* **2018**, *7*, 273. [CrossRef]

110. Karra, K.; Kontgis, C.; Statman-Weil, Z.; Mazzariello, J.C.; Mathis, M.; Brumby, S.P. Global land use/land cover with Sentinel 2 and deep learning. In Proceedings of the 2021 IEEE International Geoscience and Remote Sensing Symposium IGARSS, Brussels, Belgium, 11–16 July 2021; pp. 4704–4707.

111. de Sousa, C.; Fatoyinbo, L.; Neigh, C.; Boucka, F.; Angoue, V.; Larsen, T. Cloud-computing and machine learning in support of country-level land cover and ecosystem extent mapping in Liberia and Gabon. *PLoS ONE* **2020**, *15*, e0227438. [CrossRef]

112. Yang, C.; Yu, M.; Hu, F.; Jiang, Y.; Li, Y. Utilizing cloud computing to address big geospatial data challenges. *Comput. Environ. Urban Syst.* **2017**, *61*, 120–128. [CrossRef]

113. Dubertret, F.; Le Tourneau, F.M.; Villarreal, M.L.; Norman, L.M. Monitoring Annual Land Use/Land Cover Change in the Tucson Metropolitan Area with Google Earth Engine (1986–2020). *Remote Sens.* **2022**, *14*, 2127. [CrossRef]

114. Mou, X.; Li, H.; Huang, C.; Liu, Q.; Liu, G. Application progress of Google Earth Engine in land use and land cover remote sensing information extraction. *Remote Sens. Land Resour.* **2021**, *33*, 1–10.

115. Phan, T.N.; Kuch, V.; Lehnert, L.W. Land Cover Classification using Google Earth Engine and Random Forest Classifier—The Role of Image Composition. *Remote Sens.* **2020**, *12*, 2411. [CrossRef]

116. Gorelick, N.; Hancher, M.; Dixon, M.; Ilyushchenko, S.; Thau, D.; Moore, R. Google Earth Engine: Planetary-scale geospatial analysis for everyone. *Remote Sens. Environ.* **2017**, *202*, 18–27. [CrossRef]

117. Xie, S.; Liu, L.; Zhang, X.; Yang, J.; Chen, X.; Gao, Y. Automatic land-cover mapping using landsat time-series data based on google earth engine. *Remote Sens.* **2019**, *11*, 3023. [CrossRef]

118. Xia, H.; Zhao, J.; Qin, Y.; Yang, J.; Cui, Y.; Song, H.; Ma, L.; Jin, N.; Meng, Q. Changes in water surface area during 1989–2017 in the Huai River Basin using Landsat data and Google earth engine. *Remote Sens.* **2019**, *11*, 1824. [CrossRef]

119. Kumar, L.; Mutanga, O. Google Earth Engine applications since inception: Usage, trends, and potential. *Remote Sens.* **2018**, *10*, 1509. [CrossRef]

120. Wang, L.; Diao, C.; Xian, G.; Yin, D.; Lu, Y.; Zou, S.; Erickson, T.A. A summary of the special issue on remote sensing of land change science with Google earth engine. *Remote Sens. Environ.* **2020**, *248*, 112002. [CrossRef]

121. Liu, X.; Hu, G.; Chen, Y.; Li, X.; Xu, X.; Li, S.; Pei, F.; Wang, S. High-resolution multi-temporal mapping of global urban land using Landsat images based on the Google Earth Engine Platform. *Remote Sens. Environ.* **2018**, *209*, 227–239. [CrossRef]

122. Zhang, X.; Liu, L.; Wu, C.; Chen, X.; Gao, Y.; Xie, S.; Zhang, B. Development of a global 30 m impervious surface map using multisource and multitemporal remote sensing datasets with the Google Earth Engine platform. *Earth Syst. Sci. Data* **2020**, *12*, 1625–1648. [CrossRef]

123. Tang, Z.; Li, Y.; Gu, Y.; Jiang, W.; Xue, Y.; Hu, Q.; LaGrange, T.; Bishop, A.; Drahota, J.; Li, R. Assessing Nebraska playa wetland inundation status during 1985–2015 using Landsat data and Google Earth Engine. *Environ. Monit. Assess.* **2016**, *188*, 654. [CrossRef] [PubMed]

124. Floreano, I.X.; de Moraes, L.A.F. Land use/land cover (LULC) analysis (2009–2019) with Google Earth Engine and 2030 prediction using Markov-CA in the Rondônia State, Brazil. *Environ. Monit. Assess.* **2021**, *193*, 239. [CrossRef] [PubMed]

125. Ferreira, K.R.; Queiroz, G.R.; Camara, G.; Souza, R.C.; Vinhas, L.; Marujo, R.F.; Simoes, R.E.; Noronha, C.A.; Costa, R.W.; Arcanjo, J.S.; et al. Using remote sensing images and cloud services on AWS to improve land use and cover monitoring. In Proceedings of the 2020 IEEE Latin American GRSS & ISPRS Remote Sensing Conference (LAGIRS), Santiago, Chile, 22–26 March 2020; pp. 558–562.

126. Tran, B.H.; Aussenac-Gilles, N.; Comparot, C.; Trojahn, C. Semantic integration of raster data for earth observation: An RDF dataset of territorial unit versions with their land cover. *ISPRS Int. J. Geo-Inf.* **2020**, *9*, 503. [CrossRef]

127. Gutman, G.; Byrnes, R.A.; Masek, J.; Covington, S.; Justice, C.; Franks, S.; Headley, R. Towards monitoring land-cover and land-use changes at a global scale: The Global Land Survey 2005. *Photogramm. Eng. Remote Sens.* **2008**, *74*, 6–10.

128. Li, X.; Chen, G.; Liu, X.; Liang, X.; Wang, S.; Chen, Y.; Pei, F.; Xu, X. A new global land-use and land-cover change product at a 1-km resolution for 2010 to 2100 based on human–environment interactions. *Ann. Am. Assoc. Geogr.* **2017**, *107*, 1040–1059. [CrossRef]

129. Buyantuyev, A.; Wu, J. Effects of thematic resolution on landscape pattern analysis. *Landsc. Ecol.* **2007**, *22*, 7–13. [CrossRef]

130. Lechner, A.M.; Rhodes, J.R. Recent progress on spatial and thematic resolution in landscape ecology. *Curr. Landsc. Ecol. Rep.* **2016**, *1*, 98–105. [CrossRef]

131. Di Gregorio, A.; O'BRIEN, D. Overview of land-cover classifications and their interoperability. In *Remote Sensing of Land Use and Land Cover: Principles and Applications*; Giri, C.P., Ed.; CRC Press: Boca Raton, FL, USA, 2012; pp. 37–47.

132. Yang, H.; Li, S.; Chen, J.; Zhang, X.; Xu, S. The standardization and harmonization of land cover classification systems towards harmonized datasets: A review. *ISPRS Int. J. Geo-Inf.* **2017**, *6*, 154. [CrossRef]

133. Feng, C.C.; Flewelling, D.M. Assessment of semantic similarity between land use/land cover classification systems. *Comput. Environ. Urban Syst.* **2004**, *28*, 229–246. [CrossRef]

134. Herold, M.; Latham, J.S.; Di Gregorio, A.; Schmullius, C.C. Evolving standards in land cover characterization. *J. Land Use Sci.* **2006**, *1*, 157–168. [CrossRef]

135. Cruz, I.F.; Sunna, W.; Makar, N.; Bathala, S. A visual tool for ontology alignment to enable geospatial interoperability. *J. Vis. Lang. Comput.* **2007**, *18*, 230–254. [CrossRef]

136. Katharopoulos, A.; Fleuret, F. Not all samples are created equal: Deep learning with importance sampling. In Proceedings of the International Conference on Machine Learning, Stockholm, Sweden, 10–15 July 2018; pp. 2525–2534.

137. Ghorbanian, A.; Kakooei, M.; Amani, M.; Mahdavi, S.; Mohammadzadeh, A.; Hasanlou, M. Improved land cover map of Iran using Sentinel imagery within Google Earth Engine and a novel automatic workflow for land cover classification using migrated training samples. *ISPRS J. Photogramm. Remote Sens.* **2020**, *167*, 276–288. [CrossRef]

138. Traunmueller, M.; Marshall, P.; Capra, L. Crowdsourcing safety perceptions of people: Opportunities and limitations. In *International Conference on Social Informatics*; Springer: Cham, Switzerland, 2015; pp. 120–135.

139. Ji, M.; Jensen, J.R. Effectiveness of subpixel analysis in detecting and quantifying urban imperviousness from Landsat Thematic Mapper imagery. *Geocarto Int.* **1999**, *14*, 33–41. [CrossRef]

140. Powell, R.L.; Roberts, D.A.; Dennison, P.E.; Hess, L.L. Sub-pixel mapping of urban land cover using multiple endmember spectral mixture analysis: Manaus, Brazil. *Remote Sens. Environ.* **2007**, *106*, 253–267. [CrossRef]

141. MacLachlan, A.; Roberts, G.; Biggs, E.; Boruff, B. Subpixel land-cover classification for improved urban area estimates using Landsat. *Int. J. Remote Sens.* **2017**, *38*, 5763–5792. [CrossRef]
142. He, D.; Shi, Q.; Liu, X.; Zhong, Y.; Zhang, X. Deep subpixel mapping based on semantic information modulated network for urban land use mapping. *IEEE Trans. Geosci. Remote Sens.* **2021**, *59*, 10628–10646. [CrossRef]
143. Huang, Z.; Qi, H.; Kang, C.; Su, Y.; Liu, Y. An ensemble learning approach for urban land use mapping based on remote sensing imagery and social sensing data. *Remote Sens.* **2020**, *12*, 3254. [CrossRef]
144. Naushad, R.; Kaur, T.; Ghaderpour, E. Deep transfer learning for land use and land cover classification: A comparative study. *Sensors* **2021**, *21*, 8083. [CrossRef]
145. Foody, G.M. Explaining the unsuitability of the kappa coefficient in the assessment and comparison of the accuracy of thematic maps obtained by image classification. *Remote Sens. Environ.* **2020**, *239*, 111630. [CrossRef]
146. McGwire, K.C.; Fisher, P. Spatially variable thematic accuracy: Beyond the confusion matrix. In *Spatial Uncertainty in Ecology*; Springer: New York, NY, USA, 2001; pp. 308–329.
147. Pontius Jr, R.G.; Millones, M. Death to Kappa: Birth of quantity disagreement and allocation disagreement for accuracy assessment. *Int. J. Remote Sens.* **2011**, *32*, 4407–4429. [CrossRef]
148. Guo, N.; Xiong, W.; Wu, Q.; Jing, N. An efficient tile-pyramids building method for fast visualization of massive geospatial raster datasets. *Adv. Electr. Comput. Eng.* **2016**, *16*, 3–9. [CrossRef]
149. Malik, K.; Robertson, C.; Roberts, S.A.; Remmel, T.K.; Long, J.A. Computer vision models for comparing spatial patterns: Understanding spatial scale. *Int. J. Geogr. Inf. Sci.* **2022**, *36*, 1–35. [CrossRef]
150. Förstner, W.; Bonn, U. Computer Vision and Remote Sensing-Lessons Learned. *Fritsch Dieter (Hg.) Photogramm. Week* **2009**, 241–249.
151. Wilkinson, G.G. Recent developments in remote sensing technology and the importance of computer vision analysis techniques. In *Machine Vision and Advanced Image Processing in Remote Sensing*; Kanellopoulos, I., Wilkinson, G.G., Moons, T., Eds.; Springer: Berlin/Heidelberg, Germany, 1999; pp. 5–11.
152. Chen, W.; Li, X.; He, H.; Wang, L. A review of fine-scale land use and land cover classification in open-pit mining areas by remote sensing techniques. *Remote Sens.* **2017**, *10*, 15. [CrossRef]
153. Liu, X.; He, J.; Yao, Y.; Zhang, J.; Liang, H.; Wang, H.; Hong, Y. Classifying urban land use by integrating remote sensing and social media data. *Int. J. Geogr. Inf. Sci.* **2017**, *31*, 1675–1696. [CrossRef]
154. Heydari, S.S.; Mountrakis, G. Meta-analysis of deep neural networks in remote sensing: A comparative study of mono-temporal classification to support vector machines. *ISPRS J. Photogramm. Remote Sens.* **2019**, *152*, 192–210. [CrossRef]
155. Yang, X.; Chen, Z.; Li, B.; Peng, D.; Chen, P.; Zhang, B. A fast and precise method for large-scale land-use mapping based on deep learning. In Proceedings of the IGARSS 2019—2019 IEEE International Geoscience and Remote Sensing Symposium, Yokohama, Japan, 28 July–2 August 2019; pp. 5913–5916.

land

Article

Multi-Criteria GIS-Based Analysis for Mapping Suitable Sites for Onshore Wind Farms in Southeast France

Mohammed Ifkirne [1], Houssam El Bouhi [2], Siham Acharki [3], Quoc Bao Pham [4,*], Abdelouahed Farah [5] and Nguyen Thi Thuy Linh [4]

[1] Faculty of Geography and Planning, University of Strasbourg, 3, rue de l'Argonne, 67000 Strasbourg, France
[2] Mathematics and Computer Science Research Training Unit, 7, rue René Descartes, 67084 Strasbourg, France
[3] Department of Earth Sciences, Faculty of Sciences and Techniques of Tangier, University Abdelmalek Essaadi, Tetouan 93000, Morocco
[4] Institute of Applied Technology, Thu Dau Mot University, Thu Dau Mot 75000, Binh Duong Province, Vietnam
[5] Remote Sensing Laboratory (2GRNT), Department of Geology, Geoscience, Geotourism, Natural Hazards, Faculty of Sciences Semlalia, University of Cadi Ayyad, BP 2390, Marrakesh 40000, Morocco
* Correspondence: phambaoquoc@tdmu.edu.vn

Abstract: Wind energy is critical to traditional energy sources replacement in France and throughout the world. Wind energy generation in France is quite unevenly spread across the country. Despite its considerable wind potential, the research region is among the least productive. The region is a very complicated location where socio-environmental, technological, and topographical restrictions intersect, which is why energy production planning studies in this area have been delayed. In this research, the methodology used for identifying appropriate sites for future wind farms in this region combines GIS with MCDA approaches such as AHP. Six determining factors are selected: the average wind speed, which has a weight of 38%; the protected areas, which have a relative weight of 26%; the distance to electrical substations and road networks, both of which have a significant influence on relative weights of 13%; and finally, the slope and elevation, which have weights of 5% and 3%, respectively. Only one alternative was investigated (suitable and unsuitable). The spatial database was generated using ArcGIS and QGIS software; the AHP was computed using Excel; and several treatments, such as raster data categorization and weighted overlay, were automated using the Python programming language. The regions identified for wind turbines installation are defined by a total of 962,612 pixels, which cover a total of 651 km^2 and represent around 6.98% of the research area. The theoretical wind potential calculation results suggest that for at least one site with an area bigger than 400 ha, the energy output ranges between 182.60 and 280.20 MW. The planned sites appear to be suitable; each site can support an average installed capacity of 45 MW. This energy benefit will fulfill the region's population's transportation, heating, and electrical demands.

Keywords: spatial energy planning; France; GIS; MCDA-AHP; suitability map; onshore wind farms

Citation: Ifkirne, M.; El Bouhi, H.; Acharki, S.; Pham, Q.B.; Farah, A.; Linh, N.T.T. Multi-Criteria GIS-Based Analysis for Mapping Suitable Sites for Onshore Wind Farms in Southeast France. *Land* **2022**, *11*, 1839. https://doi.org/10.3390/land11101839

Academic Editor: Chuanrong Zhang

Received: 14 September 2022
Accepted: 14 October 2022
Published: 19 October 2022

1. Introduction

Increased energy consumption in developed and developing countries as a result of prolonged economic growth [1] may lead to fast resource depletion, environmental degradation, biodiversity loss, and climate change [2–4]. Therefore, governments are required to focus their efforts on reducing greenhouse gas emissions and other environmental, social, and economic problems [5–7], as well as converting energy supply to green energy production methods.

Indeed, France generates electricity, heat, and transportation using a variety of energy sources. This energy mix includes nuclear, fossil fuels, and renewable sources. According to the Minister of Environmental Transition's key energy numbers for 2021/2020, nuclear accounts for 40% of energy, oil for 28%, and natural gas for 16%. However, in France, renewable energy accounts for just 13% of total energy consumption [8]. Due to its support for activities concerned with the energy and ecological transition, as well as its obligations to reduce the dangers associated with global warming, France would want to see the 30% renewable energy target reached by 2030 [8].

Previous investigations have demonstrated that wind power is one of the most promising renewable energy sources [9–12]. It is becoming increasingly popular worldwide due to its several benefits, including simple access to efficient multi-megawatt wind turbines [13]. Furthermore, wind energy sources will supply more power than any other form of energy source by 2050 in the European Union's renewable energy decarbonization scenario [14]. In addition, due to variables such as the availability of stronger and longer-lasting winds and land for installation, wind energy has recently become an important component of France's increasing renewable energy sector. Currently, there are already 11,625 onshore wind turbines in France. Thus, wind power has increased its proportion of the country's energy output from 2.2% 10 years ago to 7.9% in 2020, up from 6.1% in 2019 (https://www.revolution-energetique.com/, accessed on 2 July 2022). In addition, most wind turbines are unevenly spread across the Hauts-de-France and Grand-Est regions (as illustrated in Figure 1). Nonetheless, some regions, such as Aquitaine, Auvergne-Rhône-Alpes, and Provence-Alpes-Côte d'Azur, have insufficient infrastructures.

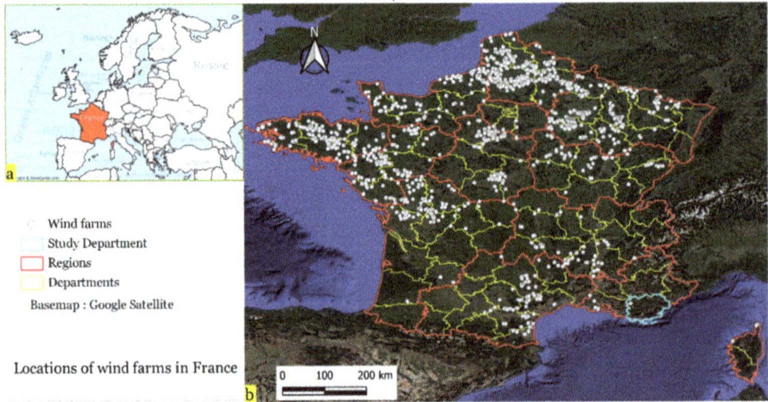

Figure 1. France's situation in Europe (**a**), Wind turbine geographical distribution in France (**b**) [15].

Many studies have reported significant environmental, social, economic, political, legal, and technological issues associated with wind farms sitting around the world [16]. To address these limits, geographic information systems (GIS) and hierarchical multi-criteria analysis (AHP) methodologies have been frequently adopted. These methods have recently been applied in various studies, including the identification of suitable sites for sitting solar farms [17–21] and suitable sites for marine wind farms [16,22–27].

Unfortunately, no study has been conducted in France on the application of multicriteria analysis approaches combined with GIS. However, there are studies that have been conducted on territories in Europe such as Greece [28,29] or southern Spain [30].

Different scenarios were considered. The result is a site suitability index map ranging from inadequate to highly suitable, which seems impractical for a sensitivity analysis of the overall site suitability index. Uncertainty is often inherent in such cases and leads to decision-making problems and inconsistency between decision makers' preferences [31]. All this research has highlighted the capabilities of GIS-based multi-criteria analysis approaches to site selection for onshore wind farms while considering regulations, legislation, and other constraints.

The results vary from one study area to another depending on the area of the study area, its topography, its natural resources (wind, temperature, etc.), and the criteria chosen for the study and the weights assigned to the criteria.

We conducted this study with the aim of identifying suitable sites for the planning of future wind farm construction projects in an area that is extremely complex due to the existing constraints in the region, considering environmental or topographical factors. The proposed methodological approach can be applied to any region of the world by adapting the characteristics considered. The implementation of the proposed methodology could facilitate the achievement of national objectives in the energy sector and encourage energy interdependence between many geographical areas in France.

Thus, our study area choice is influenced by previously mentioned reasons, such as the lack of wind farms in this region, which has a high population and high-power consumption (including heating), as well as environmental and relief limits that make it difficult to find suitable sites. The research aims are as follows: (1) to promote the use of GIS-based multi-criteria analysis methodologies in decision-making processes; (2) to contribute to the country's growth by providing cartographic and documentary materials related to wind projects; (3) to provide users with a Python code template that combines each component of the multi-criteria analysis technique for choosing potential onshore wind farm locations. By replacing the criteria, this code can be used for any research that involves decision making. Some of the pre-processing activities must be performed using GIS software such as QGIS or ArcGIS because they are not included in the source code.

It is hoped that this research will contribute to France's efforts in spatial energy planning. Effective wind farm siting options, as outlined in this study, could help the state meet its energy goals and policies.

2. Study Area

The research area is in southeast France (Figure 2a), which is part of the Provence-Alpes-Côte d'Azur region and covers 80% of the Var department. Moreover, it is bounded to the west by the Bouches-du-Rhône department, to the north by the Alpes-de-Haute-Provence department, to the east by the Alpes-Maritimes department, and to the south by the Mediterranean Sea (Figure 2b). It has an area of 11,208 km^2 with a perimeter of around 424 km (Figure 2c).

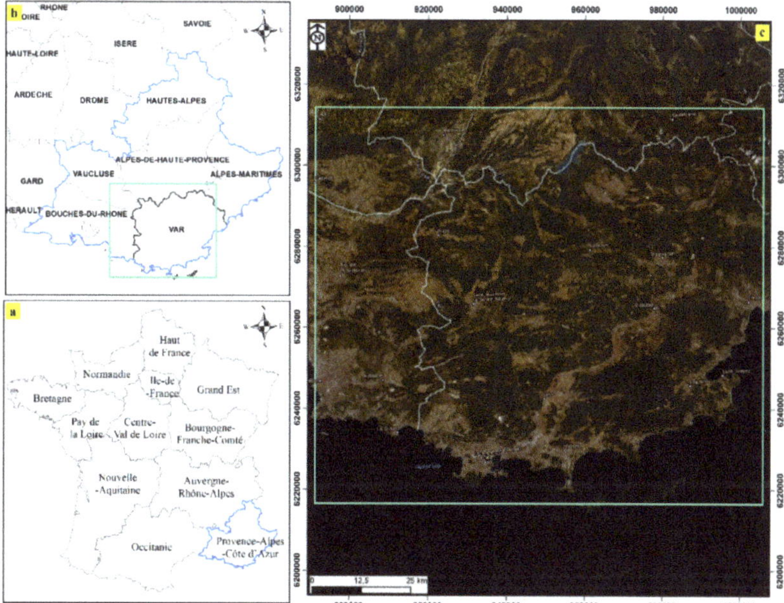

Figure 2. Study area's geographical location (**a**), on a national scale, (**b**) on regional and departmental scales, and (**c**) on a local scale.

The study area is known for its Mediterranean climate, of the Trewartha Cs or Köppen Csa type on the coast. Although it is a maritime climate, the annual temperature range is between 11 and 14 °C (Figure 3). The average annual rainfall is between 45 and 95 mm (Figure 3). The dominant winds are the Mistral (especially in Provence) and the Tramontane (especially in Languedoc) whose power comes from the channeling effect of the surrounding massifs to the north and west (Alps, Pyrenees, and Massif Central). Generally, these winds dry the air and clear the sky, and their intensity is very variable from one place to another, depending strongly on the sheltering or accelerating effect of the neighboring massifs [32]. In recent years, the average annual wind speed has been between 5.2 and 5.9 m/s (Figure 3). Interannual mean temperature (°C) and precipitation (mm) data were collected from the Climatic Research Unit Time Series (CRUTS) database at the University of East Anglia (CRU TS v. 4.01, https://www.cru.uea.ac.uk/, accessed on 10 July 2022, Harris et al., (2014)). Wind speed data were downloaded from the "Power Data Access" site via the link (https://www.uea.ac.uk/web/groups-and-centres/climatic-research-unit/data, accessed on 15 July 2022).

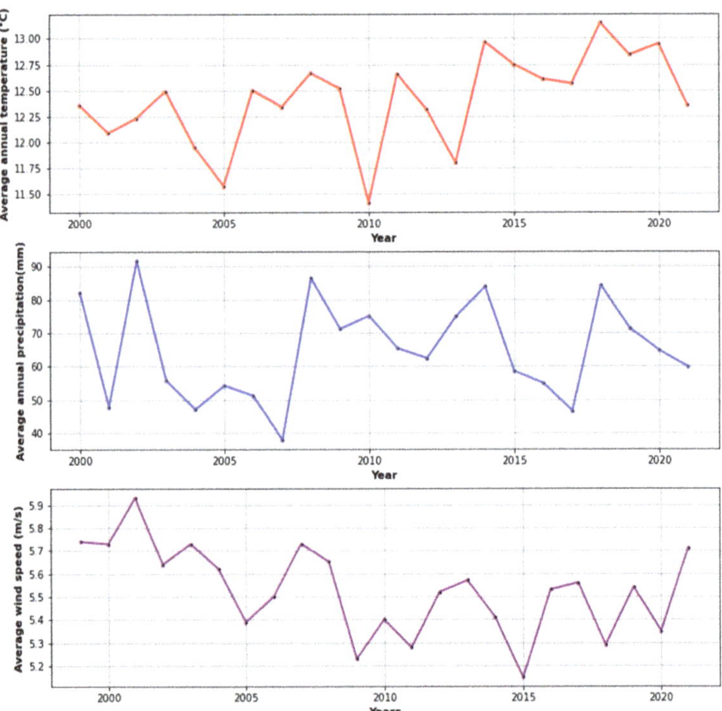

Figure 3. Average annual variability of climate data (precipitation, temperature, and wind) in the study area over the last 20 years.

France's southeast is one of the country's most heavily populated regions. According to INSEE's 2019 census statistics (Figure 4), the research's area population distribution is heterogeneous, reaching around 5200 inhabitants/km^2 in the south (along the coastline), particularly in important towns such as Marseille, Toulon, and Aix-en-province. In contrast, population density in the region's north is modest, with values lower than 100 inhabitants/km^2. Consequently, the high population density in southeastern France leads to high energy and electricity usage. However, with only one wind farm, as indicated in Figure 1, satisfying electricity needs using wind is not possible, prompting us to identify potential places for further wind project execution.

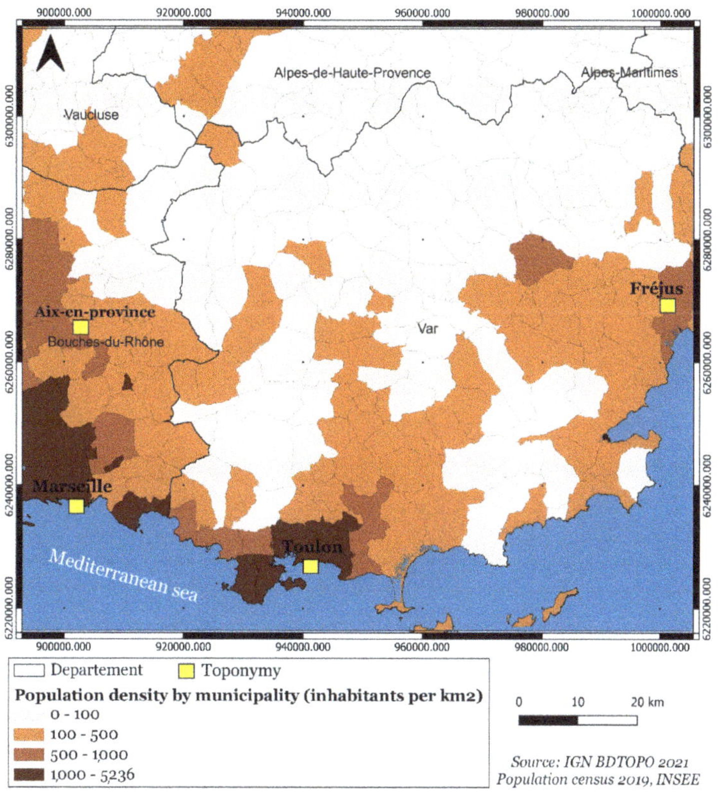

Figure 4. Research area's population density (inhabitants/km^2).

3. Methodology

To map suitable sites for wind farm construction in southeast France, we adopted the methodology presented in Figure 5.

3.1. Data Collection

This research's approach mobilizes the whole set of data that determines onshore wind project placement planning. The data collected covers socioeconomic, environmental, and technical parameters (Table 1).

Table 1. Data collection and their sources.

Data	File Format	Source
Wind speed	Grid	Global Wind Atlas 3.0 [33,34]
Digital elevation (STRM)	Grid	U.S. Geological Survey available at https://earthexplorer.usgs.gov, accessed on 15 July 2022
Protected area	Shapefile	BD TOPO IGN [35]
Road network	Shapefile	BD TOPO IGN [35]
Electrical Substation	Shapefile	BD TOPO IGN [35]

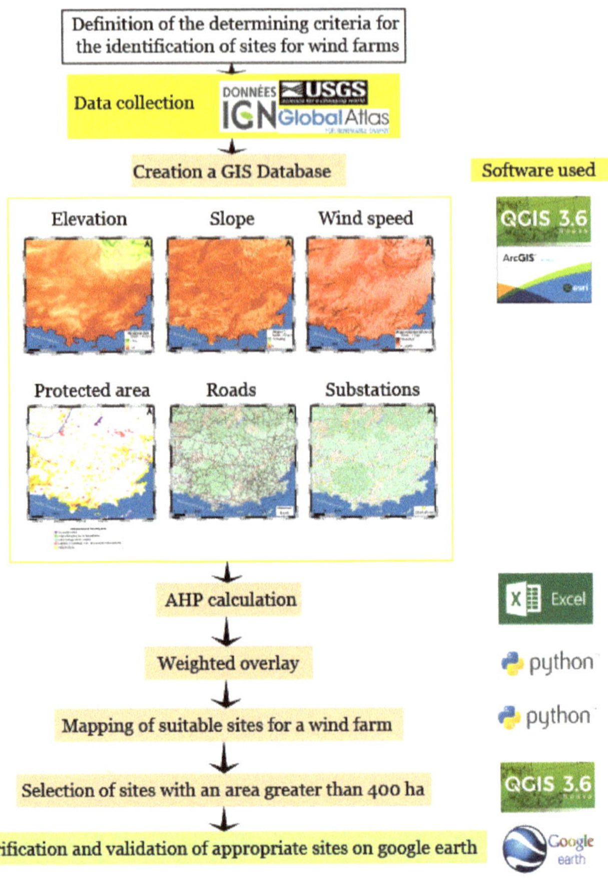

Figure 5. Study flowchart illustrating modeling strategy.

The average wind speed raster data with a spatial resolution of 300 m was obtained from the "Global wind speed" website (https://globalwindatlas.info/, accessed on 15 July 2022). These data are based on ten years of hourly measurements recorded at a height of 100 m (2001–2010). Subsequently, the digital terrain model (DTM) retrieved from the USGS website was employed to generate a mosaic of two DTM rasters covering the whole research region with a spatial resolution of 30 m. Moreover, the IGN 2021 topo database [35] was used to collect information on protected areas (urbanized areas, industrial or commercial areas, infrastructures and equipment, continental waters), road networks (departmental, national, highways, railroads), and electrical substations. All data are resampled in 26 m.

3.2. GIS-Based Spatial Database Creation

Elevation is an important criterion; however, in numerous studies, high altitudes have not been indicated for wind projects [36–38]. The researchers mentioned have proposed that locations below 1000 m be considered extremely appropriate for wind projects.

This is due to access issues and a lack of basic infrastructure in higher places. Thus, to reduce the high expenses connected with construction, regions lower than 1000 m in height appear to be the most efficient and appropriate. As shown in Figure 6a, elevations above 1000 m account for less than 30% of the research region's total area.

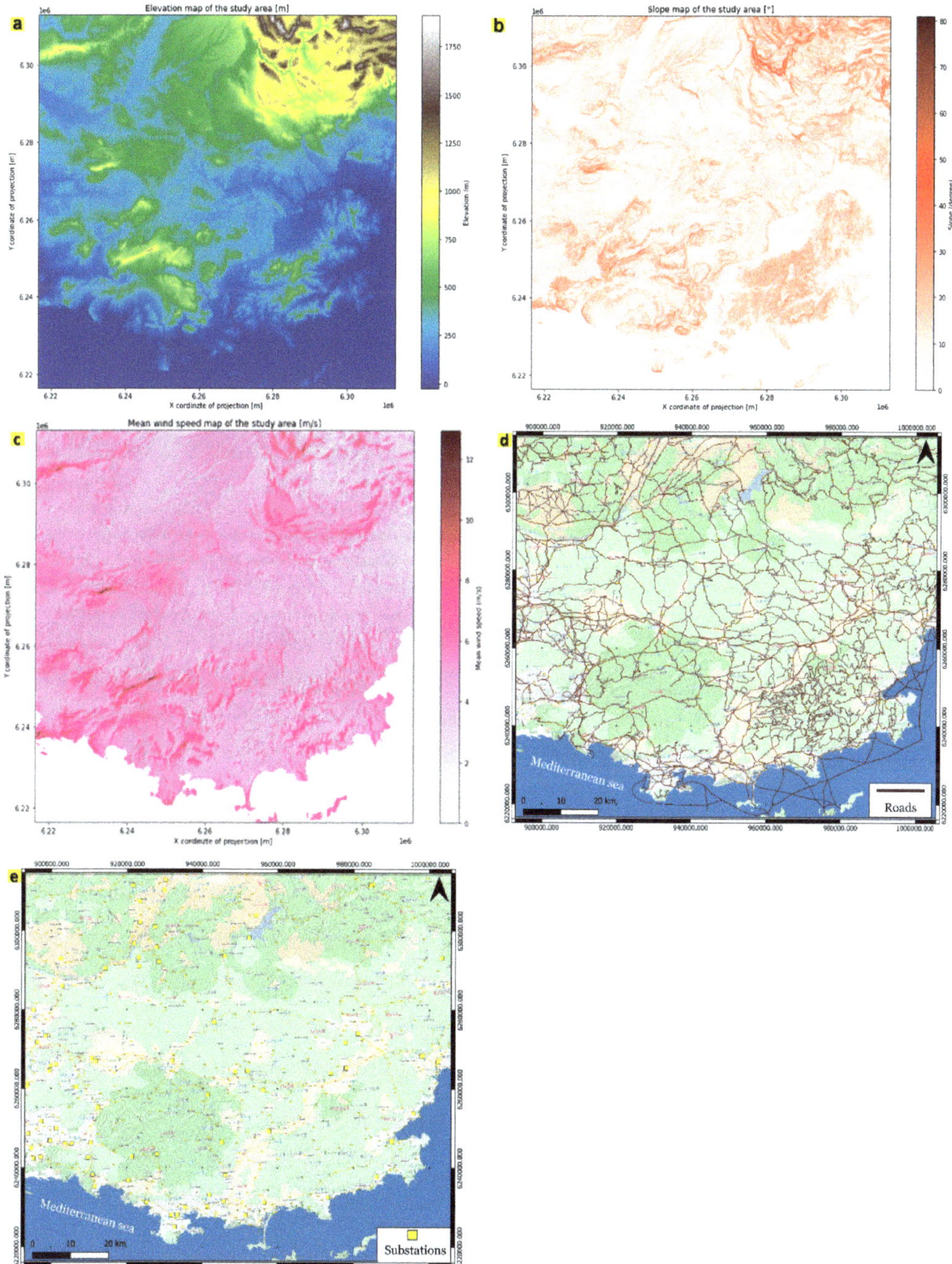

Figure 6. (**a**) Elevation map (m); (**b**) average wind speed map (m/s); (**c**) slope map (°); (**d**) road network; (**e**) locations of power plants and substations in the study area.

Wind speed: Average wind speed has been the most essential and weighted parameter in wind farm location evaluation studies, as reported in most previous studies [24,28,39,40].

This parameter is directly related to the project's profitability [41]. In our wind farm siting analysis, areas, sites having an average annual wind speed of less than 5 m/s at a height of 100 m above mean sea level were considered inappropriate for wind farm sitings, as recommended by [28,42]. Nonetheless, several studies suggest that an annual average wind speed of more than 6 m/s is required for a functional wind farm installation [43,44]. Conversely, extremely high wind speeds can damage the wind turbines and the project execution in general.

Slope: A slope map in degrees is produced by combining two SRTM rasters acquired from the USGS website. This criterion can be applied to exclude areas with steep slopes of greater than 15 degrees and high relief. These are typically inaccessible and so unsuitable for wind turbines. The highest slopes, as illustrated in Figure 6c, are in the research's area northeast, towards the province of Alpes de Haute, and also surround the shoreline in the south.

Indeed, our choice of 15% (maximum limit of suitable slopes) has already been defined by research works [45,46]; others have adopted a constraint of 25% [47], while some [48,49] have raised the constraint threshold to 30%. In addition, some researchers have considered areas with slopes greater than 10% as infeasible areas for wind turbine installation [50,51]. Selection of land having a slope of less than 15% is planned to facilitate crane and truck accessibility to sites and to reduce installation and maintenance costs due to turbulence.

Protected area: Wind farm construction is controlled by various laws, most notably the French energy code, the urban planning code, and the environmental code. Any prospective wind project needs to evaluate its environmental impact by including parameters such as landscape impact, biodiversity, noise, and dangers to nearby inhabitants. The protected areas in this study, which include urban areas, wetlands, biodiversity parks, and water surfaces (Figure 7), were gathered from the IGN's BD 2021.

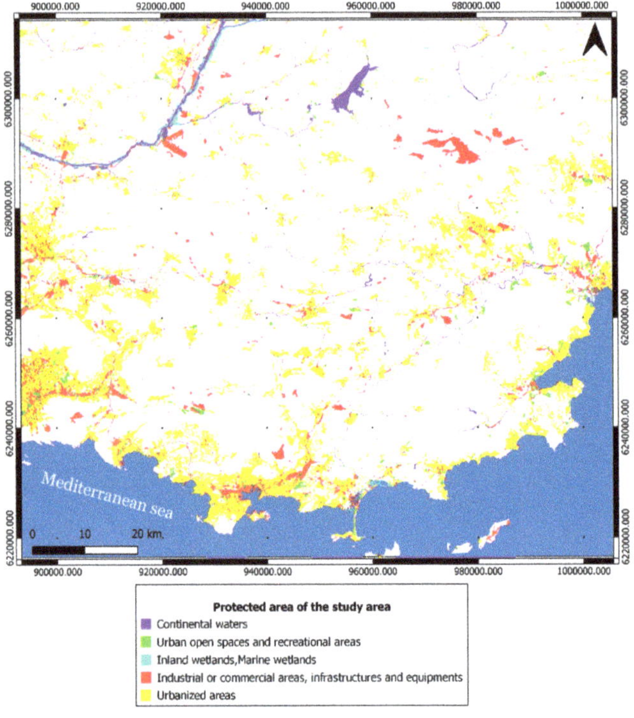

Figure 7. Protected areas of the study area.

The study area is characterized by the presence of forests, pastures, beautiful landscapes, biodiversity parks (fauna and flora), NATURA 2000 protected sites [52,53], and the most important large urban agglomerations. These areas of environmental interest were not absolutely excluded according to the literature but also according to the national legislation (the minimum distances were determined after the decision approving the environmental conditions ("DAEC")). To avoid the destruction of these spaces and the negative impact of wind farms on the nature of these areas, a minimum distance of 2000 m is required [54,55].

Road network: This was generated using data from the IGN 2021 database (Figure 6d). The road's proximity is a critical parameter in various studies. It is particularly relevant for studies related to the search for suitable sites for a large project implementation requiring massive equipment to keep transportation costs, as well as construction and maintenance costs, low [24,39,40,56].

Electrical substations: Close proximity to electrical substations minimizes wire costs, prevents power losses, and simplifies installation and maintenance processes [57]. Figure 6e represents electrical substations in the research area.

3.3. MCDM Using an AHP Approach

Suitable site selection for implementation of a sensitive project such as a wind farm is always difficult since it requires a combination of various parameters and criteria defining the project location. Therefore, decision-making solutions to overcome these obstacles have been developed by integrating all of these determining criteria. Generally, the multi-criteria decision-making (MCDM) approach is always applied to address problems with many stakeholders, criteria, and objectives [58]. Moreover, this approach has been widely applied in various fields, including the energy sector to plan renewable energy projects [59–62]. The analytic hierarchy process (AHP) is a well-known MCDA approach that was initially proposed by [63] and has subsequently been greatly improved.

The methods for weighting the criteria in the MCDA are diverse. Some of these methods include AHP, fuzzy measures [64], Analytic Network Process (ANP) [65], Swara [66], entropy [67], Dematel [68], and standard deviation [69]. Although these methods are quite limited, AHP is one of the most essential and widely used methods in MCDA. The AHP method is similar to Swara's in that the expert's opinion specifies the importance and prioritization of alternatives. As for the entropy method, there are two different views of this method. According to some studies, entropy is reliable and effective [70]. However, from another point of view, entropy results do not always take into account the importance of the indices [71]. Dematel is similar to the Swara method, except that the Dematel approach is used to solve extremely complex problems. In the Dematel decision process, the expert opinion is used to develop the pairwise comparison matrix, and it has three main characteristics. The attributes are compensatory and independent of each other. Qualitative attributes are transformed into quantitative attributes [66]. The Swara and Dematel methods have been widely used in MCDA problems, especially in the renewable energy sector [72–74]. In this study, the AHP method was employed to address site selection problems for several reasons:

It is commonly used for its ease of design and implementation. It is highly compatible with GIS, which is widely used for planning and spatial analysis of site selection problems. The consideration of the consistency and inconsistency of alternatives is one of the main advantages of this method [60].

AHP can be combined with other methods of multicriteria analysis, genetic algorithms, neural networks, etc. [60]. It also takes into account quantitative and qualitative criteria to interpret the problem [75].

The AHP method can apply various sensitivity analyses to the criteria. AHP facilitates the decision-making process, using pairwise comparison between criteria [60]. For site selection problems, in which the main objective is to select the best locations, simple approaches such as AHP are satisfactory, and more complex approaches such as Fuzzy-AHP do not necessarily lead to distinct results [76].

AHP is a structured decision support method that is primarily focused on sophisticated computations with matrix algebra [77,78]. Through this approach, a decomposition of a complicated decision-making issue into a top-down hierarchical structure can be carried out in most cases. In recent years, as geographic information system technologies have improved, GIS integration with MCDA approaches has become increasingly popular. This integration is adaptable and suited to the qualitative and quantitative investigation of multi-criteria issues with a geographical component.

In this research, we developed the decision process required for the usage of the AHP approach. This approach is provided in four steps, each of which requires clear problem identification or study's objective.

Step 1: Deconstruct the decision-making problem and explain its main characteristics or components (criteria, sub-criteria, options, etc.). Then, using a limited number of levels, create a linear hierarchy of concerns (Figure 8). Each level has a set number of selection criteria. The aim is expressed at the most fundamental level. Subsequently, the second and third layers comprise the criterion and sub-criteria. The bottom of the hierarchy is allotted to alternatives.

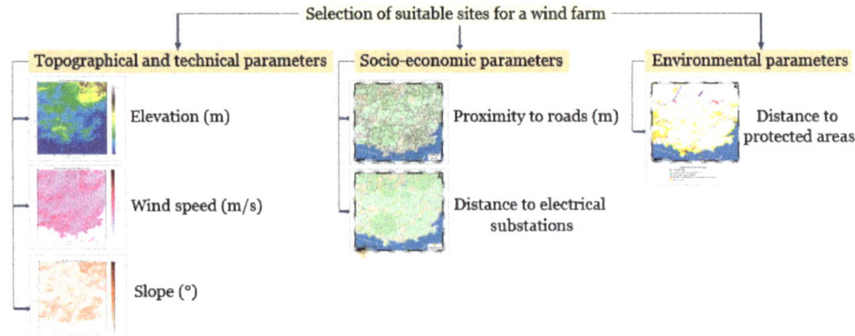

Figure 8. Hierarchical structure of wind farm-related factors and site selection criteria.

Step 2: Design the judgment matrix and pairwise comparison matrices for each criterion. Based on the Saaty scale (Table 2), the pairwise comparisons are grouped into a matrix using the following criteria:

$$
A = [aij] = \begin{array}{c} C1 \\ C2 \\ \vdots \\ Cn \end{array} \begin{pmatrix} C1 & C2 & \cdots & Cn \\ 1 & a12 & \cdots & a1n \\ 1/a12 & 1 & \cdots & a2n \\ \vdots & \cdots & 1 & \vdots \\ 1/a1n & 1/a2n & \cdots & 1 \end{pmatrix} \tag{1}
$$

In relation to the comparisons of two criteria C1 and C2, we designate an important value of the evaluation element "*a*". We place the "*a*" value in the cell column "*i*" and line "*j*" of an important criterion. Then, we need to place the value ratio "1/*a*" in the cell considered less important of the comparison. C1, C2, and *Cn* are the comparison criteria in row "*i*" and column "*j*", which correspond to the comparison values *Ci* and *Cj*. The entries *aij* are often taken from the ratio scale (1/9-9) [79]. The matrix's element semantic description is provided in Table 3.

Table 2. Saaty's comparison scale.

Rating Scale	Definition	Description
1	Equal importance	Two requirements are of equal values
3	Moderate importance on one over another	Experience slightly favors one requirement over another
5	Essential of strong importance	Experience strongly favors one requirement over another
7	Very strong importance	A requirement is strongly favored, and its dominance is demonstrated in practice
9	Extreme importance	the evidence favoring one over another is of the highest possible order of affirmation
2, 4, 6, 8	Intermediate values between the two adjacent judgement	When compromise is needed

Table 3. Pairwise comparison matrix.

Criteria	(1)	(2)	(3)	(4)	(5)	(6)
Slope (1)	1	2	1/6	1/5	1/4	1/4
Elevation (2)	1/2	1	1/7	1/6	1/5	1/5
Wind speed (3)	6	7	1	2	4	4
Distance to protected areas (4)	5	6	1/2	1	3	3
Distance from power stations (5)	4	5	1/4	1/3	1	1
Distance to Roads (6)	4	5	1/4	1/3	1	1
Total	20.5	26	2.31	4.03	9.45	9.45

Using the evaluations provided in the previous step, each hierarchy element's relative relevance was determined. Furthermore, the eigenvector problem is addressed to establish each matrix's element priority.

First, compute the sum of each jth column value as follows:

$$Sum(i) = \sum_{i=1}^{n} aij \tag{2}$$

Subsequently, a normalized comparison matrix $n \times n$ aij^* is generated, in which each aij in the matrix is divided by the sum of its jth column, as expressed in Equation (3):

$$aij^* = \frac{aij}{sum(i)} \tag{3}$$

The weights' ith criterion is then computed as follows:

$$Wi = \frac{\sum_{j=1}^{1} aij}{n} \text{ for all k} = 1, 2, 3, \ldots, n \tag{4}$$

Step 3: The individual criteria weights are calculated using the eigenvalue procedure's pairwise comparison matrices. The eigenvalue λmax is calculated by multiplying each column value by the criteria weight as follows:

$$ai = [\prod_{i=1}^{n} wiaij] = [dij]n - n \tag{5}$$

Then, using the following equation, we determine the weighted sum value Sw by adding the sum of each preceding matrix's rows ai:

$$Swi = \sum_{j=1}^{n} dij \tag{6}$$

Eventually, for each row, the ratio between the weighted value sum Sw and the weighting criterion is calculated as follows:

$$Ratio\ i = \frac{Swi}{wi} \tag{7}$$

By averaging the *ratio i* we obtain the highest eigenvalue max.

Step 4: Calculate the consistency ratio CR (Equation (8)). The final criteria weights are validated using this ratio. Discrepancies in the comparison matrix are identified at this stage:

$$CR = \frac{CI}{RI} \tag{8}$$

where the consistency index CI is calculated as follows:

$$CI = \frac{\gamma max - n}{n - 1} \tag{9}$$

The value of the RI varies with the size of the matrix. Table 4 shows the RI values according to the number of criteria chosen.

Table 4. Random Consistency Index (*RI*), [80].

n	2	3	4	5	6	7	8	9	10	11
RI	0	0.58	0.90	1.12	1.24	1.32	1.41	1.45	1.49	1.51

The RC should be lower than 10% to determine that the pairwise comparison evaluations are consistent. If this is not the case, the matrix should be updated, and the element values re-evaluated.

The weighted findings (Table 5) indicate the most important parameters in wind farm development. The wind's existence is the greatest driver of wind energy, with a relative weight of 38% (Table 4), followed in second place by protected regions with a relative weight of 26%. However, distances to power plants and road networks have a significant influence, with relative weights of 13% each, while slope and elevation have the lowest relative weights of 5% and 3%, respectively. Despite their low weights, these criteria should be considered in all wind projects to avoid or minimize potential negative impacts. This weighting choice is based on our study area's good knowledge.

Table 5. Evaluation criterion weighting.

	(1)	(2)	(3)	(4)	(5)	(6)	Weight %
Slope (1)	0.05	0.08	0.07	0.05	0.03	0.03	5.03
Elevation (2)	0.02	0.04	0.06	0.04	0.02	0.02	3.47
Wind speed (3)	0.29	0.27	0.43	0.50	0.42	0.42	38.96
Distance to protected areas (4)	0.24	0.23	0.22	0.25	0.32	0.32	26.24
Distance from power stations (5)	0.20	0.19	0.11	0.08	0.11	0.11	13.15
Distance to Roads (6)	0.20	0.19	0.11	0.08	0.11	0.11	13.1

Consistency measure = 6.26, CR = 0.04, CI = 0.05.

After completing all of the AHP calculation processes, the following step is to normalize the criteria (Table 6). The vector data are then converted to a raster format, and the matrices are reclassified into two groups (adequate: code 1; and inadequate: code 0).

Table 6. Standardization table for selected criteria.

Criteria	Suitable: Score 1	Unsuitable: Score 0
Slope	<15 degrees	>15 degrees
Elevation	<1000 m	>1000 m
Wind speed	>5 m/s	<5 m/s
Distance to protected areas	>2000 m	<2000 m
Distance from power stations	<1500 m	>1500 m
Distance to roads	<2000 m	>2000 m

3.4. Weighted Superposition

The weighted overlay tool is one of the most frequently used methods for solving multi-criteria problems, such as site selection and suitability models. For instance, users can use this functionality to combine several spatial layers with varied weights to produce a final result. Each raster layer is assigned a weight in the suitability analysis. The raster layer values are re-ranked on a scale (two classes in our case). In this study, the weighted overlay analysis was utilized to identify the most suitable and appropriate sites for future wind farm siting based on the AHP-derived weights assigned to each evaluated parameter. According to Equation (10), all selected criteria in raster format that have been reclassified to equal size (number of columns equal to the number of rows) (Figure 9) are combined into a single raster layer (Figure 10). Weighted overlay is defined as follows:

$$WOA = \sum_{i=1}^{n} Wi * Ri \tag{10}$$

where Wi is the weight of a specific choice criterion, Ri is the criterion's matrix layer, and n is the number of decision criteria.

In total, 962,612 pixels define the wind turbine installation sites. The research area has a total size of 9319 km^2. The detected locations cover an area of 650,725,712 m^2, or 651 km^2. This accounts for roughly 6.98% of the study area's surface.

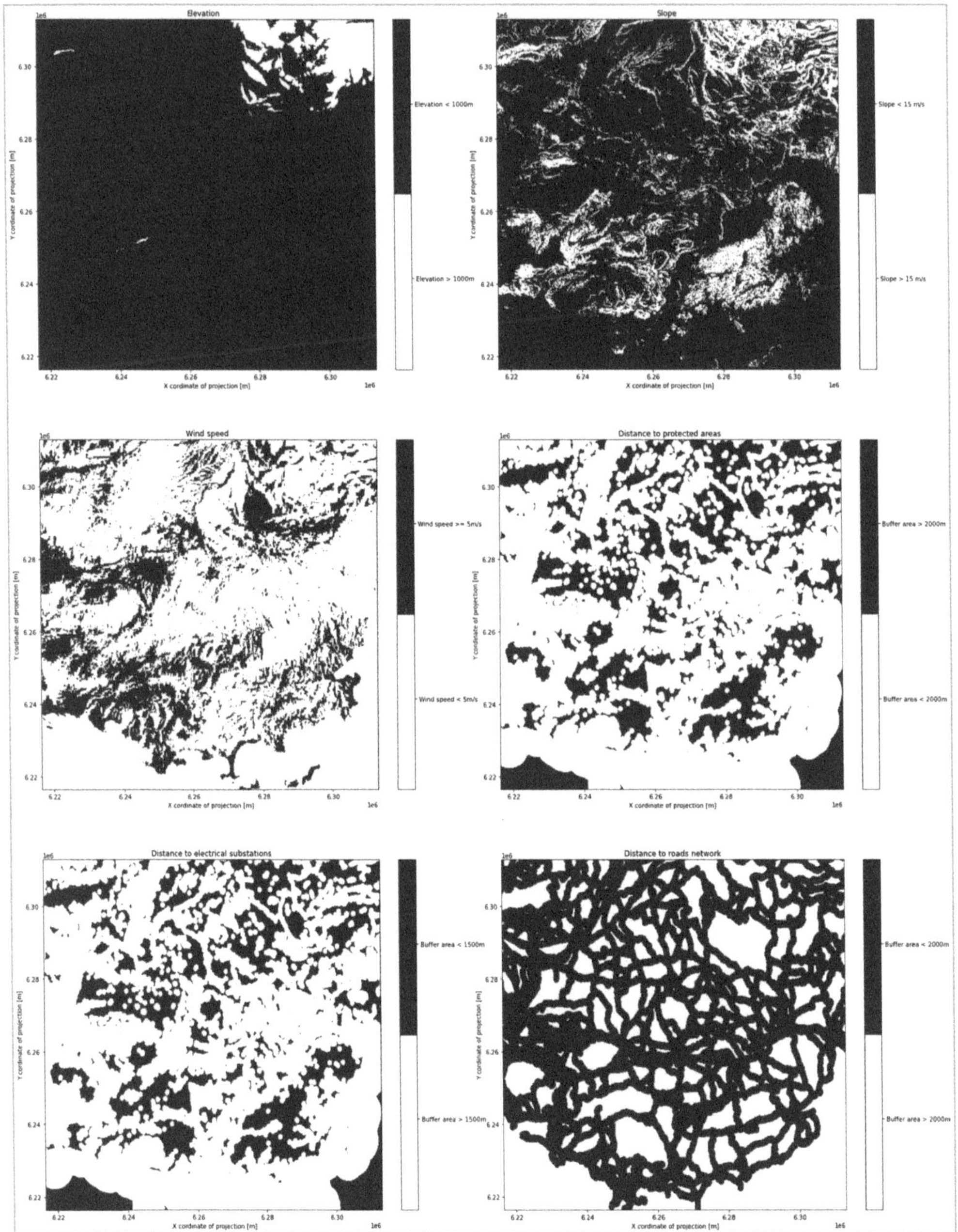

Figure 9. Each weighted criterion's reclassified rasters (in blue: appropriate, in white: inappropriate).

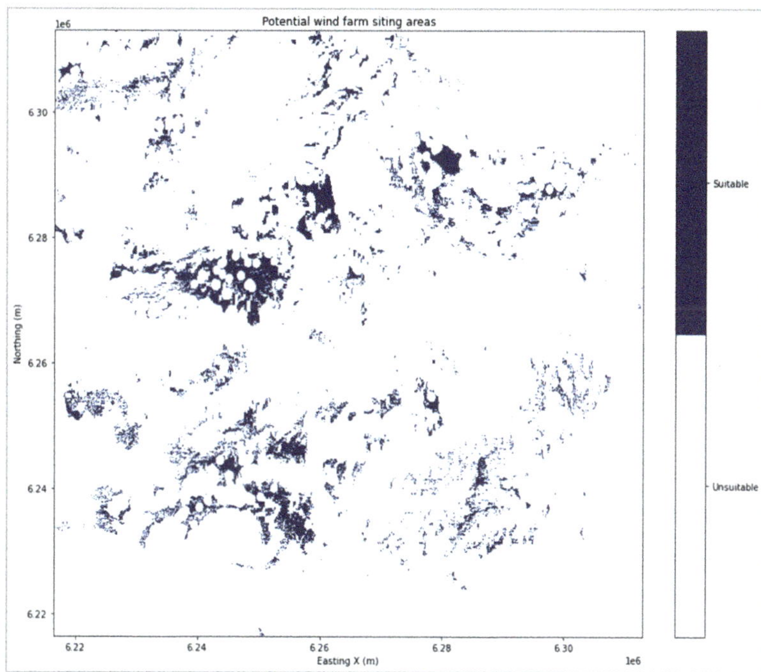

Figure 10. Reclassified raster's weighted overlay map.

4. Results

AHP factor weights were computed using technical, environmental, and economic requirements for wind turbines in France. Factor weights used to evaluate appropriate sites for wind farm installation are shown in Figure 11. As can be seen in Figure 9, wind speed is the most important factor, with a weight of 38%. It is followed by the respect for buffer zones around protected areas (urban areas, wetlands, biodiversity parks, etc.) with a weight of 26%, and the proximity of electrical substations and the road network with a weight of 13% each. Slope and elevation are ranked last, with weights of 5% and 4%, respectively. It should be noted that the eigenvalue max (max = 6.26) is calculated after computing criteria's weights. *CI* and *CR* values are 0.05 and 0.04, respectively. The *CR* value is 10%, suggesting that the research was satisfactory.

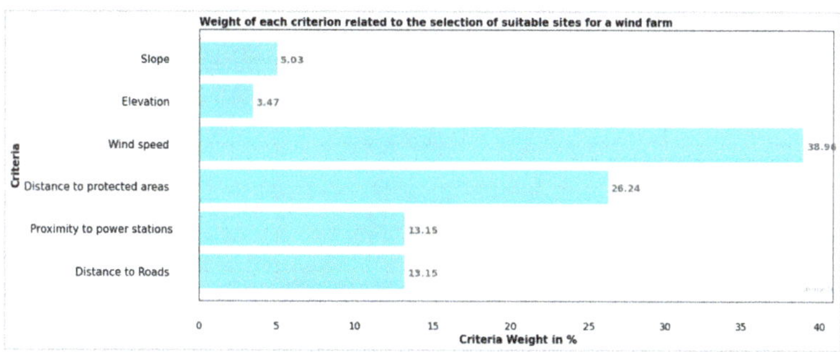

Figure 11. Decision criteria priority weights for selecting suitable sites for future wind projects.

Figure 12 depicts the appropriate location distribution for planning future onshore wind farms in various research regions' departments. Figure 10 only shows locations greater in size than 400 hectares and the road network and the electricity substation's locations. Furthermore, calculating the eligible site's surface shows that 74.62%, or 35,127.92 hectares, is in the Var department, which controls more than 80% of the study area's surface. Only 10%, or 4962 hectares, of Alpes-de-Haute-Provence department is suitable for future wind farm development, compared to 13.70%, or 6449 hectares, in Bouches-de-Rhones. In the Alpes-Maritimes, however, 1.45%, or 535 hectares, is protected (Figure 13).

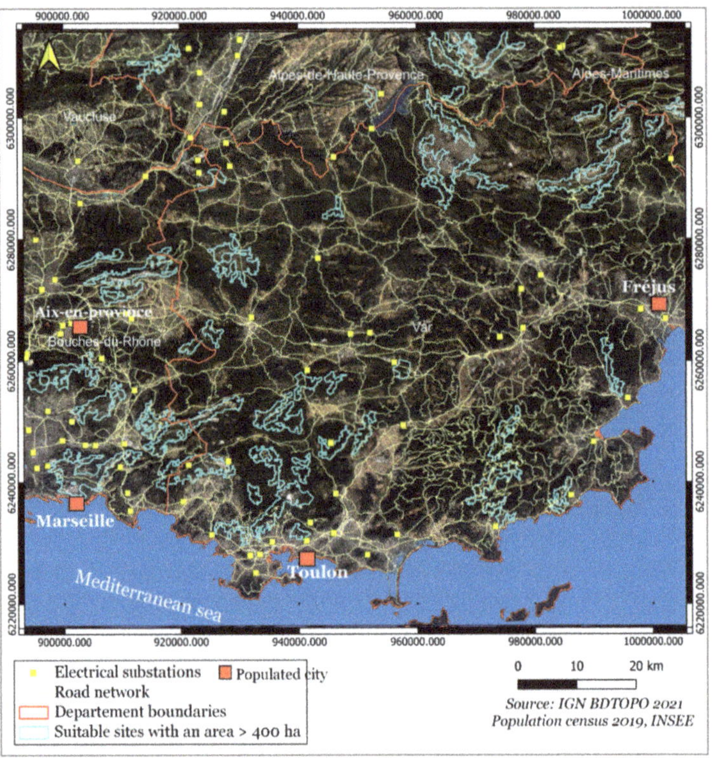

Figure 12. Maps showing potential locations for future onshore wind farms in southern France's Provence-Alpes-Côte d'Azur region.

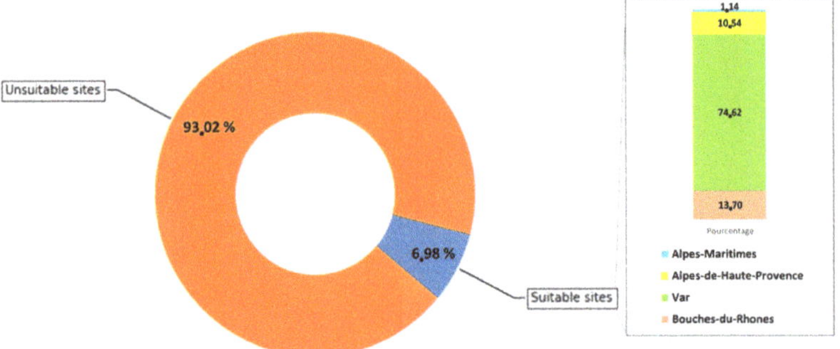

Figure 13. Suitable site percentage distribution (%) for future wind project implementation by department.

Figure 14 demonstrates that 1121 hectares, or 12.07% of the area suitable for future wind farm construction, have an average wind speed greater than 5 m/s, which is required for wind turbine development. Furthermore, 11.24% of the area, or 1044 hectares, is located lower than 1500 m above sea level, while just 8% is on slopes less than 15 degrees; 5.92% is next to roadways, 2.34% is near electrical substations, and 7.91%, or 735 hectares, is outside of protected areas.

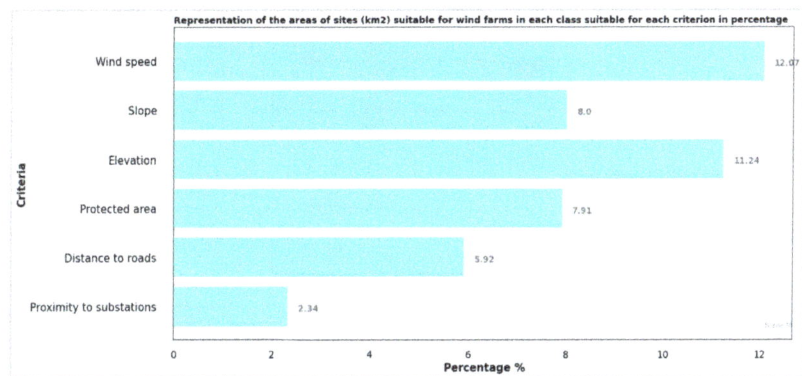

Figure 14. Representation of the percentage regions in the "appropriate = score 1" class for each criterion for potential wind farm locations.

On Google Earth imagery, the selected appropriate site locations for wind farm development were projected (Figure 15). Four locations were recommended, and their selection was based on their unique characteristics (area, location, elevation, slope, wind speed, accessibility, closeness to electrical substations, etc.) as well as their proximity to populous regions while respecting buffer zones relative to protected areas. Onshore wind turbines in France typically have a power range of 1.8 to 3 MW, with rotor diameters ranging from 80 to 110 m and total heights ranging from 80 to 155 m. In fact, a 2 MW wind turbine generates 4200 MWh per year, which is roughly equivalent to the average electricity consumption of over 800 French households [8]. France is classified by the International Electrotechnical Commission (IEC) as having strong winds with high average turbulence intensity. Some wind turbine types that are easily useable in the French market have been chosen in accordance with IEC design criteria. Table 7 contains detailed information about the wind turbine types and their attributes.

Table 7. Theoretical potential of wind energy on highly suitable land.

Manufacturer	Wind Turbine Model	Rotor Diameter (m)	Capacity (MW)	$7\,d \times 5\,d$ Area (Km^2)	Area Factor (MW/Km^2)	Theoretical Wind Power Potential (MW)
Vesta	V110-2.0	110	2.0	0.424	4.72	182.60
GE	1.6 to 82.5 WT	82.5	1.6	0.238	6.72	259.98
Vent Inox	93 RD + 80 HH	93	2.0	0.303	6.60	255.26
ReGen Powertech	VENSYS-77	77	1.5	0.207	7.25	280.20

Surface in km^2 of the selected sites very suitable (Figure 11) = 38.67 km^2.

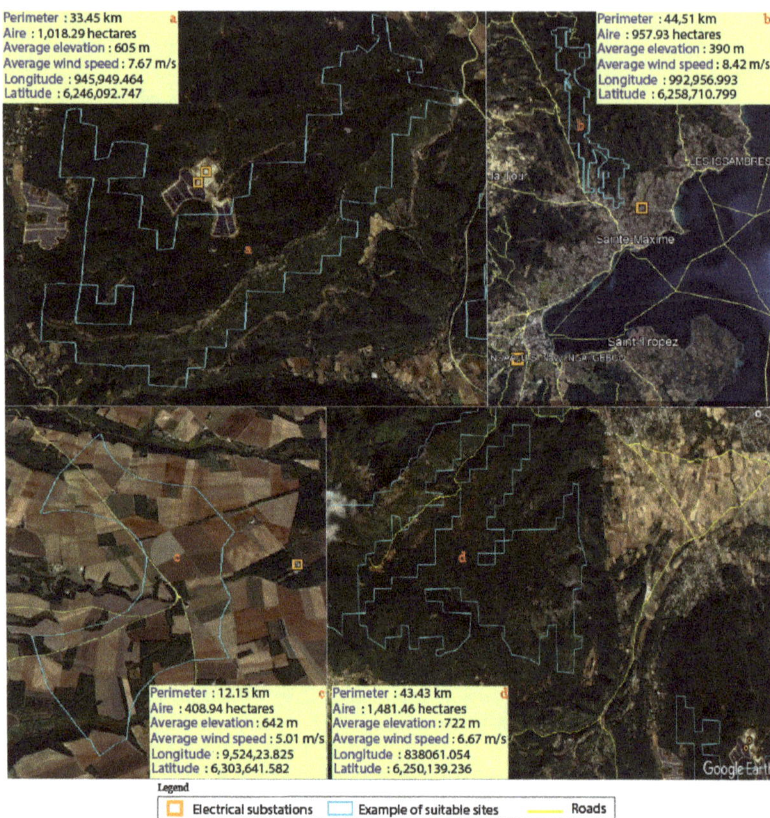

Figure 15. Example of the most suitable sites selected and their characteristics (geographical coordinates of their centroids, area in hectares, perimeter in km, average wind speed, and average altitude) for the development of onshore wind farms.

Theoretical wind power potential may be evaluated using Equation (11) based on wind turbine output capacity, rotor diameter, and total area of appropriate land [81–83].

$$TWPP = TA * AF \tag{11}$$

$TWPP$ is theoretical wind power potential (MW), TA is the total area of the four appropriate locations (km^2) (Figure 13), and AF is the area factor (MW/km^2). Our computations were performed on wind turbines that were situated 7d × 5d apart, where d is the rotor diameter.

Based on the theoretical wind potential calculations, the four proposed sites for future wind turbine installations may generate between 182.60 and 280.20 MW of electricity. This energy benefit will suit the study region's population demands in terms of power consumption, heating, and transportation.

5. Discussion

In 2021, the wind sector in France grew in relevance, accounting for 7% of the country's net power generation. Furthermore, wind power now accounts for 7.7% of total consumption [32]. More than half of France's wind farms are concentrated in two regions: Hauts-de-France and Grand Est (Figure 16), with an almost complete absence in the country's southeast.

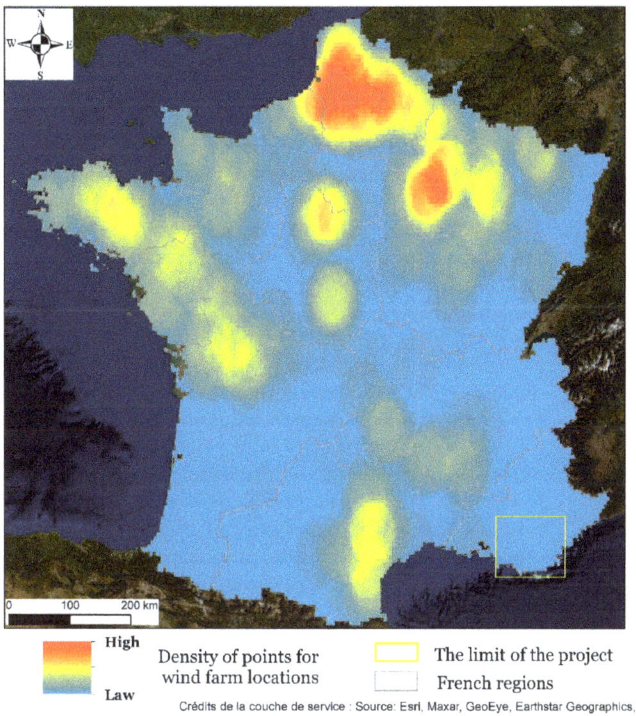

Figure 16. Wind farm point density in France's spatial distribution.

Wind power production (wind farms), according to the International Energy Agency, is very unevenly distributed among areas. Despite the study region possessing the country's most populated cities, it has the lowest energy productivity (201 GWh in 2021). However, the study region (Provence-Alpes-Côte d'Azur, region code: 13) ranks second to last in terms of wind energy generation [60], with 77 wind turbines and a very low installed capacity of 99 MW, compared to demand (Figure 17).

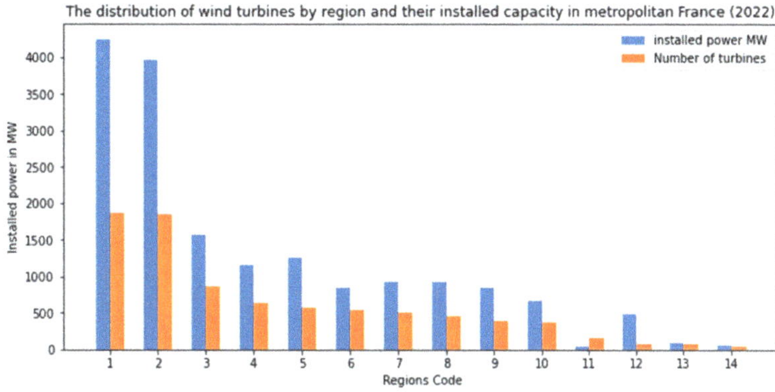

Figure 17. Wind turbine distribution by region and their installed capacity in metropolitan France (1: Hauts-de-France; 2: Grand Est; 3: Occitanie; 4: Nouvelle-Aquitaine; 5: Centre-Val de Loire; 6: Bretagne; 7: Pays de la Loire 8: Bourgogne-Franche-Comté; 9: Normandie; 10: Auvergne-Rhone-Alpes; 11 Guadeloupe; 12: Auvergne-Rhone-Alpes; 13: Provence-Alpes-Cote d'Azure; 14: ile-de-France).

The study area's onshore potential is greatly limited by certain constraints, including the requirement to avoid exclusion zones imposed by environmental protection areas, historical perimeters, and the requirement to build more than 500 m from homes, as well as habitat dispersion, which reduces the percentage of territory eligible for wind power. This research was conducted for all of these reasons. In addition, it may assist various governmental agencies, policymakers, researchers, and investors in planning and developing wind energy projects in this difficult location.

Regarding criteria, the Ministers of Ecological Transition, Territorial Cohesion, and Energy Transition previously investigated a set of documents and reports on technical requirements, regulations, and environmental and urban planning issues related to wind turbine development in France, which served as the foundation for our criteria and constraints. The thresholds, on the other hand, must be closely tied to certain location characteristics. As a result, the criteria suggested in this study were used with considerable caution. They are currently being explored by wind planning professionals. Practical experience in the subject field is also advantageous for the assessment of visual findings [30].

In addition to the French regulations concerning the determining criteria for land use planning of future wind farm projects, a consultation of confidential reports and a discussion with experts and former researchers in the field of wind farm planning was carried out. The values for each criterion were selected according to French legislation. As no studies have been published on this topic in France, we also based our selection on research undertaken in Europe (e.g., Greece) as mentioned above. Other means can be used to define the important criteria, such as filling in questionnaires by experts in the field. Interviews with experts could also be an effective solution for the determination of criteria.

Furthermore, restriction criteria for onshore wind farm planning in France are well defined; only one scenario is required in this case: average wind speed greater than 5 m/s, altitude less than 1500 m, slope less than 15%, proximity to roads (2 km) and electrical substations (1.5 km), and at least 2 km from protected areas.

Unsurprisingly, as shown in Figure 18a, there is a substantial correlation between areas of high average wind speed and selected site locations as suitable for wind farms. Indeed, the project limits were previously chosen based on a wind speed map in France, which is a region where the average yearly wind speed (at least 50 m above ground level) exceeds 5 m/s. These locations are all accessible by national, regional, or occasionally freeway roads, and the majority of them are near electricity substations (Figure 18d). Furthermore, 30% of the locations are at high altitudes (over 1500 m) (Figure 18c). This is due to the high wind speed in these high-altitude areas, as well as high weight assigned to the wind parameter, and the low weight assigned to the elevation parameter. The selected sites are more than 2 km away from the protected regions (Figure 18e). Since the Southern Alps surround the northern and northwestern parts of the research region, various sites are on steep slopes. Nonetheless, we were able to choose really good locations on moderate slopes (Figure 15). Other criteria, such as acceptance of these installations by populations and associations; administrative procedures and their validation by local authorities; energy demand in these territories; and pre-existing installation replacement, can all have an impact on decisions to install wind turbines. Indeed, the criteria required for wind farm construction, namely minimum average wind speed required, minimum acreage required, closeness of highways and electrical substations, and distance from protected areas, are all gathered at the identified sites. Some research employed the same choice criteria with various limitation levels based on the area and state restrictions. For example, [30] used the same criteria but with different limitation values because the study was conducted in Spain. Nonetheless, three situations were investigated, each with distinct limitation values and weights for the criterion. This scenario-based method is useful when there is little information or rigid wind-level limitations. However, [61] conducted a study to identify viable places for developing onshore wind projects in a rural zone using only four characteristics (urban area or habitat, vegetation, slope, and wind speed). Furthermore, they investigated three scenarios. The multi-criteria analysis technique used, however, is fuzzy rather than AHP. In summary, the multi-criteria technique used in the research varied, as did the number of scenarios examined, the number of criteria addressed, and weights assigned to the criteria. According to the findings of all studies on decision making for project implementation or multi-factor problem solving using GIS-based multi-criteria analysis approaches [62,84,85], when no restriction values are well defined by the state or agreed upon by experts in the field in question, it is recommended that several alternatives be implemented.

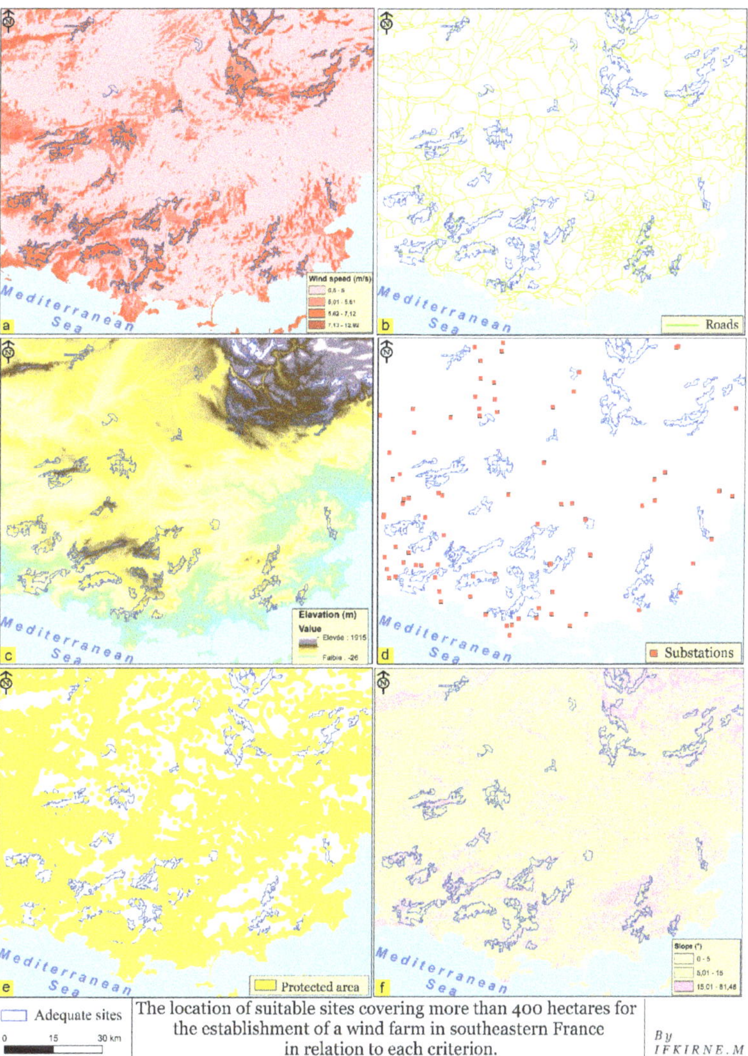

Figure 18. Suitable site locations for future wind energy projects about each decision parameter: (**a**): average wind speed; (**b**): accessibility to roads; (**c**): elevation; (**d**): proximity to substations; (**e**): buffer to protected areas; (**f**): slope.

The number and total area of suitable sites vary by region. For example, in our case, only 7% of the total area of the study area is suitable for the development of future wind projects. Even though it is a difficult location with various environmental, topographical, and urban restrictions, research and planning for wind farms in this area with great wind potential are still on hold. However, based on the theoretical wind potential calculation equation (Equation (10)), each of the sites depicted in Figure 18 produces an average of 45 MW, which is more than adequate to fulfill the demands of the local population.

6. Conclusions

This research aimed to offer a method for identifying potential sites for future wind energy projects based on geographic information systems (GIS) and multi-criteria decision-making (AHP), as well as to contribute to the literature on renewable energy planning. To the best of the author's knowledge, this is the first research of its kind in France. Thus, this research was carried out in a region of France's southeast that has high wind energy potential but is also the least productive. Another reason for choosing this region was to overcome the numerous constraints that limited the region's energy output. For wind farm siting, six criteria were adopted, practically addressing in full the economic, technical, and socio-environmental challenges associated with these facilities and uses. Most of the criteria were based on worldwide literature, in addition to French wind turbine legislation.

According to the findings, many sites were identified as suitable for wind farms. Visual and manual analyses were performed on these sites to choose those with an area larger than 400 ha, a high average wind speed, accessibility by roads, proximity to electrical substations, and a distance from protected areas. Four sites with an average installed capacity of 45 MW were selected and must be confirmed by the appropriate state authorities. The decision tool provided in this article may be utilized in any part of the world by adapting it to the specific characteristics of each territory, as well as the distinct needs and policies.

Although the results presented in this paper are specific to France, the methodology presented provides an interesting reference model that can be transposed and adapted with relative ease. This assumes that the different constraints and criteria are adapted to the specific needs of energy planners and to the particularities of each study area.

Despite the quality and reliability of the IGN database used, the methodology followed (including the choice of the MCDA method), and the analysis performed in this work, certain aspects are to be recommended for future projects, such as taking into account the knowledge of the study area and the regulations put in place by the government concerned, and the relevant choice of determination (decision) criteria and their weightings, which often vary between experts' opinions and from one country to another.

The spatial resolution of the data used is also an important element, especially for topographic data (DEM) and wind data.

In addition, a validation of the identified suitable sites could lead to a more robust and real interpretation of the results, either through aerial photography (drone) or a field visit to these sites.

Future studies could consider extending the proposed method to investigate the theoretical energy potential of wind generation in order to benefit from their complementarity and overcome the inherent intermittency of renewable energies. This study can also be the starting point for a project to install wind turbines or solar panels in the study area by simply replacing the wind variable with the temperature variable. Furthermore, this study can also contribute to the creation of new investments and, consequently, new jobs in the region.

Author Contributions: M.I.: Original project, Conceptualization, Methodology, Software, Writing—Original version, Visualization, Reviewing and Editing. H.E.B.: Conceptualization, Methodology, Software. S.A.: Formal analysis, original version, Conceptualization, Writing, Supervision, Revision. Q.B.P.: Formal analysis, original version, Conceptualization, Writing, Supervision, Revision. A.F. and N.T.T.L.: Formal analysis, Revision, Writing. All authors have read and agreed to the published version of the manuscript.

Funding: The authors have no relevant funding to acknowledge.

Data Availability Statement: The data that support the findings of this study are available upon reasonable request.

Acknowledgments: The authors would like to thank the many data authors: The National Institute of Geographic and Forestry Information (IGN) for the BD TOPO®, a current database containing the following data: Administration (boundaries and administrative units); Addresses (postal addresses); Buildings (constructions); Hydrography (water-related features); Places (place or locality with a place name and describing a natural space or inhabited place); Land use (vegetation, foreshore); Services and activities (utilities, energy storage and transport, industrial places and sites); Transportation (road, rail or air infrastructures); Regulated areas (most areas are subject to specific regulations). Transport (road, rail or air infrastructure); Restricted areas (most areas are subject to specific regulations). The National Institute of Statistics and Economic Studies (INSEE). The Global Wind Atlas 3.0, a free web-based application developed, owned, and operated by the Technical University of Denmark (DTU) in partnership with the World Bank Group, using data provided by Vortex, with funding provided by the Energy Sector Management Assistance Program (ESMAP). For more information: https://globalwindatlas.info.

References

1. Criqui, P.; Kouvaritakis, N. World energy projections to 2030. *Int. J. Glob. Energy Issues* **2000**, *14*, 116. [CrossRef]
2. Foley, J.A.; DeFries, R.; Asner, G.P.; Barford, C.; Bonan, G.; Carpenter, S.R.; Chapin, F.S.; Coe, M.T.; Daily, G.C.; Gibbs, H.K.; et al. Global consequences of land use. *Science* **2005**, *309*, 570–574. [CrossRef] [PubMed]
3. Ceballos, G.; Ehrlich, P.R.; Dirzo, R. Biological annihilation via the ongoing sixth mass extinction signaled by vertebrate population losses and declines. *Proc. Natl. Acad. Sci. USA* **2017**, *114*, E6089–E6096. [CrossRef] [PubMed]
4. Steffen, W.; Richardson, K.; Rockström, J.; Cornell, S.E.; Fetzer, I.; Bennett, E.M.; Biggs, R.; Carpenter, S.R.; De Vries, W.; De Wit, C.A.; et al. Planetary boundaries: Guiding human development on a changing planet. *Science* **2015**, *347*, 1259855. [CrossRef]
5. Giamalaki, M.; Tsoutsos, T. Sustainable siting of solar power installations in Mediterranean using a GIS/AHP approach. *Renew. Energy* **2019**, *141*, 64–75. [CrossRef]
6. Shorabeh, S.N.; Firozjaei, M.K.; Nematollahi, O.; Firozjaei, H.K.; Jelokhani-Niaraki, M. A risk-based multi-criteria spatial decision analysis for solar power plant site selection in different climates: A case study in Iran. *Renew. Energy* **2019**, *143*, 958–973. [CrossRef]
7. Han, Y.; Tan, S.; Zhu, C.; Liu, Y. Research on the emission reduction effects of carbon trading mechanism on power industry: Plant-level evidence from China. *Int. J. Clim. Change Strateg. Manag.* **2022**. *ahead-of-print*. [CrossRef]
8. Ministère de la Transition Écologique et de la Cohésion des Territoires et Ministère de la Transition Énergétique (MTECT & MTE). Les Énergies Renouvelables. 2022. Available online: https://www.ecologie.gouv.fr/energies-renouvelables (accessed on 10 July 2022).
9. Hepbasli, A.; Ozgener, O. A review on the development of wind energy in Turkey. *Renew. Sustain. Energy Rev.* **2004**, *8*, 257–276. [CrossRef]
10. Kamdar, I.; Ali, S.; Taweekun, J.; Ali, H.M. Wind Farm Site Selection Using WAsP Tool for Application in the Tropical Region. *Sustainability* **2021**, *13*, 13718. [CrossRef]
11. Shu, Z.; Li, Q.; Chan, P.W. Statistical analysis of wind characteristics and wind energy potential in Hong Kong. *Energy Convers. Manag.* **2015**, *101*, 644–657. [CrossRef]
12. Østergaard, P.A.; Duic, N.; Noorollahi, Y.; Kalogirou, S.A. Recent advances in renewable energy technology for the energy transition. *Renew. Energy* **2021**, *179*, 877–884. [CrossRef]
13. Baseer, M.; Rehman, S.; Meyer, J.; Alam, M. GIS-based site suitability analysis for wind farm development in Saudi Arabia. *Energy* **2017**, *141*, 1166–1176. [CrossRef]
14. Réseau de Trasnport et d'Électricité (RTE), Futurs Énergétiques 2050: Principaux Résultats. 2021. Available online: https://assets.rte-france.com/prod/public/2021-10/Futurs-Energetiques-2050-principaux-resultats_0.pdf (accessed on 10 July 2022).
15. Réseau de Transport et D'Éléctricité (RTE) & Référentiel Éolien Terrestre National (RETN). Carte des Éoliennes en Service au Sein de la France Métropolitaine. 2021. Available online: https://www.data.gouv.fr/fr/datasets/retn-referentiel-eolien-terrestre-national-carte-des-eoliennes-en-service-au-sein-de-la-france-metropolitaine-hors-corse/ (accessed on 16 July 2022).
16. Taoufik, M.; Fekri, A. GIS-based multi-criteria analysis of offshore wind farm development in Morocco. *Energy Convers. Manag. X* **2021**, *11*, 100103. [CrossRef]
17. Amjad, F.; Shah, L.A. Identification and assessment of sites for solar farms development using GIS and density based clustering technique- A case of Pakistan. *Renew. Energy* **2020**, *155*, 761–769. [CrossRef]

18. Barzehkar, M.; Parnell, K.E.; Dinan, N.M.; Brodie, G. Decision support tools for wind and solar farm site selection in Isfahan Province, Iran. *Clean Technol. Environ. Policy* **2020**, *23*, 1179–1195. [CrossRef]
19. Elboshy, B.; Alwetaishi, M.; Aly, R.M.H.; Zalhaf, A.S. A suitability mapping for the PV solar farms in Egypt based on GIS-AHP to optimize multi-criteria feasibility. *Ain Shams Eng. J.* **2022**, *13*, 101618. [CrossRef]
20. Tercan, E.; Eymen, A.; Urfalı, T.; Saracoglu, B.O. A sustainable framework for spatial planning of photovoltaic solar farms using GIS and multi-criteria assessment approach in Central Anatolia, Turkey. *Land Use Policy* **2021**, *102*, 105272. [CrossRef]
21. Wilson, T.N.E.; Camille, K.M. Identification des sites favorables à l'installation des centrales solaires photovoltaïques à l'aide de l'analyse multicritères et des SIG: Cas de l'arrondissement de Bélabo, Cameroun. *Int. J. Innov. Appl. Stud.* **2019**, *26*, 938–952. Available online: https://www.proquest.com/openview/c0a8941925ffca23139ea86c22156205/1?pq-origsite=gscholar&cbl=2031961 (accessed on 17 September 2022).
22. Koc, A.; Turk, S.; Şahin, G. Multi-criteria of wind-solar site selection problem using a GIS-AHP-based approach with an application in Igdir Province/Turkey. *Environ. Sci. Pollut. Res.* **2019**, *26*, 32298–32310. [CrossRef]
23. Mahdy, M.; Bahaj, A.S. Multi criteria decision analysis for offshore wind energy potential in Egypt. *Renew. Energy* **2018**, *118*, 278–289. [CrossRef]
24. Moradi, S.; Yousefi, H.; Noorollahi, Y.; Rosso, D. Multi-criteria decision support system for wind farm site selection and sensitivity analysis: Case study of Alborz Province, Iran. *Energy Strat. Rev.* **2020**, *29*, 100478. [CrossRef]
25. Ayodele, T.; Ogunjuyigbe, A.; Odigie, O.; Munda, J. A multi-criteria GIS based model for wind farm site selection using interval type-2 fuzzy analytic hierarchy process: The case study of Nigeria. *Appl. Energy* **2018**, *228*, 1853–1869. [CrossRef]
26. Diemuodeke, E.; Addo, A.; Oko, C.; Mulugetta, Y.; Ojapah, M. Optimal mapping of hybrid renewable energy systems for locations using multi-criteria decision-making algorithm. *Renew. Energy* **2018**, *134*, 461–477. [CrossRef]
27. Rehman, S.; Baseer, M.; Alhems, L. GIS-based multi-criteria wind farm site selection methodology. *FME Trans.* **2020**, *48*, 855–867. [CrossRef]
28. Spyridonidou, S.; Vagiona, D.G. A comparative analysis of decision-making methods on site suitability for on- and offshore wind farms: The case of regional unit of Euboea, Greece. *Circ. Econ. Sustain.* **2021**, 1–14. [CrossRef]
29. Feloni, E.; Karandinaki, E. GIS-based MCDM Approach for Wind Farm Site Selection-A Case Study. *J. Energy Power Technol.* **2021**, *3*, 39. [CrossRef]
30. Díaz-Cuevas, P. GIS-based methodology for evaluating the wind-energy potential of territories: A case study from Anda-lusia (Spain). *Energies* **2018**, *11*, 2789. [CrossRef]
31. Malczewski, J. *GIS and Multicriteria Decision Analysis*; John Wiley & Sons: Hoboken, NJ, USA, 1999.
32. Wikipédia. Climat de la France. 2017. Available online: https://fr.wikipedia.org/wiki/Climat_de_la_France (accessed on 22 July 2022).
33. NASA Prediction of Worldwide Energy Resources. The POWER Project. Available online: https://power.larc.nasa.gov/ (accessed on 15 September 2022).
34. Denmark. The Global Wind Atlas 3.0, T.U.o. Editor, 2019.
35. IGN. BD TOPO®. 2021. Available online: https://geoservices.ign.fr/bdtopo (accessed on 17 July 2022).
36. Yousefi, H.; Motlagh, S.G.; Montazeri, M. Multi-Criteria Decision-Making System for Wind Farm Site-Selection Using Geographic Information System (GIS): Case Study of Semnan Province, Iran. *Sustainability* **2022**, *14*, 7640. [CrossRef]
37. Atici, K.B.; Simsek, A.B.; Ulucan, A.; Tosun, M.U. A GIS-based Multiple Criteria Decision Analysis approach for wind power plant site selection. *Util. Policy* **2015**, *37*, 86–96. [CrossRef]
38. Konstantinos, I.; Georgios, T.; Garyfalos, A. A Decision Support System methodology for selecting wind farm installation locations using AHP and TOPSIS: Case study in Eastern Macedonia and Thrace region, Greece. *Energy Policy* **2019**, *132*, 232–246. [CrossRef]
39. Gorsevski, P.V.; Cathcart, S.C.; Mirzaei, G.; Jamali, M.M.; Ye, X.; Gomezdelcampo, E. A group-based spatial decision support system for wind farm site selection in Northwest Ohio. *Energy Policy* **2013**, *55*, 374–385. [CrossRef]
40. Simao, A.; Densham, P.J.; Haklay, M.M. Web-based GIS for collaborative planning and public participation: An application to the strategic planning of wind farm sites. *J. Environ. Manag.* **2009**, *90*, 2027–2040. [CrossRef] [PubMed]
41. Sunak, Y.; Madlener, R. The impact of wind farm visibility on property values: A spatial difference-in-differences analysis. *Energy Econ.* **2016**, *55*, 79–91. [CrossRef]
42. Schallenberg-Rodríguez, J.; Montesdeoca, N.G. Spatial planning to estimate the offshore wind energy potential in coastal regions and islands. Practical case: The Canary Islands. *Energy* **2018**, *143*, 91–103. [CrossRef]
43. Holland, G.B.; Provenzano, J.J. *The Hydrogen Age: Empowering a Clean-Energy Future*; Gibbs Smith: Layton, UT, USA, 2007.
44. Opalek, C. *Wind Power Fraud*; Lulu.com: Morrisville, NV, USA, 2010.
45. Noorollahi, Y.; Yousefi, H.; Mohammadi, M. Multi-criteria decision support system for wind farm site selection using GIS. *Sustain. Energy Technol. Assess.* **2016**, *13*, 38–50. [CrossRef]
46. Villacreses, G.; Gaona, G.; Martínez-Gómez, J.; Jijón, D.J. Wind farms suitability location using geographical information system (GIS), based on multi-criteria decision making (MCDM) methods: The case of continental Ecua-dor. *Renew. Energy* **2017**, *109*, 275–286. [CrossRef]
47. Latinopoulos, D.; Kechagia, K. A GIS-based multi-criteria evaluation for wind farm site selection. A regional scale application in Greece. *Renew. Energy* **2015**, *78*, 550–560. [CrossRef]

48. Tegou, L.-I.; Polatidis, H.; Haralambopoulos, D.A. Environmental management framework for wind farm siting: Methodology and case study. *J. Environ. Manag.* **2010**, *91*, 2134–2147. [CrossRef]

49. Höfer, T.; Sunak, Y.; Siddique, H.; Madlener, R. Wind farm siting using a spatial Analytic Hierarchy Process ap-proach: A case study of the Städteregion Aachen. *Appl. Energy* **2016**, *163*, 222–243. [CrossRef]

50. Sotiropoulou, K.F.; Vavatsikos, A.P. Onshore wind farms GIS-Assisted suitability analysis using PROMETHEE II. *Energy Policy* **2021**, *158*, 112531. [CrossRef]

51. Watson, J.J.; Hudson, M.D. Regional Scale wind farm and solar farm suitability assessment using GIS-assisted multi-criteria evaluation. *Landsc. Urban Plan.* **2015**, *138*, 20–31. [CrossRef]

52. Greek Legislation. *Law 4432/B*; Government Gazette: Athens, Greece, 2017.

53. European Commission. Natura 2000—Environment. Available online: https://ec.europa.eu/environment/nature/natura2000/index_en.htm (accessed on 13 September 2022).

54. Gharaibeh, A.A.; Al-Shboul, D.A.; Al-Rawabdeh, A.M.; Jaradat, R.A. Establishing Regional Power Sustainability and Feasibility Using Wind Farm Land-Use Optimization. *Land* **2021**, *10*, 442. [CrossRef]

55. Zahedi, R.; Ghorbani, M.; Daneshgar, S.; Gitifar, S.; Qezelbigloo, S. Measuring Iran's western regional wind power potential using GIS. *J. Clean. Prod.* **2022**, *330*, 129883. [CrossRef]

56. Janke, J.R. Multicriteria GIS modeling of wind and solar farms in Colorado. *Renew. Energy* **2010**, *35*, 2228–2234. [CrossRef]

57. Gavériaux, L.; Laverrière, G.; Wang, T.; Maslov, N.; Claramunt, C. GIS-based multi-criteria analysis for offshore wind turbine deployment in Hong Kong. *Ann. GIS* **2019**, *25*, 207–218. [CrossRef]

58. Kumar, A.; Samadder, S.R. A review on technological options of waste to energy for effective management of municipal solid waste. *Waste Manag.* **2017**, *69*, 407–422. [CrossRef]

59. Algarín, C.R. An Analytic Hierarchy Process Based Approach for Evaluating Renewable Energy Sources. 2017. Available online: http://hdl.handle.net/11159/1258 (accessed on 9 September 2022).

60. Nedjar, S. Répartition des Éoliennes en France: Découvrez le Classement. Hello Watt. 2022. Available online: https://www.hellowatt.fr/blog/etude-eoliennes-terrestres-france/ (accessed on 20 July 2022).

61. Uzar, Ş.M. Suitable map analysis for wind energy projects using remote sensing and GIS: A case study in Turkey. *Environ. Monit. Assess.* **2019**, *191*, 459. [CrossRef]

62. Abdelouhed, F.; Ahmed, A.; Abdellah, A.; Yassine, B.; Mohammed, I. Using GIS and remote sensing for the mapping of potential groundwater zones in fractured environments in the CHAOUIA-Morocco area. *Remote Sens. Appl. Soc. Environ.* **2021**, *23*, 100571. [CrossRef]

63. Saaty, T.L. *Optimization in Integers and Related External Problems*; McGraw-Hill: New York, NY, USA, 1970.

64. Ishii, K.; Sugeno, M. A model of human evaluation process using fuzzy measure. *Int. J. Man-Mach. Stud.* **1985**, *22*, 19–38. [CrossRef]

65. Saaty, T.L.; Vargas, L.G. The Analytic Network Process. In *Decision Making with the Analytic Network Process*; Springer: Boston, MA, USA, 2013; pp. 1–40. Available online: https://link.springer.com/chapter/10.1007/978-1-4614-7279-7_1 (accessed on 12 September 2022).

66. Alinezhad, A.; Khalili, J. *New Methods and Applications in Multiple Attribute Decision Making (MADM)*; Springer: Cham, Switzerland, 2019; Volume 277. Available online: https://link.springer.com/book/10.1007/978-3-030-15009-9 (accessed on 12 September 2022).

67. Zhu, Y.; Tian, D.; Yan, F. Effectiveness of Entropy Weight Method in Decision-Making. Mathematical Problems in Engineering. 2020. Available online: https://www.hindawi.com/journals/mpe/2020/3564835/ (accessed on 12 September 2022).

68. Si, S.L.; Vous, X.Y.; Liu, H.C.; Zhang, P. DEMATEL Technique: Une revue systématique de la littérature de pointe sur les méthodologies et les applications. *Math. Probl. Ing.* **2018**, *2018*, 3696457.

69. Chang, Y.S. Cartes de contrôle multivariiées CUSUM et EWMA pour les populations asymétriques utilisant des écarts-types pondérés. *Commun. Stat. Simul. Calcul.* **2007**, *36*, 921–936. [CrossRef]

70. Lu, X.; Li, L.Y.; Lei, K.; Wang, L.; Zhai, Y.; Zhai, M. Water quality assessment of Wei River, China using fuzzy synthetic evaluation. *Environ. Earth Sci.* **2010**, *60*, 1693–1699. [CrossRef]

71. Cui, Y.; Feng, P.; Jin, J.; Liu, L. Water resources carrying capacity evaluation and diagnosis based on set pair analysis and improved the entropy weight method. *Entropy* **2018**, *20*, 359. [CrossRef]

72. Ahmadi, M.H.; Hosseini Dehshiri, S.S.; Hosseini Dehshiri, S.J.; Mostafaeipour, A.; Almutairi, K.; Ao, H.X.; Rezaei, M.; Techato, K. A Thorough Economic Evaluation by Implementing Solar/Wind Energies for Hydrogen Production: A Case Study. *Sustainability* **2022**, *14*, 1177. [CrossRef]

73. Tanackov, I.; Badi, I.; Stević, Ž.; Pamučar, D.; Zavadskas, E.K.; Bausys, R. A Novel Hybrid Interval Rough SWARA–Interval Rough ARAS Model for Evaluation Strategies of Cleaner Production. *Sustainability* **2022**, *14*, 4343. [CrossRef]

74. Meng, R.; Zhang, L.; Zang, H.; Jin, S. Evaluation of Environmental and Economic Integrated Benefits of Photovoltaic Poverty Alleviation Technology in the Sanjiangyuan Region of Qinghai Province. *Sustainability* **2021**, *13*, 13236. [CrossRef]

75. Koundinya, S.; Chattopadhyay, D.; Ramanathan, R. Incorporating qualitative objectives in integrated resource planning: Application of analytic hierarchy process and compromise programming. *Energy Sources* **1995**, *17*, 565–581. [CrossRef]

76. Mosadeghi, R.; Warnken, J.; Tomlinson, R.; Mirfenderesk, H. Comparison of Fuzzy-AHP and AHP in a spatial multi-criteria decision making model for urban land-use planning. *Comput. Environ. Urban Syst.* **2015**, *49*, 54–65. [CrossRef]

77. Aydi, A.; Zairi, M.; Ben Dhia, H. Minimization of environmental risk of landfill site using fuzzy logic, analytical hierarchy process, weighted linear combination methodology in a geographic information system environment. *Environ. Earth Sci.* **2012**, *68*, 1375–1389. [CrossRef]
78. Shahabi, H.; Keihanfard, S.; Bin Ahmad, B.; Amiri, M.J.T. Evaluating Boolean, AHP and WLC methods for the selection of waste landfill sites using GIS and satellite images. *Environ. Earth Sci.* **2013**, *71*, 4221–4233. [CrossRef]
79. Saaty, T. *The Analytic Hierarchy Process (AHP) for Decision Making*; Kube, Japan, 1980; pp. 1–69. Available online: http://www.cashflow88.com/decisiones/saaty1.pdf (accessed on 15 September 2022).
80. Donegan, H.A.; Dodd, F.J. A note on Saaty's random indexes. *Math. Comput. Model.* **1991**, *15*, 135–137. [CrossRef]
81. Anwarzai, M.A.; Nagasaka, K. Utility-scale implementable potential of wind and solar energies for Afghanistan using GIS multi-criteria decision analysis. *Renew. Sustain. Energy Rev.* **2017**, *71*, 150–160. [CrossRef]
82. Bina, S.M.; Jalilinasrabady, S.; Fujii, H.; Farabi-Asl, H. A comprehensive approach for wind power plant potential as-sessment, application to northwestern Iran. *Energy* **2018**, *164*, 344–358. [CrossRef]
83. Saraswat, S.; Digalwar, A.K.; Yadav, S.; Kumar, G. MCDM and GIS based modelling technique for assessment of solar and wind farm locations in India. *Renew. Energy* **2021**, *169*, 865–884. [CrossRef]
84. Abdelouhed, F.; Ahmed, A.; Abdellah, A.; Yassine, B.; Mohammed, I. GIS and remote sensing coupled with analytical hierarchy process (AHP) for the selection of appropriate sites for landfills: A case study in the province of Ouarzazate, Morocco. *J. Eng. Appl. Sci.* **2022**, *69*, 19. [CrossRef]
85. Akay, H. Towards Linking the Sustainable Development Goals and a Novel Proposed Snow Avalanche Susceptibility Mapping. *Water Resour. Manag.* **2022**, *36*, 1–18. [CrossRef]

 *land*

Article

Modelling Floodplain Vegetation Response to Groundwater Variability Using the ArcSWAT Hydrological Model, MODIS NDVI Data, and Machine Learning

Newton Muhury [1,2,*], Armando A. Apan [1,2,3], Tek N. Marasani [2] and Gebiaw T. Ayele [4,*]

[1] School of Civil Engineering and Surveying, University of Southern Queensland, Toowoomba, QLD 4350, Australia
[2] Institute for Life Sciences and the Environment, University of Southern Queensland, Toowoomba, QLD 4350, Australia
[3] Institute of Environmental Science and Meteorology, University of the Philippines Diliman, Quezon City 1101, Philippines
[4] Australian Rivers Institute and School of Engineering and Built Environment, Griffith University, Nathan, QLD 4111, Australia
* Correspondence: newton.muhury@usq.edu.au (N.M.); gebiaw.ayele@griffithuni.edu.au or gebeyaw21@gmail.com (G.T.A.)

Citation: Muhury, N.; Apan, A.A.; Marasani, T.N.; Ayele, G.T. Modelling Floodplain Vegetation Response to Groundwater Variability Using the ArcSWAT Hydrological Model, MODIS NDVI Data, and Machine Learning. *Land* **2022**, *11*, 2154. https://doi.org/10.3390/land11122154

Academic Editor: Chuanrong Zhang

Received: 9 October 2022
Accepted: 22 November 2022
Published: 29 November 2022

Abstract: This study modelled the relationships between vegetation response and available water below the soil surface using Terra's moderate resolution imaging spectroradiometer (MODIS), Normalised Difference Vegetation Index (NDVI), and soil water content (SWC). The Soil & Water Assessment Tool (SWAT) interface known as ArcSWAT was used in ArcGIS for the groundwater analysis. The SWAT model was calibrated and validated in SWAT-CUP software using 10 years (2001–2010) of monthly streamflow data. The average Nash-Sutcliffe efficiency during the calibration and validation was 0.54 and 0.51, respectively, indicating that the model performances were good. Nineteen years (2002–2020) of monthly MODIS NDVI data for three different types of vegetation (forest, shrub, and grass) and soil water content for 43 sub-basins were analysed using the WEKA, machine learning tool with a selection of two supervised machine learning algorithms, i.e., support vector machine (SVM) and random forest (RF). The modelling results show that different types of vegetation response and soil water content vary in the dry and wet seasons. For example, the model generated high positive relationships (r = 0.76, 0.73, and 0.81) between the measured and predicted NDVI values of all vegetation in the sub-basin against the groundwater flow (GW), soil water content (SWC), and combination of these two variables, respectively, during the dry season. However, these relationships were reduced by 36.8% (r = 0.48) and 13.6% (r = 0.63) against GW and SWC, respectively, in the wet season. Our models also predicted that vegetation in the top location (upper part) of the sub-basin is highly responsive to GW and SWC (r = 0.78, and 0.70) during the dry season. Although the rainfall pattern is highly variable in the study area, the summer rainfall is very effective for the growth of the grass vegetation type. The results predicted that the growth of vegetation in the top-point location is highly dependent on groundwater flow in both the dry and wet seasons, and any instability or long-term drought can negatively affect these floodplain vegetation communities. This study has enriched our knowledge of vegetation responses to groundwater in each season, which will facilitate better floodplain vegetation management.

Keywords: ArcSWAT; machine learning; floodplain vegetation; MODIS NDVI; groundwater

1. Introduction

Floodplain vegetation plays an important role in catchment hydrology and energy flow. Floodplain vegetation distribution is directly influenced by several factors, including rainfall, temperature, and groundwater [1]. Rainfall, temperature, and groundwater are highly variable in arid and semi-arid regions [2]. The annual rainfall in arid regions is much less than the annual potential evapotranspiration and surface water flows (i.e., surface runoff), which provides limited water supply for vegetation systems [3]. Therefore, groundwater becomes the only water source in arid regions affecting the spatial and temporal distribution of soil water content (SWC) which, in turn, affects the growth of vegetation [4]. An accurate understanding of the distribution of SWC in arid regions is important since water deficit is gradually becoming one of the major factors limiting agricultural productivity and ecological development [5]. As one of the driest continents in the world, Australia has been facing severe droughts over the last 50 years, noticeably in the south-eastern part of the country [6]. This area will become drier in the coming decades due to increasing annual average temperatures and decreasing rainfall [7]. Therefore, understanding the vegetation response to SWC is critical for sustainable ecosystem improvements in arid regions [8].

SWC can be estimated using both direct and indirect methods. The direct method, such as the oven drying technique, is widely used because of its reliability and simplicity [9]; however, the direct method is labour-intensive, time-consuming, and costly for continuous application in large catchments. On the other hand, hydrological simulation and remote sensing techniques can be used for the same purpose at a catchment or global scale [10]. SWC can also be estimated for previous years using remote sensing techniques, which is not possible to obtain from experimental measurements [10]. Therefore, model-simulated results can fulfil temporal and spatial data requirements and improve SWC and vegetation response relationship studies.

The SWC also influences vegetation productivity and water stress [11,12]. The amount of soil water availability in drought regions for vegetation intake affects the length of the growing period [13]. However, groundwater is the main source of water for vegetation growth in arid regions [14]. Any changes in the groundwater tables decrease the accessibility of the dependent vegetation and may create water stress [15]. Moreover, water stress can trigger a longer growing period and photosynthesis reduction, thereby resulting in reduced productivity and increased vegetation mortality [12]. The reduction in accessible soil water availability under a changing climate may exaggerate ecological droughts during the plantation season [16]. Researchers have identified that the change in groundwater depth affects the vegetation physiology and dynamics [17,18]. Another study also focused on individual vegetation responses by examining the leaf, tree, canopy, and population [19]. However, according to our knowledge, accessible water in soil and vegetation response modelling is still lacking. This research focuses on SWC that is accessible to floodplain vegetation and understanding their relationship in a seasonal context.

The Soil and Water Assessment Tool (SWAT) is a physically based and semi-distributed hydrological model widely used for quantitative hydrological modelling [20,21]. Many researchers have used SWAT for evaluating soil water at the catchment scale [22–24]. Previous studies have shown that changes in the water balance components, specifically soil water storage, evapotranspiration, land use/land cover dynamics, and water yield, are more sensitive under wet climate and heterogeneous soils [25,26]. The SWAT model has also been successfully applied in the U.S. to estimate SWC for drought monitoring and predicting crop production [27]. However, the SWAT application in the Australian region is limited [28]. In our study, a SWAT model was used to estimate SWC for the Burrinjuck Dam sub-catchment within the Murrumbidgee River catchment. The suitability of the model simulation for long-term SWC datasets was assessed using a combination of physically measured and remotely sensed data. This type of simulation helps to correlate with long-term historical vegetation data.

The Normalised Difference Vegetation Index (NDVI), which can be derived from remote sensing, is frequently applied for studies on vegetation dynamics over large scales [29–32]. Researchers used NDVI to understand the relationships between terrestrial vegetation and climate [31]. Several studies found a linear relationship between NDVI and climate variables in arid regions [33–35]. Relationships also were investigated for NDVI and groundwater levels and groundwater flow discharge [36–38]. However, none of these previous studies analysed the relationship between NDVI and hydrological model simulated SWC in an arid region.

This study aims to analyse and model the relationships between seasonal SWC variability and floodplain vegetation responses using MODIS-derived NDVI data and machine learning algorithms for 20 years (2001–2020). The specific objectives of this study are the following: (a) to understand the relationship between different types of vegetation responses (NDVI) and groundwater variables as simulated by the SWAT model at the basin level; (b) to assess the correlation between the vegetation response (as measured by NDVI) and SWAT-simulated variables at different positions (top and bottom) within the sub-basin; and (c) to model seasonal vegetation responses to groundwater variables at the basin level using the WEKA machine learning tool developed by the University of Waikato, New Zealand [39,40].

The WEKA tool is a collection of machine learning algorithms for data mining activities that supports data pre-processing, clustering, classification, regression, and visualization [41]. This software can be run under the General Public License (GNU) with a selected classifier compared to other data mining tools [42].

The results of this study provide qualitative information on catchment hydrology and water resources on temporal and spatial dimensions at the sub-catchment level. A calibrated model at this scale can be used for various analyses such as sedimentation, water pollution, and future stream flow prediction. This study also contributes to developing sustainable water resource management for the dry and wet season in an efficient way. The modelling results may be used to improve domestic agricultural production by selecting appropriate crops and plants that can grow commercially in similar regions. An understanding of seasonal vegetation water requirements from this study can be implemented to review the floodplain water management policies for better water management.

2. Materials and Methods

2.1. Study Area

The study area resides within the Upper Murrumbidgee catchment (Figure 1) in the south-east of the Murray Darling Basin (MDB), in south-eastern Australia. The Burrinjuck Dam catchment area size is 13,000 km^2 (approx.) which is one-seventh that of the Murrumbidgee River catchment [43]. The latitude and longitude of the study area are 34.53° S–35.14° S and 148.31° E–148.55° E. The Burrinjuck Dam is situated within the upper catchment of the Murrumbidgee River basin, which was built (1910–1927) to develop an irrigation project after the devastating drought in 1902. The Murrumbidgee River rises at an altitude of around 1500 m in Kosciuszko National Park and flows approximately 316 km before entering Burrinjuck Reservoir at an altitude of 370 m (approx.). The topography of the Burrinjuck Dam area consists of gentle and moderate slopes and the elevation varies from 370 to 934 m [44]. The upper mountainous section of the Murrumbidgee River flow is regulated by dams for hydroelectric power generation and water supply [45]. The main land use in this part is forest and pasture. However, this area also contributes to agricultural production by growing wheat and cereals [46]. Having a diverse climate in the upper and lower Murrumbidgee, the mean annual rainfall varies 350 mm in the Riverina plains and 1700 mm in the Snowy Mountains [47]. According to the Köppen-Geiger climate classification system, the climate of the study area is temperate, without a dry season mostly hot summer with average 22 °C temperature in the hottest months [48]. The Burrinjuck Dam and surrounding area contribute to the maximum river flow by adding 24% of the total rainfall as runoff [49]. The climate has enriched the Burrinjuck reserve possesses a high diversity of vegetation types and ecosystems.

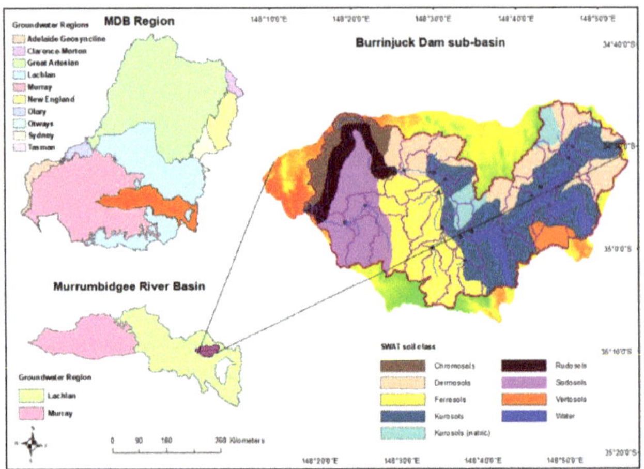

Figure 1. Study area of the Burrinjuck Dam sub-catchment of the Murrumbidgee River catchment within the Murray-Darling Basin region.

2.2. Methods

Figure 2 presents an overview of the research methods applied in this study. The SWC and groundwater flow (GW) were simulated in ArcSWAT. The datasets used in this study were obtained from various local and international data portals, such as the Australian Bureau of Meteorology (BOM) and U.S. Geological Survey (USGS). We used the ArcGIS tool [50] and Microsoft Excel [51] for spatial and attribute data pre-processing and formatted the data to apply in the ArcSWAT hydrological model. We analysed the model output data using the WEKA machine learning tool [52] with different vegetation responses as measured by MODIS NDVI values. Different machine learning algorithms have been applied to model the relationships between vegetation types, and their location within the sub-basin and seasonal groundwater variability.

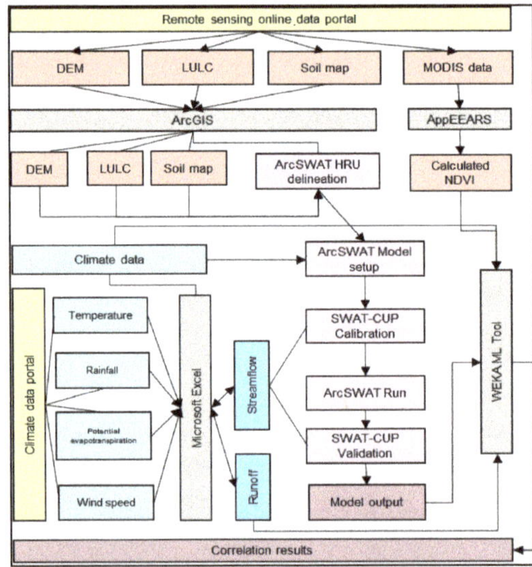

Figure 2. An overview of the research methodology for vegetation responses and groundwater variables modelling using machine learning algorithms.

2.3. Hydrological Model Setup

An ArcSWAT interface of the SWAT2012 model was used in this study [21]. We installed compatible ArcGIS version 10.6 on a desktop to run SWAT2012 from the user interface. The SWAT model is a continuous physically based distributed parameter model that operates on a daily time-step. This model is capable of simulating catchment hydrology, land use impact on water, sediments, plant growing, agricultural-chemical yields, etc., within agricultural watersheds [21,53]. SWAT divides the watershed into multiple sub-basins based on spatial characteristics. These sub-basins are further subdivided into hydrological response units (HRUs) that consist of unique land use, soils, and slope characteristics [54]. Each HRU is simulated for SWC, groundwater flow, nutrient cycles, sedimentation, crop growth, and management practices [44]. The simulated results from the HRUs represent the sub-basin scale. SWAT [53] simulates the hydrological cycle based on the following daily water balance equation:

$$SW_t = SW_0 \sum_{i=0}^{t} \left(R_{day} - Q_{surf} - E_a - W_{seep} - Q_{gw} \right)_i \tag{1}$$

where SW_t is the ultimate water content in (mm), SW_0 is the amount of water content on the first soil of the day i (mm), t is time (days), R_{day} is the amount of rainfall on day i (mm), Q_{surf} is the amount of surface runoff on specific day i (mm), E_a is the amount of evapotranspiration on day i (mm), W_{seep} is the amount of water percolated into the vadose zone from the soil profile on day i (mm), and Q_{gw} is the amount of return flow on day i (mm).

The SWAT model was delineated from a 30 m resolution digital elevation model (DEM) (Figure 3). A threshold drainage area of 1342 km^2 was selected based on the DEM and Murrumbidgee River network to divide the watershed into 43 sub-basins, which were later categorised into 350 HRUs depending on land cover and land use, soil types, and slope. The model was run for 20 years of data, starting from 2001 and ending in 2020. The SWC data for Australia was obtained from the Australian Water Resource Assessment Landscape water balance model (AWRA-L), which was calibrated against the streamflow data. It is not best practice to use data from a different model simulation to run a hydrological model as it may not provide good modelling results. To avoid this confusion, the model was calibrated and validated against observed streamflow data instead of SWC.

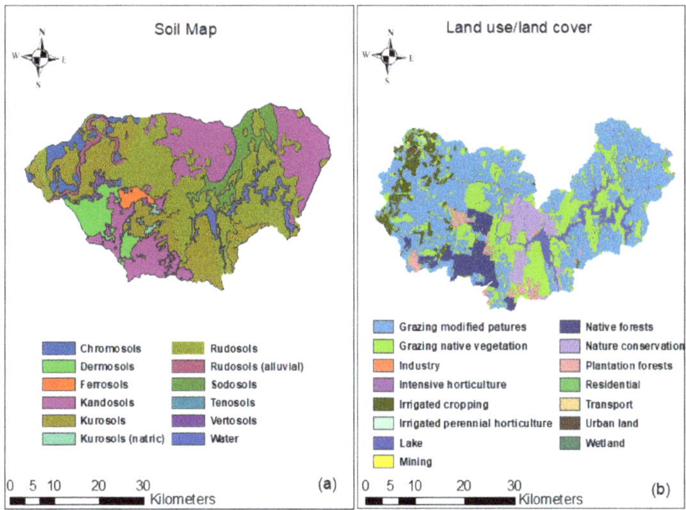

Figure 3. *Cont.*

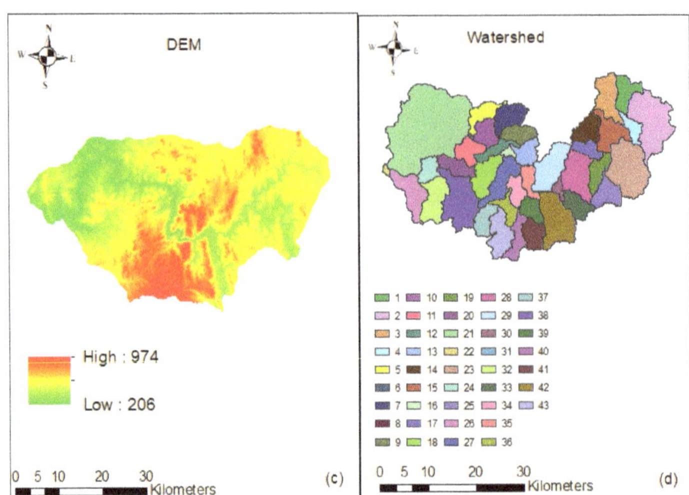

Figure 3. In the above figure four images are captured: (**a**) Study area soil map, (**b**) Land use/land cover map, (**c**) DEM, and (**d**) Delineated watershed.

2.3.1. Data Preparation

A combination of climatological and land properties data were required to develop a semi-distributed model using the ArcSWAT interface (Appendix A). Some data such as DEM, soil, land use, and weather data are mandatory to run the dynamics of the watershed; however, streamflow, reservoir information, sediment, water quality, chemical, and pesticide data are non-mandatory. The data used in this study and their sources are listed in Table 1.

Table 1. The datasets used in this study including their descriptions and sources.

Data	Frequency	Description	Source
Precipitation	Daily	Station gauged, temporal	Bureau of Meteorology
Temperature	Daily	Station gauged, temporal	Bureau of Meteorology
Evapotranspiration	Daily	Satellite-derived, 0.05 degree (approximately 5 × 5 km)	Bureau of Meteorology
Wind speed	Hourly	Station gauged, temporal	Bureau of Meteorology
Runoff	Daily	Satellite-derived, 0.05 degree (approximately 5 × 5 km)	Bureau of Meteorology
Streamflow (discharge)	Daily	Station gauged, temporal	NSW Office of Water
MODIS NDVI	16-Day	250 m spatial resolution	U.S. Geological Survey
DEM	-	30 m spatial resolution	U.S. Geological Survey
Soil Map	-	250 m spatial resolution	Digital Atlas of Australian Soil
Land cover/land use map	-	50 m spatial resolution	NSW Office of Environment and Heritage

2.3.2. Study Period

The study period (2001–2020) was selected to include a long-term drought (2001–2006) and flooding (2007–2010) phases. Both dry and wet phases were included in the study to ensure any long-term change in the vegetation condition was identified in the NDVI data. The annual data were divided into two seasons: (i) dry and (ii) wet, which were categorised based on rainfall and temperature anomalies. The average dry season (Oct–Mar) and wet season (Apr–Sep) rainfall are 52.4 mm, 66.45 mm and 70.74 mm, 73.91 mm in the drought and flooding periods, respectively.

2.3.3. DEM

The sub-basin parameters (gradient and length of the slope) and stream network characteristics (slope, width, and length of the channel) were obtained from the DEM file. For this study, we used a 30 m resolution DEM downloaded from the Shuttle Rader Topography Mission (STRM) using the USGS data portal [55]. DEM for the Burrinjuck Dam study area was masked for the SWAT application (Figure 3c).

2.3.4. Land Use/Land Cover Data

The land use data for the study area used in the ArcSWAT HRU delineation was developed by the NSW Office of Environment and Heritage. These satellite imagery data were derived for the period of 2001 to 2005 and verified with Google Earth and a field survey of specific land cover types. The raster files were processed in ArcGIS to reclassify for the SWAT model (Figure 3b).

2.3.5. Soil Data

The SWAT model requires soil information of the basin area including a database table of soil texture, pH number, available water content, hydraulic conductivity, bulk density, and organic carbon content for each soil type [44,56]. The soil map of the study area was downloaded from the Digital Atlas of Australian Soil [8] (Figure 3a). A 'usersoil' database table was prepared for this study from the available soil information and lookup tables, and then replaced the default 'usersoil' table in the SWAT database.

2.3.6. Climate Data

The climate data we used in this study included daily rainfall, temperature (maximum and minimum), wind speed, solar radiation, and relative humidity. They were obtained from the Australian Bureau of Meteorology [57]. The climate data was obtained for a period of 21 years (from 2000 to 2020) in daily time series format. These data were processed using the Microsoft Excel tool to fill 0.2 of the missing data by the linear interpolation method [58].

2.3.7. Sensitivity Analysis and Hydrological Model Calibration

We applied sensitivity analysis following the guidelines explained in the previous studies [59], using the SWAT Calibration and Uncertainty Programs (SWAT-CUP). The SWAT-CUP has five algorithm options for model calibration (SUFI-2, PSO, GLUE, ParaSol, and MCMC), 11 functions (mult, sum, R2, chi2, NS, br2, ssqr, PBIAS, KGE, RSR, MNS) and integrated features such as plot visualisation [60]. The sensitivity analysis was done using SUFI-2, considering the one-at-a-time method of 15 parameters related to the processes of streamflow, recharge, evapotranspiration, percolation, and infiltration from the list to identify the most sensitive ones for the model simulations at the Burrinjuck Dam. According to previous studies [61], the Curve Number for moisture condition II (CN2) and the coefficient of water percolation to the deep aquifer (RCHRG_DP) were identified as the two most important sensitive parameters. Based on the literature review, among the two sensitive parameters, CN2 was chosen for the model calibration of this study. However, some other parameters such as the surface runoff lag coefficient (SURLAG) and Manning's roughness coefficient (CH_N2) were also analysed, which were not as sensitive as in the previous modelling done by Saha and Zeleke [44]. The fact is that the previous study was done in the Yass River gauging station, which was upstream of the Burrinjuck Dam basin, while the present study focuses on the whole basin. Acquiring knowledge from several previous studies that applied the SWAT model close to the study area helps parameter selection for sensitivity analysis. Thirteen parameters were chosen to do sensitivity analysis (Table 2) based on previous SWAT model applications in the Kyeamba Creek basin [28] and Yass River basin [44]. The difference in basin scale could interfere in the sensitivity analysis. Therefore, the parameters used for calibration in this study are not necessarily the same proposed by Saha [44].

Table 2. The table below shows the number of parameters applied, their definitions, and ranking in the SWAT-CUP simulation.

Parameter Definition	Value Range	Unit	Method	Par.inputfile	Ranking
Initial SCS runoff curve number for moisture condition	35–89	%	r	CN2	1
Effective hydraulic conductivity in the main channel alluvium	0–500	mm/h	v	CH_K2.rte	13
Manning's *n* value for the main channel	0–0.3	—	v	CH_N2.rte	12
Base flow alpha factor	0–1	days	v	ALPHA_BF.gw	5
Groundwater delay	30–500	days	v	GW_DELAY.gw	10
Groundwater "revap" coefficient	0.02–0.2	—	v	GW_REVAP.gw	11
Threshold depth of water in the shallow aquifer for return flow to occur	0–5000	mm H2O	v	GWQMN.gw	3
Threshold depth of water in the shallow aquifer required for "revap" to occur	0–1	mm H2O	v	REVAPMN.gw	8
Soil evaporation compensation factor	0–0.65	-	v	ESCO.bsn	2
Average slope length	10–150	m	r	SLSUBBSN.hru	9
Surface runoff lag coefficient	0.05–24	—	v	SURLAG.bsn	15
Available water capacity of the soil layer	−0.5–0.5	mm H2O/mm	r	SOL_AWC.sol	4
Depth from the soil surface to layer bottom	−0.5–0.5	mm	r	SOL_Z.sol	6
Peak rate adjustment factor for sediment routing	1–2	-	r	ADJ_PKR.bsn	14
Maximum canopy storage	0–100	mm H2O	v	CANMX.hru	7

In this study, we used the sequential uncertainty fitting algorithm (SUFI-2) and selected the Nash–Sutcliffe model efficiency (NS) coefficient as a target function for calibration procedures. In the calibration process, SUFI-2 captures the uncertainties of the model run. The six parameters applied in the calibration process were selected from the sensitivity analysis table based on their ranking (Table 2). A researcher [61] found that the calibration process and uncertainties are closely related, and identifying these relationships are important. In the SUFI-2 interface, the input parameter uncertainty is expressed as ranges, whereas the output parameter's uncertainties are expressed from the 95 PPU (95% probability distribution), which is calculated using Latin American hypercube sampling from the cumulative distribution of an output variable at 2.5% and 97.5%. The adjustment between the simulation results and observed data can be done by the *p-factor* (the fraction of measured data bracketed by the 95PPU band) and the *R-factor* (ratio of the average width of the 95PPU band and the standard deviation of the measured variable) known as statistical indices [61]. The *p-factor* value > 0.7 and *R-factor* value <1.5 are desirable for streamflow discharge depending on the situation [62].

The SWAT model was calibrated (2004–2007) and validated (2008–2010) with a warm-up period of three years (2000–2002). The calibration and validation processes were done in monthly timestep at two different points within the watershed, starting from the upstream of the streamflow station (Yass station) and then to the downstream station (Burrinjuck Dam station).

2.3.8. Hydrological Model Performance Evaluation

In this study, we assessed model calibration performance using the coefficient of determination (R^2), Nash-Sutcliffe efficiencies (NSE), and percent bias (PBIAS) quantitative statistics, which were used in previous studies [56,63,64]. Moreover, we applied 15 parameters in the SWAT-CUP simulation and ranked them following the model performance acceptance guidelines documented by Arnold et al., [21], which are presented in Table 2.

The Nash–Sutcliffe simulation efficiency (NSE) coefficient is a dimensionless statistic, indicating the accuracy of simulated versus observed data against the 1:1 line [65]. NSE is the most widely used statistical indicator for hydrological model performance, in which the NSE value 1 represents observed and simulated values as the same, while negative NSE value means simulations are extremely poor. NSE is defined as:

$$NSE = 1 - \frac{\sum_{i=1}^{n}(Q_{obs},i - Q_{sim},i)^2}{\sum_{i=1}^{n}(Q_{obs},i - \overline{Q_{obs}})^2}$$

(2)

where n is the number of time steps, Q_{obs}, i is the observed flow at time step i (daily here), Q_{obs} is the mean of the observed flow, and Q_{sim}, i is the simulated flow. The range of NSE is $[-\infty, 1]$, where 1 represents a perfect match between the observed and simulated flow.

A hydrological model with higher R^2 is considered as a good result [66]. R^2 is defined as:

$$R^2 = \left\{ \frac{\sum_{i=1}^{n}\left(Q_i^{obs} - \overline{Q}^{sim}\right)\left(Q_i^{sim} - \overline{Q}^{sim}\right)}{\sum_{i=1}^{n}\left(Q_i^{obs} - \overline{Q}^{obs}\right)^2 \sum_{i=1}^{n}\left(Q_i^{obs} - \overline{Q}^{obs}\right)^2} \right\}^2$$

(3)

where, Q_i^{obs} and Q_i^{sim} are representing the measured and simulated data for ith observation and $\overline{Q}^{obs}$ and $\overline{Q}^{sim}$ are the mean of the measured and simulated data, respectively.

The percent bias (PBIAS) determines the average tendency to be greater or smaller simulated values than their observed data [63]. The maximum $PBIAS$ value is zero, indicating the simulation is exactly the same as the observed data. In general, a smaller $PBIAS$ value signifies accurate model simulation. $PBIAS$ is calculated as:

$$PBIAS = \frac{\sum_{i=1}^{n}\left(Q_i^{obs} - Q_i^{sim}\right) * 100}{\sum_{i=1}^{n} Q_i^{obs}}$$

(4)

where Q_i^{obs} and Q_i^{sim} are representing the measured and simulated data for the ith observation, respectively.

2.3.9. Remote Sensing Data

Moderate resolution imaging spectroradiometer (MODIS) data are available from the U.S. Geological Survey website for free of cost [55]. We used the MODIS (Terra) 16-Day Global 250 m composite product of MOD13Q1 (version V006) to identify the vegetation condition. The NDVI values were selected from the available vegetation indices in the MOD13Q1 product from imagery acquired during the period 2001 to 2020. We have selected six plots of different vegetation types (average size between 1 and 2 km^2) within the study area (such as grass, shrub, and tree). These plots were selected randomly (i.e., stratified random sampling) based on the specific vegetation type dominant in the selected plot area. We also selected point areas (500 m radius) at the bottom and top of each sub-basin (Figure 4). A total of 60 areas (point area) were calculated for 40 sub-basins (three sub-basins were too small to create a point). These plots have been converted into polygons in the Google Earth Pro and then saved as KML files, which were later processed into shapefiles in ArcGIS [50]. A pre-processing tool called the Application for Extracting and Exploring Analysis Ready Samples (AppEEARS) was selected to obtain pre-processed NDVI time-series data for those shapefiles prepared earlier.

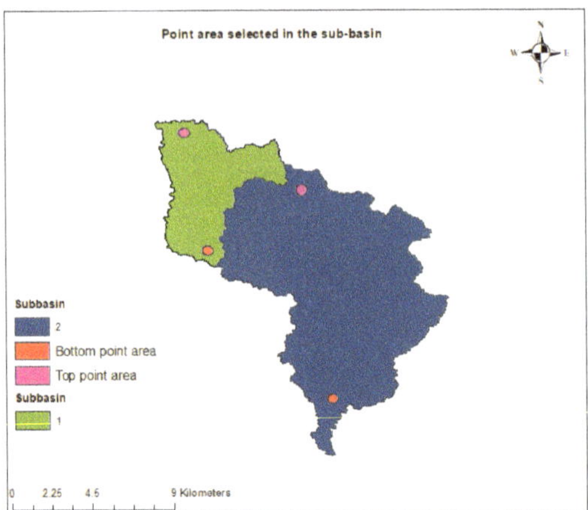

Figure 4. The vegetation NDVI was also calculated for the point area with a radius of 500 m selected both from the top and bottom location within all sub-basins. The above figure only shows the point locations of sub-basin 1 and 2.

2.3.10. Normalised Difference Vegetation Index (NDVI)

The NDVI data were processed using the AppEEARS tool [67]. The MODIS sensor captures a range of broad spectrum of reflected sunlight from tree leaves. The healthy vegetation mostly absorbs light from the red spectrum and reflects light from the near-infrared (NIR) spectrum. NDVI utilises the contrast of strong reflectance in the near-infrared region and the strongly absorbed reflectance in the red wavelength region. The NDVI calculation was performed applying the difference between the red and near-infrared bands and normalising it over the sum of the red and near-infrared bands (Equation (5)).

$$\text{NDVI} = \frac{(Near\ Infrared - Visible\ red\ light)}{(Near\ Infrared + Visible\ red\ light)} \tag{5}$$

Three types of vegetation indices were obtained using the Google Earth map and U.S. Geological Survey website. Firstly, the plots were selected for forest type vegetation within the watershed in Google Earth Pro and saved into KML files. These KML files were then processed in ArcGIS to convert into shapefiles and later used to obtain 20 years (2001–2020) of NDVI data from USGS. These similar steps were followed to obtain NDVI data for shrub and grass type vegetation within the watershed. We also calculated NDVI for each of the 43 sub-basins for the same period (2001–2020).

2.3.11. Machine Learning Algorithms for Data Analysis

A machine learning (ML) algorithm is a set of computational codes that can process a large amount of data in a complex way [68]. It is also known as data-driven methods that build models based on evidence obtained from a sample data set. The algorithms read and processed data to learn the maximum possible patterns about the data [49]. In this study, we applied the Waikato Environment for Knowledge Analysis (WEKA) tool, developed by the University of Waikato, New Zealand [39,40]. Firstly, the WEKA tool was set up to run a random forest model using 43 different datasets. These datasets included the combination of SWC, groundwater flow towards stream, and different types of vegetation responses (NDVI values). Each dataset was initially set for linear regression to find the collinear and non-collinear variables. Secondly, the machine learning tool was prepared to run a support vector machine (SVM) model using the same datasets.

The performance of all models was assessed in two ways: (a) using a 10-fold cross-validation, which is a leave-one-out approach, and (b) using the 80 and 20 per cent split-sample method. These two approaches were performed to compute the root mean square error (RMSE) and correlation coefficient (r) between the SWAT output variables (SW and GW) and predicted vegetation response (NDVI value) of each model. We selected models with higher correlation coefficient (r) values and smaller RMSEs to analyse the relationship against soil water content (SWC) and groundwater flow (GW). We also analysed these relationships based on rainfall intensity such as dry season (October to March) for less intensity and wet season (April to September) for high intensity.

3. Results

3.1. Hydrological Model Calibration and Validation

Table 2 shows the sensitivity ranking of the different model parameters and their ranges applied during the calibration. The model was calibrated and validated at two different stations (Figure 5), for which the results are listed in Table 3. The results explained that manual calibration performed better than auto-calibration. The 0.51 NSE value for the manual calibration performance parameters can be marked as 'satisfactory' for the SWAT model developed in the study area. The model in the study area was able to simulate about 51% of the variance on observed streamflow data.

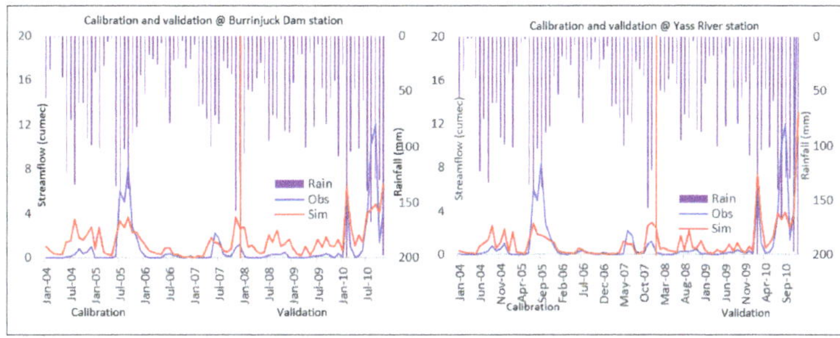

Figure 5. In this study, the SWAT-CUP tool was applied for the model calibration and validation at two different locations based on the available station, (i) Burrinjuck Dam, and (ii) Yass River station.

Table 3. The table below shows the number of parameters applied, their definitions, and ranking in the SWAT-CUP simulation.

Scenario	NSE	R^2	PBIAS
Default	0.25	0.36	73.2
Manual calibration	0.51	0.72	54.2
SUFI-2	0.41	0.55	68.2

The statistical indicators reflected a regression between observed and simulated streamflow for those two points with NSE 0.51, PBIAS 54.2, R2 0.72, *p*-factor 0.63 and NSE 0.54, PBIAS 58.6, R2 0.73, and *p*-factor 0.68, respectively. The hydrographs show that the observed and simulated values have a noticeable difference in the plots. Additionally, the model slightly overestimated the low flow during the calibration and validation periods.

3.2. Relationships of Vegetation Responses and Groundwater

The average monthly SWC and groundwater data were presented in Table 4. The average correlation coefficient of different vegetation types and SWAT model output variables over the study period in shown in Figure 6. The different correlation patterns of vegetation types of responses and SWC (SWC) suggested that vegetations were influenced considerably by SWC. The linear regression results show that shrub vegetation NDVI is highly correlated (R2 = 0.82) to SWC than forest and grass type vegetation NDVI (R2 = 0.78, and R2 = 0.72, respectively). However, grass type vegetation response is higher

(R2 = 0.59) to groundwater (GW) compared to forest vegetation (R2 = 0.24) and shrub vegetation (R2 = 0.25).

Table 4. ArcSWAT produced simulated soil water content (SWC) and groundwater flow (GW) data presented as average monthly for the study area.

Variable	January	February	March	April	May	June	July	August	September	October	November	December
SWC	86.28	98.54	93.18	96.25	112.64	130.79	131.11	129.71	122.23	106.23	100.48	78.14
GW	6.07	3.72	5.10	4.59	4.60	9.13	21.15	29.00	28.73	24.57	15.01	10.96

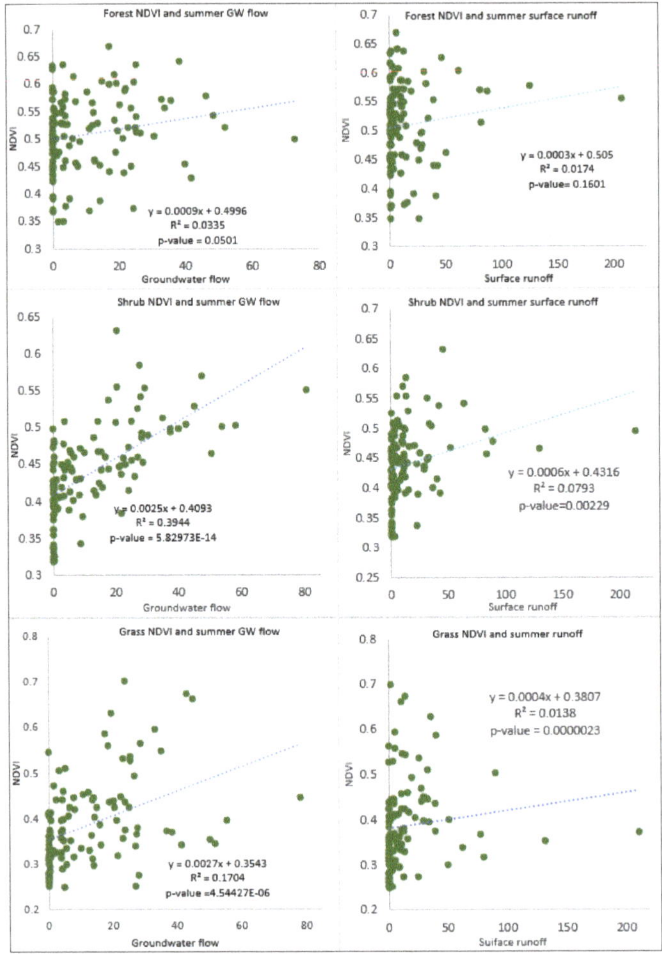

Figure 6. The forest, shrub, and grass type vegetation NDVI datasets are plotted against model-simulated surface runoff and groundwater flow (GW) to calculate the co-efficient of determination (R2).

The WEKA modelling results show that sub-basin NDVI (including all vegetation types within the sub-basin no 28) was highly responsive (r = 0.78) compared with forest NDVI (r = 0.61) when the ML algorithms were applied against SWC and GW (Table 5). Similarly, sub-basin NDVI (including all vegetation types within the sub-basin no 19 and

28) was highly responsive (r = 0.76 and r = 0.74 respectively) than shrub and grass type vegetation (r = 0.67 and r = 0.56 respectively) (Tables 6 and 7).

Table 5. The WEKA-generated modelling results for forest, sub-basin, top-point, and bottom-point NDVI against SWAT-simulated variables, soil water content (SWC), and groundwater flow (GW). The r represents the correlation coefficient.

Sub-Basin		GW			SWC			SWC and GW		
# 28		r	RMSE	RRSE	r	RMSE	RRSE	r	RMSE	RRSE
FOREST	SVM	0.373	0.064	91%	0.592	0.055	79%	0.610	0.055	78%
	RF	0.219	0.076	110.42%	0.446	0.067	91%	0.540	0.060	85%
SB_NDVI	SVM	0.597	0.075	80%	0.710	0.066	70%	0.781	0.059	62%
	RF	0.484	0.088	94%	0.604	0.079	84%	0.736	0.064	68%
TP_NDVI	SVM	0.471	0.072	89%	0.624	0.063	78%	0.660	0.061	75%
	RF	0.407	0.080	98%	0.624	0.063	78%	0.631	0.064	79%
BP_NDVI	SVM	0.267	0.072	96%	0.513	0.064	85%	0.521	0.063	85%
	RF	0.132	0.085	113%	0.330	0.078	104%	0.434	0.070	93%

Table 6. The WEKA machine learning produced modelling results for vegetation NDVI from shrub, sub-basin, top point, and bottom point against the SWAT variables soil water content (SWC) and groundwater flow (GW). The r represents the correlation coefficient in the below results.

Sub-Basin		GW			SWC			SWC and GW		
# 19		r	RMSE	RRSE	r	RMSE	RRSE	r	RMSE	RRSE
SHRUB	SVM	0.533	0.059	82%	0.681	0.051	70%	0.671	0.052	72%
	RF	0.596	0.056	77.96%	0.625	0.055	74%	0.626	0.054	74%
SB_NDVI	SVM	0.579	0.073	82%	0.689	0.064	72%	0.759	0.058	65%
	RF	0.462	0.084	94%	0.577	0.076	85%	0.685	0.066	74%
TP_NDVI	SVM	0.674	0.078	74%	0.697	0.075	71%	0.812	0.061	58%
	RF	0.609	0.087	82%	0.571	0.090	86%	0.772	0.067	64%
BP_NDVI	SVM	0.247	0.082	97%	0.456	0.075	89%	0.451	0.075	89%
	RF	0.041	0.098	117%	0.267	0.091	108%	0.363	0.082	97%

Table 7. The WEKA machine learning modelling results for grass type vegetation NDVI (sub-basin combined, vegetation located at the top point, and vegetation located at the bottom point) against SWAT variables. The correlation coefficient (r) for the random forest and support vector machine algorithms are listed in the below table.

Sub-Basin		GW			SWC			SWC and GW		
# 23		r	RMSE	RRSE	r	RMSE	RRSE	r	RMSE	RRSE
GRASS	SVM	0.4642	0.1116	84.57%	0.5342	0.105	79.28%	0.5629	0.1024	76.98%
	RF	0.4876	0.1094	83.15%	0.4607	0.112	82.75%	0.4955	0.1088	80.10%
SB_NDVI	SVM	0.6004	0.1071	80.63%	0.649	0.1007	75.75%	0.7431	0.0889	66.92%
	RF	0.5369	0.1171	88.10%	0.4353	0.1299	97.78%	0.6522	0.1025	77.11%
TP_NDVI	SVM	0.6528	0.1276	75.90%	0.6729	0.1238	73.62%	0.7883	0.1035	61.55%
	RF	0.581	0.1422	84.62%	0.4665	0.1605	95.47%	0.7031	0.121	71.97%
BP_NDVI	SVM	−0.0069	0.1265	101.07%	0.1134	0.1242	99.19%	0.2045	0.1223	97.67%
	RF	−0.0646	0.1519	121.35%	0.0884	0.1438	114.89%	0.1552	0.1312	104.79%

3.3. Vegetation Responses Considering Their Location within the Watershed

The results shown in Figure 6 were calculated from the average data for 40 sub-basins. The monthly average correlation coefficient result shows that vegetation in the top-point location in a sub-basin is more sensitive (R^2 = 0.77) to SWC when compared with vegetation in the bottom point location (R^2 = 0.72). On the other hand, vegetation in the bottom point location is more correlated to groundwater (R^2 = 0.62) than vegetation in the top point location (R^2 = 0.57).

The average correlation coefficient of top-point (distant from outlet) and bottom-point (close to outlet) NDVI and SWC is shown in Figure 7. The modelling results show that vegetation in the top-point location of the sub-basin has moderate r values against GW and SWC (0.67 and 0.69 respectively) compared with vegetation in the bottom location (0.25, and 0.46 respectively). Moreover, the result shows strong correlations for the top point vegetation NDVI against these two variables (r = 0.81 and r = 0.79, respectively) (Tables 6 and 7). The negative value of r (−0.0069) shows that vegetation in the bottom location of sub-basin #23 has no response to the GW (Table 7).

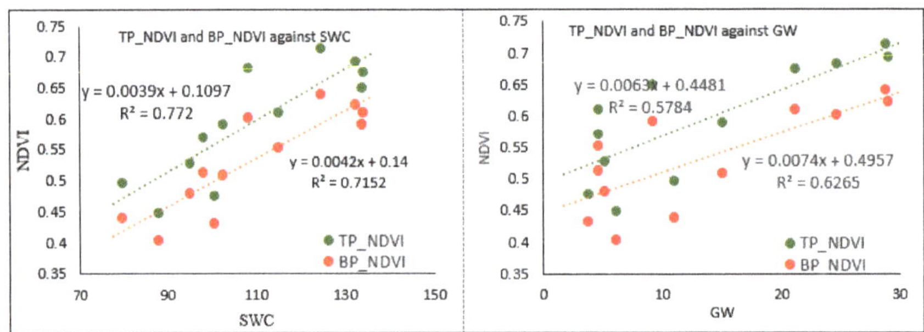

Figure 7. The NDVI collected from the top-point and bottom-point areas as vegetation response are plotted against the Soil Water Content (SWC) and groundwater flow (GW) to calculate the co-efficient of determination (R2).

3.4. Seasonal Vegetation Responses

The results of the linear correlation analysis for different vegetation types for two distinct seasons are shown in Figure 8. The correlation results show that shrub and forest vegetations are highly correlated (R^2 = 0.89 and R^2 = 0.82, respectively) to SWC during the wet season compared with grass type vegetation (R2 = 0.47). However, grass vegetation shows a better response during the dry season (R^2 = 0.52) compared with the shrub and forest (R^2 = 0.45 and R^2 = 0.43, respectively).

The vegetation responses were observed for different locations within the sub-basin (Figure 9). The regression analysis shows that vegetation in the top point and bottom point locations of the sub-basin are highly correlated to GW in the dry (R^2 = 0.79 and R^2 = 0.84, respectively) and wet season (R^2 = 0.81 and R^2 = 0.85, respectively). However, vegetation in these two locations is moderately correlated to SWC during the wet season (R^2 = 0.66 and R^2 = 0.71, respectively) than the dry season (R^2 = 0.51 and R^2 = 0.54, respectively).

The WEKA modelling results show that shrub vegetation is moderately responsive to GW and SWC (r = 0.62 and r = 0.63, respectively) in the dry season. However, forest and grass type vegetation are less responsive to GW and SWC (r = 0.52, r = 0.48, r = 0.27, and r = 0.38, respectively) in the dry season (Table 8). All three types of vegetation were less responsive to GW and SWC in the wet season.

In contrast to the sub-basin level, the vegetation NDVI is highly responsive to GW and SWC (r = 0.75 and r = 0.73, respectively) in the dry season. Furthermore, the sub-basin NDVI shows a strong relationship with SWC and GW (r = 0.81) (Table 8) in the dry season, and moderate relation (r = 0.62) in the wet season (Table 9). This result clearly indicates that the vegetation in the sub-basin is positively influenced by groundwater flow both in the dry and wet seasons.

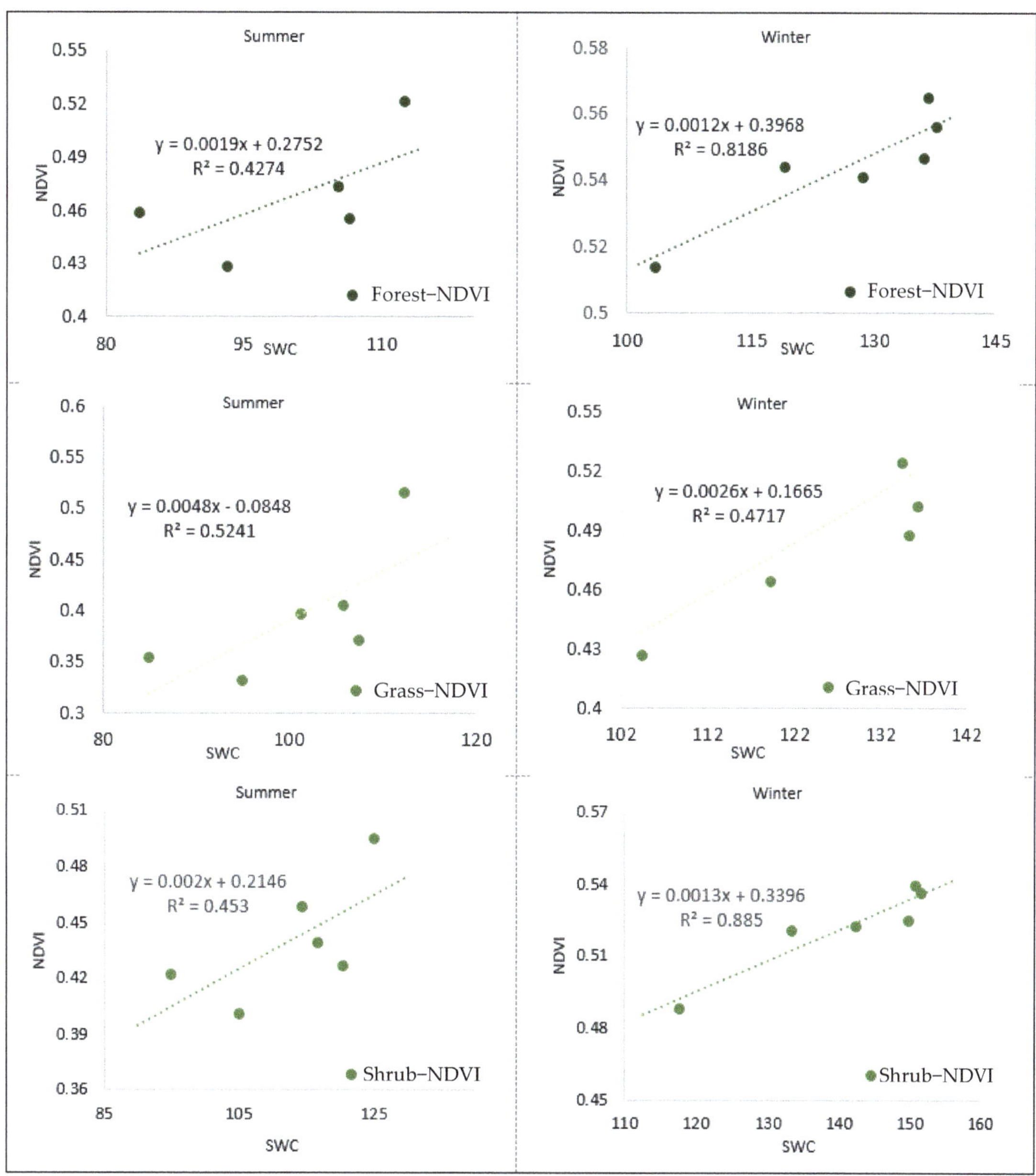

Figure 8. The vegetation responses (NDVI) against the SWC in dry and wet seasons in the study area are plotted to calculate the co-efficient of determination (R^2).

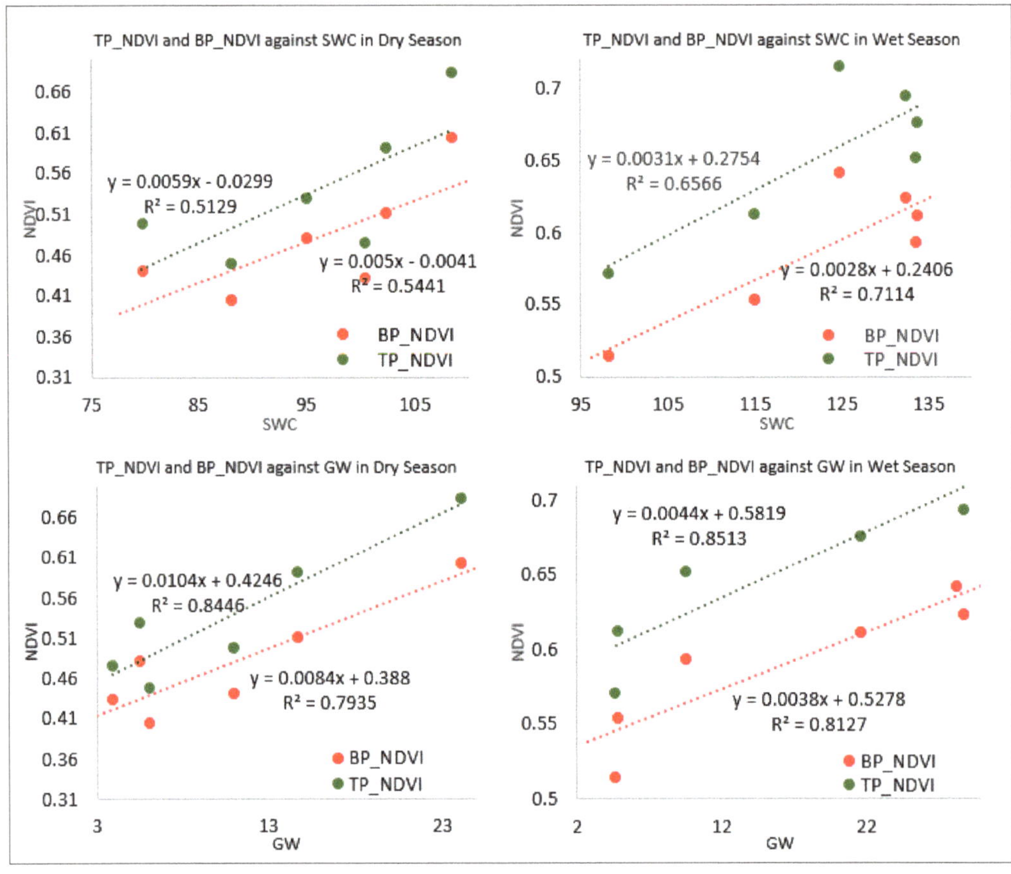

Figure 9. Seasonal vegetation responses (NDVI) from different locations (top point and bottom point) against soil water content (SWC) and groundwater flow are plotted to identify the co-efficient of determination (R^2).

Table 8. The below table shows the modelling results for different types of vegetation responses and vegetation located at different points in the sub-basin. This result shows the relationship during the dry season. The r value shows the correlation coefficient of the modelling results.

Sub-basin # 28		GW			SWC			SWC and GW		
		r	RMSE	RRSE	r	RMSE	RRSE	r	RMSE	RRSE
FOREST	SVM	0.527	0.053	0.837	0.481	0.056	0.871	0.594	0.051	0.792
	RF	0.581	0.053	0.828	0.317	0.068	1.074	0.560	0.054	0.844
SB_NDVI	SVM	**0.730**	0.058	0.674	0.570	0.071	0.815	0.782	0.054	0.625
	RF	**0.702**	0.062	0.716	0.434	0.084	0.974	0.750	0.058	0.666
TP_NDVI	SVM	0.564	0.068	0.817	0.539	0.070	0.840	0.649	0.063	0.753
	RF	0.592	0.069	0.826	0.379	0.085	1.017	0.637	0.065	0.777
BP_NDVI	SVM	0.362	0.061	0.921	0.368	0.061	0.917	0.403	0.060	0.901
	RF	0.420	0.063	0.944	0.254	0.073	1.099	0.403	0.062	0.935

Table 8. *Cont.*

Sub-basin #19		r	GW RMSE	RRSE	r	SWC RMSE	RRSE	r	SWC and GW RMSE	RRSE
SHRUB	SVM	0.629	0.048	0.777	0.631	0.048	0.799	0.666	0.046	0.766
	RF	0.627	0.048	.76.60%	0.604	0.050	0.784	0.633	0.048	0.771
SB_NDVI	SVM	0.755	0.052	0.650	0.731	0.054	0.676	0.812	0.046	0.580
	RF	0.736	0.054	0.671	0.744	0.053	0.660	0.763	0.510	0.636
TP_NDVI	SVM	0.780	0.060	0.594	0.697	0.075	0.713	0.729	0.066	0.687
	RF	0.777	0.060	0.623	0.789	0.059	0.605	0.789	0.059	0.605
BP_NDVI	SVM	0.424	0.062	0.892	0.322	0.065	0.958	0.442	0.061	0.893
	RF	0.184	0.071	1.070	0.269	0.068	1.023	0.254	0.068	1.031

Sub-basin #23		r	GW RMSE	RRSE	r	SWC RMSE	RRSE	r	SWC and GW RMSE	RRSE
GRASS	SVM	0.271	0.094	0.967	0.382	0.090	0.920	0.412	0.088	0.902
	RF	0.301	0.100	1.023	0.212	0.108	1.115	0.473	0.087	0.897
SB_NDVI	SVM	0.696	0.088	0.728	0.571	0.098	0.811	0.756	0.078	0.648
	RF	0.572	0.101	0.837	0.442	0.116	0.956	0.730	0.083	0.682
TP_NDVI	SVM	0.708	0.109	0.709	0.575	0.124	0.808	0.763	0.100	0.649
	RF	0.553	0.133	0.860	0.503	0.140	0.907	0.737	0.103	0.671
BP_NDVI	SVM	−0.128	0.116	1.025	−0.206	0.116	1.026	0.008	0.123	1.092
	RF	−0.111	0.138	1.225	−0.139	0.138	1.227	−0.202	0.118	1.043

Table 9. The below table shows the modelling results for different types of vegetation responses and vegetation located at different points in the sub-basin. This result shows the relationship during the wet season. The r value shows the correlation coefficient of the modelling results.

Sub-basin #28		r	GW RMSE	RRSE	r	SWC RMSE	RRSE	r	SWC and GW RMSE	RRSE
FOREST	SVM	0.163	0.050	98%	0.372	0.047	93%	0.356	0.048	0.934
	RF	0.230	0.055	107%	0.182	0.058	114%	0.242	0.053	1.035
SB_NDVI	SVM	0.501	0.060	86%	0.623	0.054	78%	0.710	0.049	0.699
	RF	0.530	0.066	94%	0.458	0.067	96%	0.640	0.055	0.785
TP_NDVI	SVM	0.246	0.057	96%	0.371	0.054	92%	0.358	0.055	0.927
	RF	0.361	0.058	99%	0.060	0.071	121%	0.092	0.076	1.288
BP_NDVI	SVM	0.089	0.058	99%	0.245	0.057	97%	0.203	0.057	0.981
	RF	0.159	0.063	108%	0.028	0.071	121%	0.048	0.066	1.120

Sub-basin #19		r	GW RMSE	RRSE	r	SWC RMSE	RRSE	r	SWC and GW RMSE	RRSE
SHRUB	SVM	0.346	0.045	93%	0.431	0.044	90%	0.445	0.043	0.892
	RF	0.460	0.044	90.10%	0.501	0.042	87%	0.474	0.043	0.889
SB_NDVI	SVM	0.478	0.062	87%	0.630	0.055	77%	0.623	0.056	0.778
	RF	0.568	0.060	84%	0.637	0.055	77%	0.629	0.055	0.779
TP_NDVI	SVM	0.612	0.072	79%	0.612	0.072	79%	0.749	0.060	0.658
	RF	0.640	0.072	79%	0.578	0.078	85%	0.676	0.068	0.746
BP_NDVI	SVM	−0.037	0.076	101%	0.173	0.075	99%	0.114	0.076	1.013
	RF	−0.002	0.087	116%	0.118	0.086	114%	0.142	0.079	1.052

Sub-basin #23		r	GW RMSE	RRSE	r	SWC RMSE	RRSE	r	SWC and GW RMSE	RRSE
GRASS	SVM	0.228	0.120	97%	0.350	0.117	94%	0.339	0.117	0.946
	RF	0.159	0.138	111%	0.071	0.145	117%	0.063	0.138	1.117
SB_NDVI	SVM	0.470	0.102	88%	0.519	0.099	85%	0.601	0.092	0.795
	RF	0.460	0.109	94%	0.337	0.119	102%	0.510	0.103	0.885
TP_NDVI	SVM	0.621	0.109	78%	0.567	0.115	82%	0.709	0.098	0.701
	RF	0.608	0.116	83%	0.353	0.142	102%	0.627	0.111	0.795
BP_NDVI	SVM	0.197	0.115	98%	−0.281	0.117	100%	0.173	0.117	0.995
	RF	−0.062	0.144	123%	−0.174	0.148	126%	−0.043	0.134	1.143

The vegetation in the top-point location within the sub-basin is also highly responsive to GW and SWC (r = 0.78 and r = 0.70, respectively) than vegetation in the bottom-point location (r = 0.42 and r = 0.32, respectively) in the dry season. The vegetation in the top-point location has a higher r value (r = 0.79) when correlated against GW and SWC in the dry season. However, vegetation in the top-point location has moderate responses to GW and SWC (r = 0.64 and r = 0.61, respectively), and highly responsive (r = 0.75) against these two variables together (Table 9).

4. Discussion

4.1. Relationship between Vegetation Responses (NDVI) and ArcSWAT Model Simulated Soil Water Content (SWC) and Groundwater Flow (GW) Considering Vegetation Types and Their Locations

This study presents a robust analysis of the relationships between groundwater availability and vegetation responses vigour in the floodplain zone. The hydrological model simulated different groundwater variables by calculating a range of meteorological variables, which were later analysed in relation to NDVI using different machine learning algorithms. Among random forest (RF) and support vector machine learning (SVM) algorithms, the SVM represented higher r values (r = 0.78, r = 0.75, r = 0.74 etc.) compared with RF (r = 0.73, r = 0.68, and r = 0.65 etc.) when analysed by different types of vegetation. A previous study also mentioned outperformance of random forest in terms of vegetation and water relationship modelling. Before the analysis, the SWAT model calibration was completed and produced the 0.51 NSE value. This might reflect the high volume of groundwater loss and disconnection of the deep aquifer in SWAT [10]. We found that the simulated variables (SWC and GW) and vegetation NDVI relationships vary with vegetation types when we applied data from the same sub-basin (watershed). The shrub-type vegetation is highly correlated to SWC over forest and grass vegetation; however, grass vegetation shows a high correlation to GW compared to forest and shrub vegetation [69]. The first objective of this study to understand different types of vegetation responses to SWC and groundwater is thus successful. Previous studies have found a strong correlation between different types of vegetation and SWAT-simulated SWC [32]. However, in their studies, different types of floodplain vegetation such as forest, shrub, or grass vegetation responses have not been included.

We also noticed from our study that the vegetated location within the sub-basin also impacts these relationships to SWC and GW. The vegetation located in the top point within the sub-basin, which are distant to the water outlet or stream, showed higher response to SWC (r = 0.69, 0.78 etc.). The SWC volume rate is usually high near the water outlet, and that is why the vegetation located in the bottom point zone can easily access SWC for their growth. This saturated soil enables surface and sub-surface flows and activates connectivity between soils and streams [68,70]. Moreover, vegetation located in the top point showed higher response to GW (R = 0.62) than vegetation located in the bottom point. The modelling results also showed the correlation coefficient (r) value has increased by 42% against GW for vegetation located at the top point compared to the bottom point. The correlation coefficient (r) was highly positive (0.81) for top-point vegetation when SWC and GW variables were considered together as relationship predictors. This means vegetation located in the top point can grow well when SWC and groundwater flow increases within the sub-basin.

4.2. Seasonal Variability in Each Vegetation Type

In the seasonal domain, the vegetation responses become stronger in the wet season when rainfall increases in the study area. As rainfall is the main source of water in the area of interest, the average SWC and GW values increased by 22% and 32.68%, respectively, during the wet season. Considering the inter seasonal water variability, the vegetation responses to SWC and groundwater flow varied over different types of vegetation. We found the grass vegetation response decreased by 10.6% in the wet season compared to the dry season. This variation may also be related to inter-seasonal temperature differences.

During the wet season, the average temperature in the study area is 18.45 °C (average from 2001 to 2020), which negatively impacts vegetation growth in winter months [4,70]. However, forest and shrub vegetation types are highly responsive to the sub-basin's SWC during the wet season. Therefore, forest and shrub responses were increased by 48.8% and 49.43%, respectively, in the wet season when compared to the dry season.

Similarly, we analysed vegetation responses and groundwater relationships against SWC and groundwater flow during the dry season using machine learning algorithms. The vegetation NDVI (including all vegetation in the sub-basin) against GW and SWC produced highly positive correlation coefficient values (r) (0.76, and 0.73 respectively). However, when the model was run against GW and SWC together, the r value becomes higher (0.81). The overall RF model performance was 7.3% better against runoff over the SVM classifier. The result shows that the RF classifier performs better than the SVM algorithm in the predictions. This result supports the findings of other studies where RF is widely used for crop mapping, urban studies and particularly for land use/land cover applications [71]. In this study, the WEKA model produced different r values when we applied a combined vegetation NDVI dataset at the sub-basin level. For example, the values of r between the sub-basin NDVI and GW, SWC were 0.75, 0.73, and 0.81, respectively. This means that vegetation in the sub-basin within a floodplain is highly responsive to groundwater flow and SWC during the dry season.

Not surprisingly, we found that shrub and forest type vegetation are highly responsive to GW (r = 0.63 and 0.58, respectively) compared to grass (r = 0.30). These results support that woody vegetation type is highly responsive to groundwater, while the non-woody vegetation type immediately responds to rainfall by seed or rhizome regeneration [72]. However, both shrub and forest vegetation were moderately responsive to SWC and GW (r = 0.66 and 0.59, respectively). This means tree and shrub vegetation can grow well when SWC and groundwater flow increase after the rainfall in dry season. Moreover, this study suggests the grass vegetation type is highly dependent on groundwater during the dry and winter season for their growth, and any instability or long-term drought can negatively affect these floodplain vegetation communities.

A comprehensive documentation of different types of vegetation and groundwater relationships can be prepared for efficient floodplain vegetation management based on the results of this study. Agricultural production in similar regions around the world can be increased by selecting appropriate crops based on their seasonal response to groundwater.

Authors should discuss the results and how they can be interpreted from the perspective of previous studies and of the working hypotheses. The findings and their implications should be discussed in the broadest context possible. Future research directions may also be highlighted.

5. Conclusions

The analytical results show that the vegetation system is highly dependent on groundwater hydrology during the dry season in this study area, especially shrub and grass type vegetation that are located distant from the water outlet in the HRU. This suggests that small- and medium-rooted vegetation, for instance, quince, feijoa, wheat, and oats etc., can grow well in similar floodplains globally, with possible implications for water management during the dry season.

The results of the study conclude the relationship between floodplain vegetation and catchment hydrology is two-way, and any change in the environment can directly influence the vegetation response to groundwater. For example, suitable growing temperature and available water can boost vegetation growth which, in turn, contributes to increasing the potential evapotranspiration rate. On the other hand, grass type vegetation growth helps to increase the infiltration. The hydrological simulation results suggested that rainfall dominates the study area catchment water balance, in which groundwater flow increases in the wetting period between April and September. Any changes in groundwater in the basin area can directly impact vegetation conditions, which need to be included in future studies applying LAI in the hydrological modelling. As rainfall dominates the catchment hydrology, any future changes in the rainfall pattern need to be considered carefully for better floodplain management. Measuring the field soil moisture data and applying that data for model calibration could be another option to compare model simulation to support the output results.

In summary, this study contributes scientific insight into groundwater-vegetation relationship and outlines a methodology for modelling the relationship in contrast to seasonal groundwater variations. The research outcomes can potentially support sustainable floodplain vegetation system development in arid environments. However, there are still some drawbacks. This study considered vegetation types and their distance from the streamflow while assessing their responses to the groundwater variables. There could be other factors, e.g., vegetation density and depth of root can be included in the future studies.

Further research should consider improving the modelling results applying more data for intense rainfall and drought years. Thus, the multiple regression including a time lag, temperature, or rainfall frequency as well as future climate projections may give better understanding on ecosystem hydrology.

Author Contributions: N.M.: Conceptualization, Methodology, Software, Data gathering, Formal analysis, Investigation, Validation, Writing original draft preparation, Review and Editing. A.A.A.: Supervision, Conceptualization, Validation, Review and Editing. T.N.M.: Validation, Review and Editing. G.T.A.: Software, Validation, Review and Editing, and acquisition of APC. All authors have read and agreed to the published version of the manuscript.

Funding: This research was funded by University of Southern Queensland Research Training Program. Gebiaw T Ayele funded the APC.

Data Availability Statement: The data that support the findings of this study are publicly available in Figshare which can be accessed from this link 10.6084/m9.figshare.21625994.

Acknowledgments: This research was supported by the University of Southern Queensland Research Training Program (RTP). Thanks are extended to the Australian Bureau of Meteorology, for providing the Australian meteorological dataset, to the USGS and National Aeronautics and Space Administration (NASA) for offering the MODIS NDVI product. Gebiaw T. Ayele received funding and acknowledges Griffith Graduate Research School, the Australian Rivers Institute and School of Engineering, Griffith University, Queensland, Australia.

Conflicts of Interest: The authors declare no conflict of interest.

Appendix A

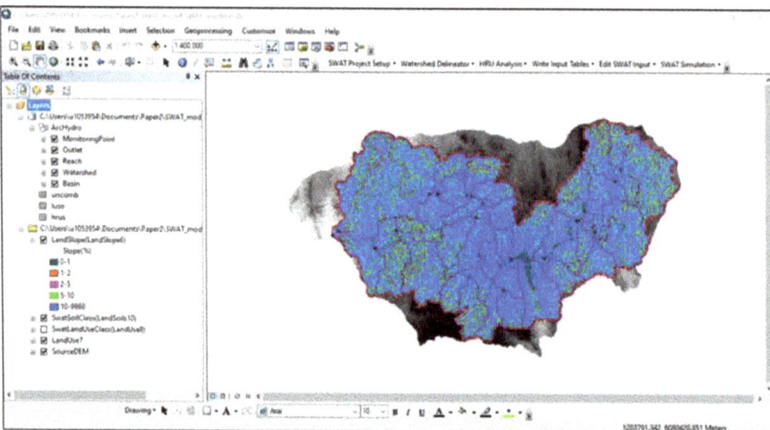

Figure A1. ArcSWAT interface shows the model set up in the Burrinjuck Dam sub-basin study area which is applied to understand the catchment hydrology.

References

1. Ponting, J.; Kelly, T.J.; Verhoef, A.; Watts, M.J.; Sizmur, T. The impact of increased flooding occurrence on the mobility of potentially toxic elements in floodplain soil—A review. *Sci. Total Environ.* **2021**, *754*, 142040. [CrossRef] [PubMed]
2. Mohammed, B.; Salah, O.; Driss, O.; Abdelghani, C. Global warming and groundwater from semi-arid areas: Essaouira region (Morocco) as an example. *SN Appl. Sci.* **2020**, *2*, 1245. [CrossRef]
3. Condon, L.E.; Atchley, A.L.; Maxwell, R.M. Evapotranspiration depletes groundwater under warming over the contiguous United States. *Nat. Commun.* **2020**, *11*, 873. [CrossRef] [PubMed]
4. Huang, F.; Zhang, D.; Chen, X. Vegetation Response to Groundwater Variation in Arid Environments: Visualization of Research Evolution, Synthesis of Response Types, and Estimation of Groundwater Threshold. *Int. J. Environ. Res. Public Health* **2019**, *16*, 1849. [CrossRef] [PubMed]
5. Cheng, Y.; Yang, W.; Zhan, H.; Jiang, Q.; Shi, M.; Wang, Y. On the Origin of Deep Soil Water Infiltration in the Arid Sandy Region of China. *Water* **2020**, *12*, 2409. [CrossRef]
6. Ma, X.; Huete, A.; Moran, S.; Ponce-Campos, G.; Eamus, D. Abrupt shifts in phenology and vegetation productivity under climate extremes. *J. Geophys. Res. Biogeosciences* **2015**, *120*, 2036–2052. [CrossRef]
7. Dai, A. Drought under global warming: A review. *WIREs Clim. Chang.* **2011**, *2*, 45–65. [CrossRef]
8. Wang, P.; Zhang, Y.; Yu, J.; Fu, G.; Ao, F. Vegetation dynamics induced by groundwater fluctuations in the lower Heihe River Basin, northwestern China. *J. Plant Ecol.* **2011**, *4*, 77–90. [CrossRef]
9. Schmugge, T.J.; Jackson, T.J.; McKim, H.L. Survey of methods for soil moisture determination. *Water Resour. Res.* **1980**, *16*, 961–979. [CrossRef]
10. Uniyal, B.; Dietrich, J.; Vasilakos, C.; Tzoraki, O. Evaluation of SWAT simulated soil moisture at catchment scale by field measurements and Landsat derived indices. *Agric. Water Manag.* **2017**, *193*, 55–70. [CrossRef]
11. Porporato, A.; Daly, E.; Rodriguez-Iturbe, I. Soil water balance and ecosystem response to climate change. *Am. Nat.* **2004**, *164*, 625–632. [CrossRef] [PubMed]
12. Tian, S.; Renzullo, L.J.; van Dijk, A.I.J.M.; Tregoning, P.; Walker, J.P. Global joint assimilation of GRACE and SMOS for improved estimation of root-zone soil moisture and vegetation response. *Hydrol. Earth Syst. Sci.* **2019**, *23*, 1067–1081. [CrossRef]
13. Leenaars, J.G.; Claessens, L.; Heuvelink, G.B.; Hengl, T.; González, M.R.; van Bussel, L.G.; Guilpart, N.; Yang, H.; Cassman, K.G. Mapping rootable depth and root zone plant-available water holding capacity of the soil of sub-Saharan Africa. *Geoderma* **2018**, *324*, 18–36. [CrossRef] [PubMed]
14. Zhu, Y.; Wu, Y.; Drake, S. A survey: Obstacles and strategies for the development of ground-water resources in arid inland river basins of Western China. *J. Arid. Environ.* **2004**, *59*, 351–367. [CrossRef]
15. Naumburg, E.; Mata-Gonzalez, R.; Hunter, R.G.; McLendon, T.; Martin, D.W. Phreatophytic Vegetation and Groundwater Fluctuations: A Review of Current Research and Application of Ecosystem Response Modeling with an Emphasis on Great Basin Vegetation. *Environ. Manag.* **2005**, *35*, 726–740. [CrossRef]
16. Schlaepfer, D.R.; Bradford, J.B.; Lauenroth, W.K.; Munson, S.M.; Tietjen, B.; Hall, S.A.; Wilson, S.D.; Duniway, M.C.; Jia, G.; Pyke, D.A.; et al. Climate change reduces extent of temperate drylands and intensifies drought in deep soils. *Nat. Commun.* **2017**, *8*, 14196. [CrossRef]
17. Tomlinson, M.; Boulton, A.J. Ecology and management of subsurface groundwater dependent ecosystems in Australia—A review. *Mar. Freshw. Res.* **2010**, *61*, 936–949. [CrossRef]
18. Zhu, Y.; Chen, Y.; Ren, L.; Lü, H.; Zhao, W.; Yuan, F.; Xu, M. Ecosystem restoration and conservation in the arid inland river basins of Northwest China: Problems and strategies. *Ecol. Eng.* **2016**, *94*, 629–637. [CrossRef]
19. Eamus, D.; Zolfaghar, S.; Villalobos-Vega, R.; Cleverly, J.; Huete, A. Groundwater-dependent ecosystems: Recent insights from satellite and field-based studies. *Hydrol. Earth Syst. Sci.* **2015**, *19*, 4229–4256. [CrossRef]
20. Adhikari, R.K.; Mohanasundaram, S.; Shrestha, S. Impacts of land-use changes on the groundwater recharge in the Ho Chi Minh city, Vietnam. *Environ. Res.* **2020**, *185*, 109–440. [CrossRef]
21. Arnold, J.G.; Moriasi, D.N.; Gassman, P.W.; Abbaspour, K.C.; White, M.J.; Srinivasan, R.; Jha, M.K. SWAT: Model use, calibration, and validation. *Trans. ASABE* **2012**, *55*, 1491–1508. [CrossRef]
22. Cuceloglu, G.; Abbaspour, K.C.; Ozturk, I. Assessing the Water-Resources Potential of Istanbul by Using a Soil and Water Assessment Tool (SWAT) Hydrological Model. *Water* **2017**, *9*, 814. [CrossRef]
23. Francesconi, W.; Srinivasan, R.; Pérez-Miñana, E.; Willcock, S.P.; Quintero, M. Using the Soil and Water Assessment Tool (SWAT) to model ecosystem services: A systematic review. *J. Hydrol.* **2016**, *535*, 625–636. [CrossRef]
24. Pisinaras, V.; Petalas, C.; Gikas, G.D.; Gemitzi, A.; Tsihrintzis, V.A. Hydrological and water quality modeling in a medium-sized basin using the Soil and Water Assessment Tool (SWAT). *Desalination* **2010**, *250*, 274–286. [CrossRef]
25. Silva-Júnior, R.O.D.; Salomão, G.N.; Tavares, A.L.; Santos, J.F.D.; Santos, D.C.; Dias, L.C.; Rocha, E.J.P.D. Response of Water Balance Components to Changes in Soil Use and Vegetation Cover Over Three Decades in the Eastern Amazon. *Front. Water* **2021**, *3*, 1. [CrossRef]
26. Yonaba, R.; Biaou, A.C.; Koïta, M.; Tazen, F.; Mounirou, L.A.; Zouré, C.O.; Queloz, P.; Karambiri, H.; Yacouba, H. A dynamic land use/land cover input helps in picturing the Sahelian paradox: Assessing variability and attribution of changes in surface runoff in a Sahelian watershed. *Sci. Total Environ.* **2021**, *757*, 143792. [CrossRef] [PubMed]

27. Narasimhan, B.; Srinivasan, R. Development and evaluation of Soil Moisture Deficit Index (SMDI) and Evapotranspiration Deficit Index (ETDI) for agricultural drought monitoring. *Agric. For. Meteorol.* **2005**, *133*, 69–88. [CrossRef]

28. Saha, P.; Zeleke, K. Assessment of streamflow and catchment water balance sensitivity to climate change for the Yass River catchment in south eastern Australia. *Environ. Earth Sci.* **2014**, *73*, 6229–6242. [CrossRef]

29. Fu, B.; Burgher, I. Riparian vegetation NDVI dynamics and its relationship with climate, surface water and groundwater. *J. Arid. Environ.* **2015**, *113*, 59–68. [CrossRef]

30. Mallick, J.; AlMesfer, M.; Singh, V.; Falqi, I.; Singh, C.; Alsubih, M.; Kahla, N. Evaluating the NDVI–Rainfall Relationship in Bisha Watershed, Saudi Arabia Using Non-Stationary Modeling Technique. *Atmosphere* **2021**, *12*, 593. [CrossRef]

31. Nouri, H.; Anderson, S.; Sutton, P.; Beecham, S.; Nagler, P.; Jarchow, C.J.; Roberts, D.A. NDVI, scale invariance and the modifiable areal unit problem: An assessment of vegetation in the Adelaide Parklands. *Sci. Total Environ.* **2017**, *584–585*, 11–18. [CrossRef]

32. Park, J.-Y.; Ahn, S.-R.; Hwang, S.-J.; Jang, C.-H.; Park, G.-A.; Kim, S.-J. Evaluation of MODIS NDVI and LST for indicating soil moisture of forest areas based on SWAT modeling. *Paddy Water Environ.* **2014**, *12*, 77–88. [CrossRef]

33. Groeneveld, D.P.; Baugh, W.M.; Sanderson, J.S.; Cooper, D.J. Annual Groundwater Evapotranspiration Mapped from Single Satellite Scenes. *J. Hydrol.* **2007**, *344*, 146–156. [CrossRef]

34. Nanzad, L.; Zhang, J.; Tuvdendorj, B.; Nabil, M.; Zhang, S.; Bai, Y. NDVI anomaly for drought monitoring and its correlation with climate factors over Mongolia from 2000 to 2016. *J. Arid. Environ.* **2019**, *164*, 69–77. [CrossRef]

35. Wen, L.; Macdonald, R.; Morrison, T.; Hameed, T.; Saintilan, N.; Ling, J. From hydrodynamic to hydrological modelling: Investigating long-term hydrological regimes of key wetlands in the Macquarie Marshes, a semi-arid lowland floodplain in Australia. *J. Hydrol.* **2013**, *500*, 45–61. [CrossRef]

36. Aguilar, C.; Zinnert, J.C.; Polo, M.J.; Young, D.R. NDVI as an indicator for changes in water availability to woody vegetation. *Ecol. Indic.* **2012**, *23*, 290–300. [CrossRef]

37. Bhanja, S.N.; Malakar, P.; Mukherjee, A.; Rodell, M.; Mitra, P.; Sarkar, S. Using Satellite-Based Vegetation Cover as Indicator of Groundwater Storage in Natural Vegetation Areas. *Geophys. Res. Lett.* **2019**, *46*, 8082–8092. [CrossRef]

38. Seeyan, S.; Merkel, B.; Abo, R. Investigation of the Relationship between Groundwater Level Fluctuation and Vegetation Cover by using NDVI for Shaqlawa Basin, Kurdistan Region—Iraq. *J. Geogr. Geol.* **2014**, *6*, 187. [CrossRef]

39. Eibe, F.; Mark, A.H.; Ian, H.W. *The WEKA Workbench. Online Appendix for "Data Mining: Practical Machine Learning Tools and Techniques"*, 4th ed.; Kaufmann, M., Ed.; University of Waikato: Burlington, MA, USA, 2016.

40. Hall, M.; Frank, E.; Holmes, G.; Pfahringer, B.; Reutemann, P.; Witten, I.H. The WEKA data mining software. *ACM SIGKDD Explor. Newsl.* **2009**, *11*, 10–18. [CrossRef]

41. Marin, D.B.; Ferraz, G.A.E.S.; Santana, L.S.; Barbosa, B.D.S.; Barata, R.A.P.; Osco, L.P.; Ramos, A.P.M.; Guimarães, P.H.S. Detecting coffee leaf rust with UAV-based vegetation indices and decision tree machine learning models. *Comput. Electron. Agric.* **2021**, *190*, 106476. [CrossRef]

42. Sharma, A.K.; Kumar, A.; Saxena, S.; Beniwal, M. Evaluating WEKA over the Open Source Web Data Mining Tools. *Int. J. Eng. Res. Technol.* **2015**, *8*, 128–132.

43. Brown, A.E.; Podger, G.M.; Davidson, A.J.; Dowling, T.I.; Zhang, L. Predicting the impact of plantation forestry on water users at local and regional scales: An example for the Murrumbidgee River Basin, Australia. *For. Ecol. Manag.* **2007**, *251*, 82–93. [CrossRef]

44. Saha, P.; Zeleke, K.; Hafeez, M. Streamflow modeling in a fluctuant climate using SWAT: Yass River catchment in south eastern Australia. *Environ. Earth Sci.* **2013**, *71*, 5241–5254. [CrossRef]

45. Wallbrink, P.J.; Olley, J.M.; Murray, A.S. *The Contribution of Subsoil to Sediment Yield in the Murrumbidgee River Basin, New South Wales, Australia*; IAHS: Wallingford, UK, 1996.

46. Verstraeten, G.; Prosser, I.P.; Fogarty, P. Predicting the spatial patterns of hillslope sediment delivery to river channels in the Murrumbidgee catchment, Australia. *J. Hydrol.* **2007**, *334*, 440–454. [CrossRef]

47. Green, D.; Petrovic, J.; Moss, P.; Burrell, M. *Water Resources and Management Overview: Murrumbidgee Catchment*; NSW Office of Water: Sydney, Australia, 2011.

48. Peel, M.; Finlayson, B.; McMahon, T. Updated world map of the Köpper-Geiger climate classification. *Hydrol. Earth Syst. Sci.* **2007**, *11*, 439–472. [CrossRef]

49. Cracknell, M.J.; Reading, A.M. Geological mapping using remote sensing data: A comparison of five machine learning algorithms, their response to variations in the spatial distribution of training data and the use of explicit spatial information. *Comput. Geosci.* **2014**, *63*, 22–33. [CrossRef]

50. ESRI. *ArcGIS Deskto2019*; Environmental Systems Research Institute: Redlands, CA, USA, 2019.

51. Microsoft Corporation. *Microsoft Excel*; Corporation, M., Ed.; Microsoft Corporation: Redmond, WA, USA, 2018.

52. Smith, T.C.; Frank, E. Introducing machine learning concepts with WEKA. In *Statistical Genomics*; Humana Press: Totowa, NJ, USA, 2016; pp. 353–378.

53. Neitsch, S.L.; Arnold, J.G.; Kiniry, J.R.; Williams, J.R. *SWAT Theoretical Documentation Version 2009. Texas Water Resources Institute Technical Report No. 406*; Texas Water Resources Institute: College Station, TX, USA, 2011.

54. Gassman, P.; Sadeghi, A.; Srinivasan, R. Applications of the SWAT Model Special Section: Overview and Insights. *J. Environ. Qual.* **2014**, *43*, 1–8. [CrossRef] [PubMed]

55. USGS. Earth Explorer. US Geological Survey. Available online: https://earthexplorer.usgs.gov/ (accessed on 28 April 2022).

56. Setegn, S.G.; Srinivasan, R.; Melesse, A.M.; Dargahi, B. SWAT Model Application and Prediction Uncertainty Analysis in the Lake Tana Basin, Ethiopia. *Hydrol. Process.* **2009**, *24*, 357–367. [CrossRef]
57. BOM. Climate Data Online. Australian Bureau of Meteorology. Available online: http://www.bom.gov.au/climate/data/index.shtml (accessed on 19 April 2022).
58. Fassò, A.; Sommer, M.; von Rohden, C. Interpolation uncertainty of atmospheric temperature profiles. *Atmos. Meas. Tech.* **2020**, *13*, 6445–6458. [CrossRef]
59. de Andrade, C.W.; Montenegro, S.M.; Montenegro, A.A.; Lima JR, D.S.; Srinivasan, R.; Jones, C.A. Soil moisture and discharge modeling in a representative watershed in northeastern Brazil using SWAT. *Ecohydrol. Hydrobiol.* **2019**, *19*, 238–251. [CrossRef]
60. Abbaspour, K.C.; Vaghefi, S.A.; Srinivasan, R. A Guideline for Successful Calibration and Uncertainty Analysis for Soil and Water Assessment: A Review of Papers from the 2016 International SWAT Conference. *Water* **2018**, *10*, 6. [CrossRef]
61. Abbaspour, K.C.; Rouholahnejad, E.; Vaghefi, S.; Srinivasan, R.; Yang, H.; Kløve, B. A continental-scale hydrology and water quality model for Europe: Calibration and uncertainty of a high-resolution large-scale SWAT model. *J. Hydrol.* **2015**, *524*, 733–752. [CrossRef]
62. Abbaspour, K.C.; Johnson, C.; Van Genuchten, M.T. Estimating uncertain flow and transport parameters using a sequential uncertainty fitting procedure. *Vadose Zone J.* **2004**, *3*, 1340–1352. [CrossRef]
63. Moriasi, D.N.; Arnold, J.G.; van Liew, M.W.; Bingner, R.L.; Harmel, R.D.; Veith, T.L. Model Evaluation Guidelines for Systematic Quantification of Accuracy in Watershed Simulations. *Trans. ASABE* **2007**, *50*, 885–900. [CrossRef]
64. Zhang, Y.; Chiew, F.H.S.; Zhang, L.; Li, H. Use of Remotely Sensed Actual Evapotranspiration to Improve Rainfall Runoff Modeling in Southeast Australia. *J. Hydrometeorol.* **2009**, *10*, 969–980. [CrossRef]
65. Nash, J.E.; Sutcliffe, J.V. River flow forecasting through conceptual models part I—A discussion of principles. *J. Hydrol.* **1970**, *10*, 282–290. [CrossRef]
66. Wu, L.; Liu, X.; Yang, Z.; Yu, Y.; Ma, X. Effects of single- and multi-site calibration strategies on hydrological model performance and parameter sensitivity of large-scale semi-arid and semi-humid watersheds. *Hydrol. Process.* **2022**, *36*, e14616. [CrossRef]
67. EarthData. Application for Extracting and Exploring Analysis Ready Samples (AρρEEARS). 2021. Available online: https://appeears.earthdatacloud.nasa.gov/ (accessed on 26 April 2022).
68. Sarkar, A. Deep Learning and the Evolution of Useful Information. *Inf. Matters* **2021**, *1*, 6. [CrossRef]
69. Jiao, L.; An, W.; Li, Z.; Gao, G.; Wang, C. Regional variation in soil water and vegetation characteristics in the Chinese Loess Plateau. *Ecol. Indic.* **2020**, *115*, 106399. [CrossRef]
70. He, B.; Chen, A.; Jiang, W.; Chen, Z. The response of vegetation growth to shifts in trend of temperature in China. *J. Geogr. Sci.* **2017**, *27*, 801–816. [CrossRef]
71. Sheykhmousa, M.; Mahdianpari, M.; Ghanbari, H.; Mohammadimanesh, F.; Ghamisi, P.; Homayouni, S. Support Vector Machine Versus Random Forest for Remote Sensing Image Classification: A Meta-Analysis and Systematic Review. *IEEE J. Sel. Top. Appl. Earth Obs. Remote Sens.* **2020**, *13*, 6308–6325. [CrossRef]
72. Sandi, S.G.; Rodriguez, J.F.; Saintilan, N.; Wen, L.; Kuczera, G.; Riccardi, G.; Saco, P.M. Resilience to drought of dryland wetlands threatened by climate change. *Sci. Rep.* **2020**, *10*, 13232. [CrossRef] [PubMed]

Review

Unveiling the Potential of Machine Learning Applications in Urban Planning Challenges

Sesil Koutra [1] and Christos S. Ioakimidis [2,*]

[1] Faculty of Architecture and Urban Planning, University of Mons, 88 Str. Havré, 7000 Mons, Belgium
[2] Inteligg P.C., Karaiskaki 28, 10554 Athens, Greece
* Correspondence: cioakim@inteligg.com

Abstract: In a digitalized era and with the rapid growth of computational skills and advancements, artificial intelligence and Machine Learning uses in various applications are gaining a rising interest from scholars and practitioners. As a fast-growing field of Artificial Intelligence, Machine Artificial Intelligence deals with smart designs, data mining and management for complex problem-solving based on experimental data on urban applications (land use and cover, configurations of the built environment and architectural design, etc.), but with few explorations and relevant studies. In this work, a comprehensive and in-depth review is presented to discuss the future opportunities and constraints in meeting the next planning portfolio against the multiple challenges in urban environments in line with Machine Learning progress. Bringing together the theoretical views with practical analyses of cases and examples, the work unveils the huge potential, but also the potential barriers of the complexity of Machine Learning to urban planning strategies.

Keywords: case-study analysis; machine learning; urban planning

1. Introduction

Digitalization is gaining rising interest in all fields of daily life, being favored by the increasing computational capacities and the emergence of efficient algorithmic processes which facilitate data mining. In line with this, Machine Learning (ML) as an intersection of informatics and statistics is a promising challenge for more evidence-based decisions [1] to fill in the gap of existing technological tools and instruments for spatiotemporal requirements. Bhavsar et al. [2] define ML as a collection of data-driven models to automate data through significant patterns, while the first attempts to develop machines to imitate living behavior dates to the 30s by Ross [3]. In 1959, Samuel approaches the concept as the *'field of study that provides computes with the ability to learn without being further programmed'* [4].

As living laboratories in a multidimensional context with tremendous environmental and social challenges, cities are being more and more implicated in these applications, especially those oriented towards meeting the complex ambitions of sustainability, resilience and climate adaptation, to cite some of them, and dealing with a noticeable mass of data ([2,3]). At the same time, rapid urbanization challenges and quality of life (QoL) degradation puts pressure on planners to channel the growth and provide monitoring strategies, while the traditional methods (e.g., surveys, etc.) are time-consuming with insufficient outcomes. Advancements in urban geography and relevant sciences, commonly geographical information systems (GIS) tools, use simulations to evaluate and analyze the complex interactions in a city with limited efficiency to simulate scenarios for future growths [4,5] and visualize spatial, demographic and other relevant data to benefit from digital innovations and patterns. These are the key transformations needed for the abovementioned roadmaps.

Combined with the technology and software advancements, ML and the field of Artificial Intelligence (AI) are being prioritized and becoming more and more essential

Citation: Koutra, S.; Ioakimidis, C.S. Unveiling the Potential of Machine Learning Applications in Urban Planning Challenges. *Land* **2023**, *12*, 83. https://doi.org/10.3390/land12010083

Academic Editor: Chuanrong Zhang

Received: 1 December 2022
Revised: 18 December 2022
Accepted: 21 December 2022
Published: 27 December 2022

for cities' operations towards smart solutions, e.g., optimization of energy performance or waste management, etc. ([6,7]). They are adopted widely for diverse tasks of the digital society while reducing the human effort [8], a recognition of the achievements in data acquisition and the practical use of algorithmic approaches [9]. Many scholars have already stressed the importance of ML for accurate predictions and correlations between spatial indicators (e.g., [10,11]).

As ML transcended the conventional techniques of modeling, a huge potential of big data management to address complex city problems is presented at the crossroads of modern urban planning challenges to make up their dynamics [12,13]. Broadly speaking, ML consist of a group of different models and patterns with the ability to minimize error using repeated processes from data collection, analysis and monitoring [14]. Based on the existing definitions, Machine Learning consists of a set of techniques to automatically detect and predict data or perform decision-making processes under an important level of uncertainty [15]. Hence, ML consists of methods leading to evidence-based processes to meet the standards and quality of a complex problem. Its rapid evolution and growth, with the parallel emergence of its potential, will equal the challenges of modern cities in several fields (mobility, energy, etc.). More and more cities are being included in this dynamic, which concerns the drivers of the urban functionalities or decision-making processes optimizing performance and leading to automation. Overall, the existing ML demonstrations on urban and spatial problems consist of spatiotemporal subjects [16]; however, their implication has not yet been fully explored, despite their large repertoire [17].

Despite the technological achievements and the progress in ML uses, data availability remains a sophisticated task and not equally distributed in every corner of the world. Lack of standards, the topics of private life and confidentiality, spatial granularities or even the lack of synergies and the nature of the heterogeneous data hinder the ML operation. On the other hand, the applications of ML algorithms on specific fields, such as geography, demonstrate the complexity of benchmarking the relevant studies due to the type of data used for the ML analyses or the missing parameters [18].

Hence, the objective of this work is to provide a critical review of the literature on ML methods and urban modelling applications at the crossroads of urban complexities as a promising research area for the years to come, associated with their advantages and opportunities for the favor of efficient data management [19]. At the same time, the aim of the work is to identify the gaps and significant constraints regarding this enterprise and to define future directions with a comprehensive benchmarking analysis and in-depth reviews to evaluate the future challenges. We also discuss the potentials and constraints of ML with a portfolio of analyzed examples to identify the future directions for ML applications sufficient to meet next generation planning strategies. Overall, the work addresses the unexplored gaps in ML's role in urban analysis in a scoping review and unveils the most prominent approaches.

The remainder of the paper is organized as follows (Figure 1). Section 2 broadly describes the taxonomy of the contents of the review for urban applications related to ML uses. In Section 3, different challenges are discussed with respect to the built environment, land uses and the transportation, to cite the most important among them. Section 4 provides a portfolio of concrete examples and cases to retrieve lessons learnt, and Section 5 concludes the review.

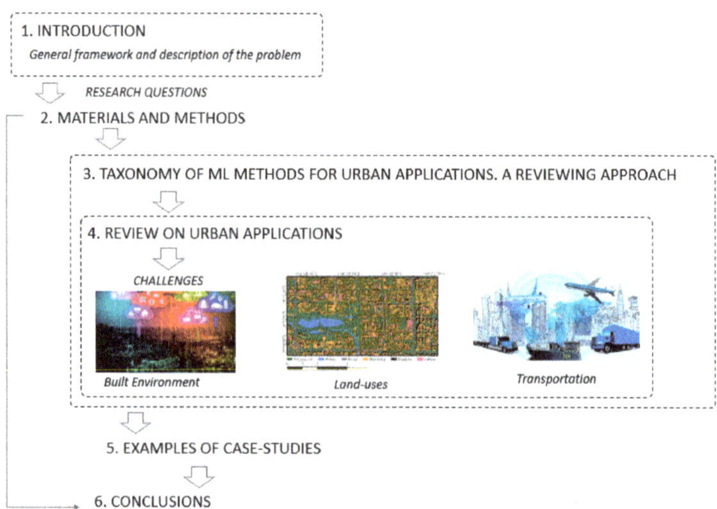

Figure 1. Paperwork plan and structure.

2. Materials and Methods

Methodologically, the chart of the study is organized in the conventional flow of a scoping review including the planning criteria for the selection of the relevant documentation, the identification and screening of the relevant scientific sources and the analysis and crossing of findings.

As keywords core to the study, ML and urban applications remain vast and inadequate for the consolidation of accurate outcomes. To identify an initial branch of sources, the Scopus platform was used to include a broad, transversal coverage to guarantee reproducible and accurate results using the bibliometric approach provided by the VOS viewer tool. More than 2000 scientific sources were detected in the Scopus platform.

For this scope, the work focuses on:

- The investigation of the spatial and temporal distribution of the selected sources with necessary filtering of the key information of title, authors, years and keywords and focusing mainly on English-language publications;
- The identification of sources by type of data provided (e.g., open or not);
- The chronological constraint from 2000 onwards;
- The identification of specific keywords and research areas (computer science, engineering, environmental and among others) (Figure 2).

While spatial data mining has been accelerated through technological advancements, the availability is not equally allocated throughout the world (Figure 5) ([20,21]). Complementary to this, Casali et al. [18] spatialized 159 related scientific documents distributed in different countries of the world and over time confirmed the discrepancies between them (Figure 3); most of the cases appeared in China and US, followed by the UK and overall only 31% were detected in Europe.

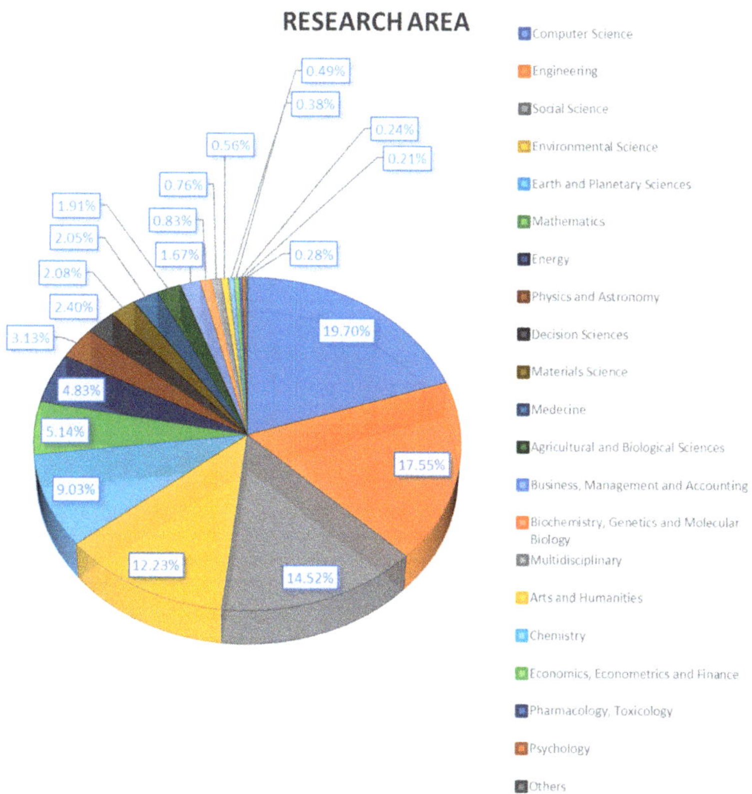

Figure 2. Machine learning on research areas (database: Scopus, authors' elaboration).

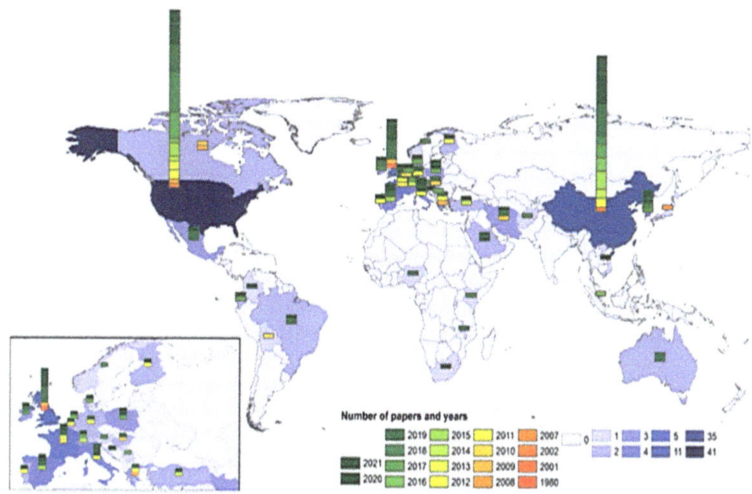

Figure 3. Spatial and temporal distribution of papers on ML related to urban applications by country and year [18].

3. Taxonomy of ML Methods for Urban Applications

Artificial Intelligence, globally, is divided into different parts, namely knowledge representation, genetic algorithms, Artificial Neural Networks (ANN), data mining, etc. The fields of urban planning and engineering are set to expand globally due to their strong, fast-growing relationship to data mining, especially in the smart cities' fields [21].

Despite the ML type and use, the quantity and type of data affect their accuracy and efficiency and enable (or not) the path towards the solution and alternative developments. In this process, Bhavsar et al. [2] underline the importance of problem definition for the appropriate application of ML methods (Figure 4). In reality, problem identification is a complex process depending on different factors, such as data mining, user skills and perceptions, etc.

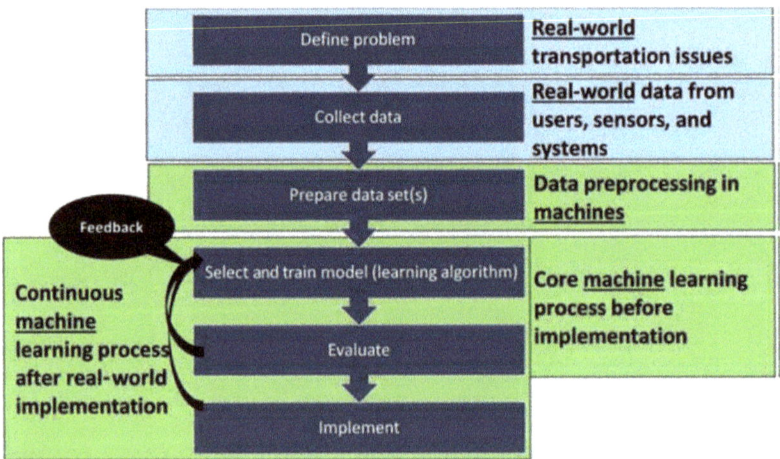

Figure 4. Machine learning algorithm steps.

Generally speaking, the ML methods are categorized based on the type of 'learning'. The most commonly known as follows [2]:

3.1. Supervised Learning

Supervised learning methods deal with a function (or an algorithm) to compute outputs based on given information and present data (e.g., the number of dwellings per ha). This information will be used for an automated process to minimize the possible risks of a prediction error, expressed as the difference between the real (data) and the computed values. Examples of this ML are the binary classifications (True or False), etc., or regression problems.

3.2. Unsupervised Learning

On the other side, unsupervised learning methods depend only on the unlabeled data and aim to identify hidden patterns of data. An example of this category is clustering, which focuses on the data grouping based on similarities or the method of association for the trends' identification concerning a specific problem.

3.3. Machine Learning Algorithms: An Overview

However, the classification and taxonomy of ML require a thorough analysis of a set of attributes when discussing urban developments. Although there are many areas of focus, ML use has a major driver on land use and cover as great support for sustainable development. Nonetheless, despite the rise of smart cities and related concepts and the advancement of big data, etc., there is little evidence regarding classification, simulation or predictions [22]; this section is an step towards the development of this ground.

Murphy [23] proposes three main types of ML methods, namely supervised (predictive) learning to identify a mapping from outputs to inputs considering a specific set of input-outputs, unsupervised learning, where only the inputs are given, and reinforcement learning, which is less commonly used and explain how to perform with the occurrence of given occasional rewards (Figure 5).

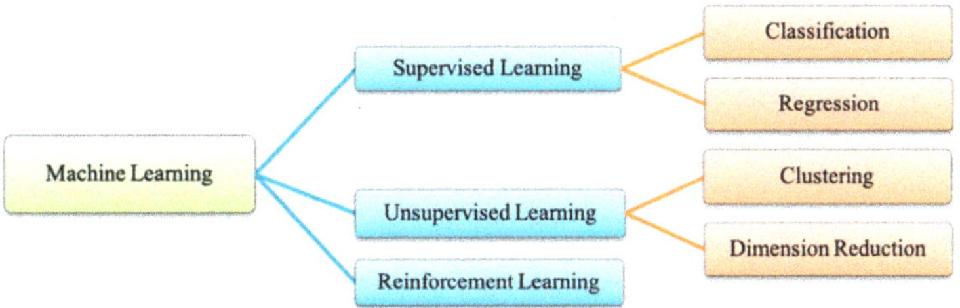

Figure 5. Taxonomy of ML common practices [23].

Emerging methods, such as convolutional neural networks (CNN), proved their efficiency in extracting features from spatial data [24], and recurrent neural networks (RNN) are promising approaches to accurate urban simulations. Examples of successful applications are found in various studies applied to road extraction from the wider perspective of both 2D optical remote sensing images and 3D point clouds commonly used for road data acquisition developed by Chen et al. [25,26]. In the same study, a comprehensive approach to the definition of morphological feature-based tools for road shape features is designed including support vector machines (SVM) ([27]). In the same study, Chen et al. provided three classifications for road area extraction based on traditional methods for identifying of road features (e.g., Lu et al. [28], Perciano et al. [29], etc.) or deep learning [30]).

Ensemble-based methods, such as random forests (RF) and similar methods, are boosted for the problem–solution studies of smart urban forms (e.g., [31–34]). On the other hand, ML methods are commonly used as a promising area to achieve smarter and more inclusive urban configurations in the tissues of modern cities [35].

3.4. Decision-Making Urban Planning Processes

Decision-making processes are fundamental in urban planning strategies, consisting, as they do, of simplified approaches to reality to enable decisions and interactions and allow planners to adjust or modify them in vitro using parametric proposals. Decision-support systems (DSS) facilitate the integration of models and enable the interactions between the diverse parameters to adjust or test solutions and evaluate the consequences leading to desirable and viable solutions. For the special case of predictive modeling, ML has been used for the identification of urban patterns and related indicators. Taking a quick look at the existing literature and the Scopus database correlations, one identifies of 585 documents for ML and the decision-making processes published in the United States and China, as presented in Figure 6.

Among the decision-making tree algorithms, the CART (Classification and Regression Tree) and ID3 (Iterative Dichotomizer 3) are the most commonly used for the land-use classifications acting as a random subset of the predicting parameters. On the other side, random forest (RF) has the particular functionality of both classification and regression analyses and handles an important volume of indicators [36]. In ML use, the DSS processes are usually represented in modeling for predictive forms to enable the decision and improve the design of the forms (Figure 7) [35].

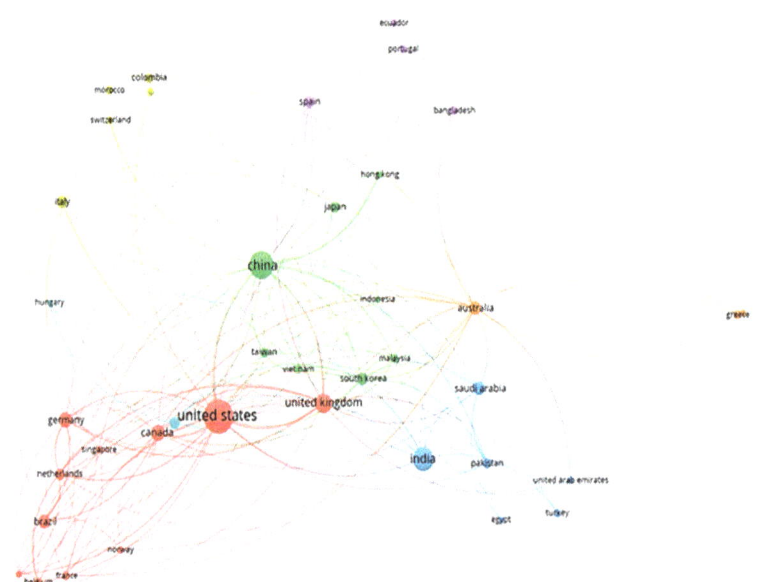

Figure 6. Correlations of ML and secision-making processes, Scopus database.

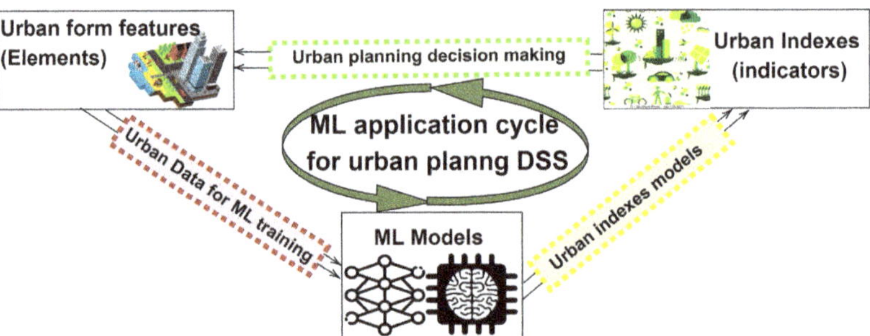

Figure 7. A cycle of ML application for urban form decision support system [35].

4. Review on Urban Applications

Within the wide access and implication of 'big data' and the increasing adoption of urban studies, numerous comprehensive insights are appearing in the literature for studying their potential for shaping the future urban environments in different sectors. The issue is the core of modern planning strategies aiming at digitalization and sustainable challenges. Specifically, to overcome the challenge of land use and cover, planners are integrating an ever-increasing complexity to qualify the dynamics and to manage the complexity [35].

Several ML algorithms have been tested for their performance on different forms of databases respecting the land-use management and/or simulation of land-use planning processes, with the more popular ones, supporting vector machines, neural networks, Markov random or GANS, experimenting on different datasets individually and in combination [37]. At the time being, there is a rising interest in ML use and applications.

Wu and Silva [38] reviewed the AI-based approaches in the projections of land use and their dynamics for spatial planning, Abdulijabbar et al. [39] related them to the mobility problems-solutions, while Yigitcanlar et al. [40] tackle the theme of 'sustainability'. Nevertheless, AI-based applications in urban contexts remain limited to supporting city planning due to the dynamic systems the urban settlements present and the increasing amount of big data mining to obtain new knowledge. AI-based tools move from static to dynamic flows to forecast urban growth and enable spatiotemporal modelling [41] with the typical example being the agent-based modelling (ABM) method for the simulation of bottom-up processes to predict future city development. Patel et al. [42] recommended the GIS integration agent-based modelling for testing informal settings' policies in real-life urban environments, while Patt [43] investigated their applications with a focus on the public space networks.

To identify some representative scientific paradigms and trends, Table 1 summarizes the AI-based approaches in urban fields and explores the linkages of different factors, while the following sections focus on particular applications.

Table 1. Examples of AI-based approaches in city planning.

Urban Theme	Scope	AI-Based Tools	Reference(s)
Polycentricity	Flow analysis and linkages, spatial simulations	Artificial neural networks, fuzzy logic, agent-based models	e.g., [31,32]
Spatial structures and dynamic analyses	Study on the functional structures of the city, mobility configurations, land-use identification	Artificial neural networks, fuzzy logic, agent-based models	e.g., [33,34]
Flows' analyses	Analysis of different types of flows in cities (e.g., energy, mobility, etc.)	Stochastic simulation models, artificial neural networks	e.g., [44–50]
Typo-morphological analysis	Analysis of urban structure, form and space	Stochastic models, Artificial neural metworks	e.g., [51–53]

4.1. Machine Learning and Built Environment

Using ML to improve data collection and management is the priority of their users towards the solutions to complex problems, such as urban themes. Their power is incorporated for building energy efficiency, allowing the analytics of its operations and the identification of solutions to issues of performance and systems' behavior to reduce the use and improve the overall energy management. A typical representation of the ML use in buildings' optimization techniques is found in Kwok and Lee's work [54], as is the use of artificial neural networks (ANN) regarding the cooling predictions and increasing the level of accuracy with the use of fixed schedules and historical data related to the occupancy. Discussing the correlation of ML and energy, the building energy simulation (BEM) has a major role in low energy configurations and the development of advanced skills and knowledge towards clean energy [55].

Schoenfeld [56] cites the optimization in:

- Forecasting energy consumption to reveal trends and predict future energy uses and assist energy planning, management or conservation to reduce the energy demand and the CO_2 emissions [57] and alternative evaluations for an optimized operation to improve demand and supply balances [58]. Nonetheless, the demand for data collection (especially via intelligent sensors/meters) is evident. To that point, Ahmad et al. [59] underline the evolution of energy metering in technological terms, while Chammas et al. [60] analyze the importance of wireless networks and IoT-based methods for energy monitoring and relevant solutions. The literature unveils a significant number of studies related to forecasting activities and prediction performances; for example, Zhaond Liu [61] with respect to a highly-accurate prediction model for the building energy load with dynamic simulations;
- Detection and prediction of faults, in which traditional models do not provide preemptive interventions;
- Seasonality modeling, i.e., correlating themes to seasonal conditions.

The interest in ML use in buildings operations is rising and the AI-based solutions for the conception (design) and operation are found in numerous scientific works (e.g., [62–65]). On the other side, AI drives smart development and enables the advanced use of technologies for buildings' operation and maintenance, examples are found in the HVAC systems or the lighting (e.g., [29,30]). Much research is being explored in AI use and integrated systems on the themes of energy efficiency, thermal comfort, etc. Tien et al. [55] visualize the transition from traditionalal to intelligent techniques (Figure 8).

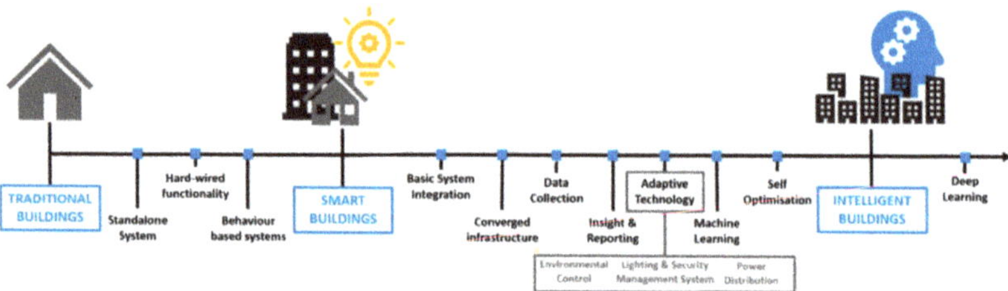

Figure 8. Evolution from traditional to smart techniques on buildings [55].

In line with construction and design, building information modeling—commonly acknowledged as BIM—is becoming the architectural norm to promote innovative and optimized designs with automated processes [66]. Nonetheless, this transition took approximately 25 years to be integrated into the market [67]; Gholizadeh et al. [68] explored the difficulties on this path, explaining that, as late as 2017, only three up to fourteen BIM functions were widespread in practical applications. The methodologies proposed are promising and incorporate both the design/conception and the construction processes for information delivery as a backbone to transcend organizational boundaries and mediate the gaps. The progress is notable considering that, in the absence of BIM and automated processes, scholars proposed stand-alone systems to represent the buildings; examples include HI-RISE for preliminary structural designs of tall buildings [69], SPEX for sizing structural cross-sections [70] or EIDOCC for the design of reinforced concrete [71]. Later on, natural language processing (NLP) was applied in design codes and regulations without further commercialization (e.g., [45,46]).

BIM enabled the automated project performance monitoring and control systems within the introduction of new concepts, e.g., 'Construction 4.0' or Digital Twins with diverse challenges from technical or conceptual standpoints, such as the integration of process information for comparison with monitored data and the need for sophisticated and complex approaches to its management (e.g., [72,73]) or ineffective production for planning and control systems [74].

Complementary to the cited approaches, Liu et al. [75] introduce the three-dimensional approach by three different methods: original building plans, field surveys and remote sensing technologies for the 3D interpretation of architectural features and extrapolation of relevant data and information, such as aerial images [76], light detection and ranging (LIDAR) data (e.g., [77,78]), satellite imagery [79] or even grouping-based stereo [80] or mono images using shadows or digital surface models [81].

4.2. Machine Learning and Land Use

Urban planning tools employ land use and cover to provide historical insights as a base for future urban development [82]. The land-use analysis uses remote sensing geographical information systems (GIS) to simulate the changes and reach a strategic decision for the designated area ([68,69]). Spatiotemporal land-use simulations as reproducible approaches for estimations and future land transitions are driving forces to support land-use policy decisions ([83–85]). These are relevant methods to this topic, commonly acknowledged with rising scientific research (e.g., [86–89]) and used is the cellular automata (CA) for the generation of urban patterns and nonlinear stochastic processes [90] and complicated interpretations of the complexity of the bottom-up model leading to the ignorance of land-use demand estimations [91].

On the other side, Machine Learning is becoming an imperative methodology to monitor ([92–94]) and forecast ([95–97]) the land-use challenges in urban areas. Undoubtedly, statistical and spatial analyses proved their popularity (e.g., [98,99]). For many years, land-use mapping and modelling of geographical, demographical and relevant data have implicated ML as a vital tool to compose models to recognize urban configurations and minimize prediction errors utilizing learning strategies and related drivers [100].

In reality, data-driven models by ML have recently been recognized as powerful means for parametric approaches to land-use distribution (e.g., [101,102]) with multiple benefits, especially in dealing with massive amounts of information and numerous variables. The ability to model sophisticated and non-linear problems [103] used to be dependent on the SVM and the random forest (RF) [104].

Land administration is a core topic in urban planning strategies, which requires multi-faced information on built and non-built-up areas, functionalities, typologies and green and public spaces, to cite some of them. Dealing with this data is a sophisticated process, which includes open-source data provided by various repositories, commercially available satellite images, aerial photographs, cadastral boundary extractions, 3D modelling, etc. Nonetheless, the first stage of the territorial analysis usually demands demographic and statistical analysis of a given population with the need for a comprehensive geospatial database to assess the existing land use and estimate the future projections and the potential possible changes. Depending on the research scope, Chaturvedi and De Vries [11] provide a relevant classification in Table 2 [11].

Table 2. Land use planning indicators with measurements, data required and applications (adapted by [11]).

Theme	Indicators	Data	Application
Urban expansion	Density, demographic profile, built/non-built	EO-based data (e.g., classified images, building footprints)	Classification and simulation (CNN, etc.)
Land restrictions	Land-use/cover, built/non-built-up spaces	A master plan, land-use regulations	Classification, and extraction of EO products (e.g., DEM), spatial logistic regression, Cellular automata
Land distribution	Policies, demographics	Census, socioeconomic data	
Zoning	Land-use distribution	Master Plan, classified images	Planned development
Land-use changes	Settlement patterns, urban growth processes, population growth	Spatiotemporal EO data	Spatial metrics, agent-based modelling

Despite the support of evidence-based algorithmic processes, few provide supportive studies on the land-use theme; examples are found in Shafizadeh-Moghadam et al. [105] of the benchmarking land-use probability models. Karimi et al. [106] detailed the use of ML in land-use changes since 2011 in cellular automata, regression models, artificial neural networks, agent-based models, true-based models and support vector machines (SVM). ML also gained wide acceptance in transportation systems but with significant limitations due to its dependency on agents' predictions. Table 3 overviews the ML models on land use based on this study.

Table 3. Machine Learning use in land-uses strategies (adapted by Karimi et al. [106]).

Machine Learning Use	Scope	Reference(s)
Cellular automata model (CA)	Land-use analysis related to transport and mobility systems (e.g., roads, railways, etc.) and population density issues	e.g., [107,108]
Artificial neural networks (ANN)	Annual population growth and land-use typologies	e.g., [83,84]
Linear regression models	Population density, land-use typology, economic centers analysis	e.g., [85,86]
Agent-based models (ABM)	Accessibility to functions and city infrastructure	e.g., [87,88]
Decision tree model (DT)	Land typologies, proximities to amenities, densities of residential, commercial	e.g., [109–115]
Support vector machines (SVM)	Land-uses typologies, built and unbuilt areas	e.g., [116–120]

Nonetheless, an important challenge on this subject is the artificial absence of data handled by classification models with the commonly acknowledged MAXENT model (maximum entropy) as an advanced ML method originated from information theory ([121,122]). The cutting-edge method that the model proposes evaluates the spatial distribution and was first adopted for the early monitoring of illegal land development in line with the probability analysis and scenario development [123] based on the maximum entropy principles mainly used for the estimation of sustainable natural habitats and the occurrence probability of species, employing one-class classification of remote sensing imageries [124]. Overall, the maximum entropy algorithms minimize the amount of information and are used in ecological modelling (e.g., [121–125]).

- An overview of the correlations of the existing works and studies of ML and urban fields to identify the authors' names per year (1.372 documents) (Figure 9).

Figure 9. Bibliometric analysis of ML and urban applications (authors per year), Scopus database.

- An overview of the correlations of the existing works and studies of ML and urban fields to identify the number of citations per country (1.372 documents) (Figure 10).

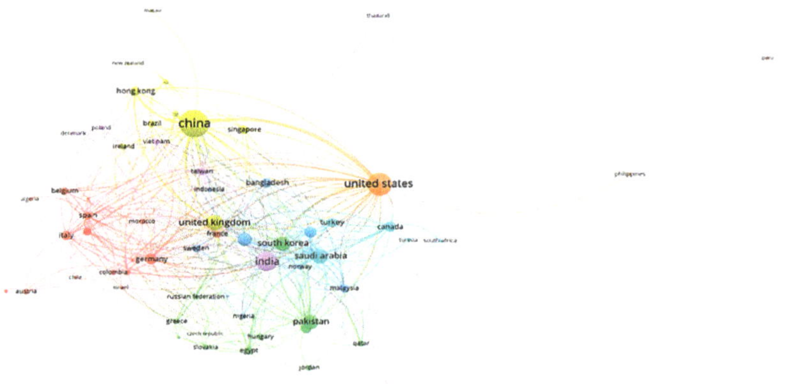

Figure 10. Bibliometric analysis of ML and urban applications (citations per country), Scopus database.

5. Examples of Case-Studies

Two typical examples of the existing case-studies portfolio of urban applications and ML implications are provided in this section, proving the importance of ML solutions to urban planning problems.

5.1. Shanghai Urban Drainage Masterplanning

Dealing with the challenges of rapid urbanization and urban growth (15 million people, a number that has tripled since 1990) to the impermeable area of the territory, Shanghai faces the threat of green space reduction and, consequently, of the rise of stormwater runoff. Considering the previous experiences of urban flooding in recent years, the city proposes in its master plan for a horizon of 2035 an upgraded drainage system. Drawing upon its expertise and successful practices, ARUP, in a collaboration with the Shanghai Urban Construction Design and Research Institute, tailored a strategy to review the traditional approaches of drainage in the city being concentrated on the sensitive and integrated urban design and decentralized infrastructure by applying remote sensing imagery and ML technologies after a comprehensive territorial analysis of the elements concerning the urban fabric of the city (Figure 11) [126].

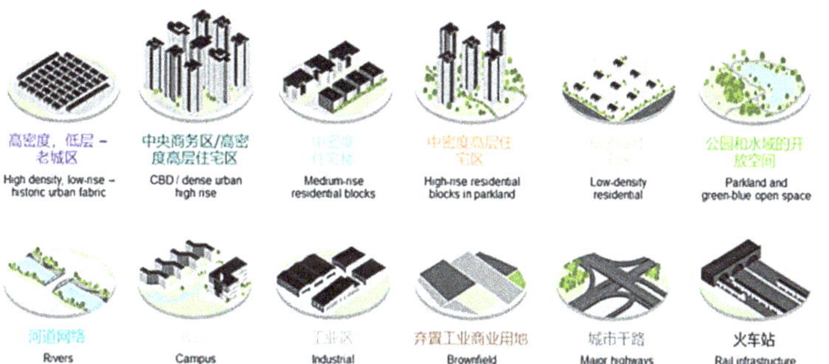

Figure 11. Land characterization of urban typologies of Shanghai.

In 2017, the design of the drainage masterplan had the specific objective of providing climate change adaptability and flood management strategies for the highly populated city center with three core objectives: reducing flooding, restoring clean water and delivering solutions to enhance the QoL in the city and the wider environment. The strategy meets the masterplan ambitions based on the following principles [127]:

- **Integrated:** considering the existing strategies in line with the 'Sponge City' and four elements: (a) a critical overarching system of governance mechanisms for collaboration and synergies; (b) green spaces to promote nature-based solutions (NBS); (c) blue equipment for flood defenses and relevant infrastructure; and (d) 'grey' equipment for drainage treatment (e.g., pumps, etc.);
- **Adaptive:** development of flexible approaches for risk management and uncertainties;
- **Smart:** integration of intelligent and digitalized models for optimization and data treatment of sophisticated scenarios based on planning strategies alongside the stormwater conditions.

Within the ML use, the concept was based on machine learning, artificial intelligence and open-source observations to identify urban typologies and NBS to address the diverse challenges proposing integrated models to improve stormwater management. Based on this approach, remote sensing tools were used for the city scanning and the classification of 12 categories required flooding protection to provide the necessary approaches for targeted water management, which facilitated the decision-making processes around the nature-based solutions beyond economics (Figure 12).

5.2. MassMotion Pedestrian Simulation

Another interesting application developed by ARUP (and commercialized through Oasys, 2014) is the MassMotion simulation for pedestrian movements in a city within ML use and the BIM tools use in a human-oriented design and approach. The goal of this application is to simulate people's movements and optimize the users' journeys in the city in an 'intelligent' way using real-time data, developing at the same time the ability to develop and test multiple alternatives and scenarios in graphical structures and allowing optimal decision in city planning.

Along with a 3D interface, the tool replicates the city models considering a series of actors (agents) and their interactions with the possible planning strategy and evaluates their roles in diverse simulation scenarios. Each of these agents has a specific origin and destination and an outset of microsimulations is performed with the ability to consider alternative solutions adapted to the specific problem [128].

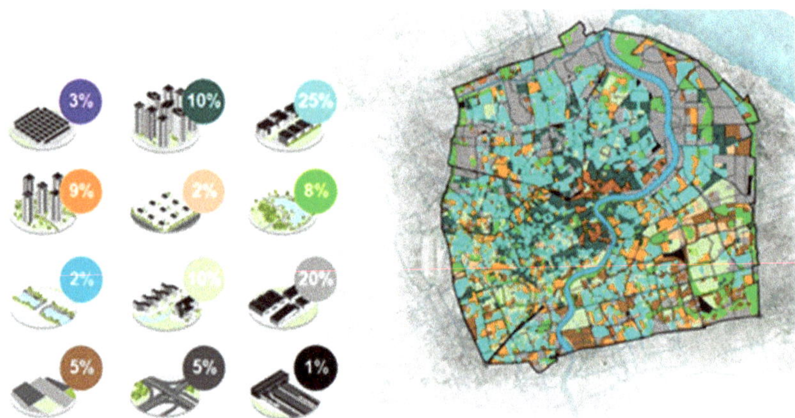

Figure 12. Machine Learning use the master planning area in different urban configurations.

Montjoy [129] characterized MassMotion as one of the world's most advanced simulation tools at the conception phase with strong visualization capabilities. As a stand-alone software, MassMotion illustrates the flows and densities in peak hours, considering crucial factors such as the speed, direction, etc., and creates geometrical models and designs for the planning of the studied transportation hubs; for example, those of the new Trondheim Central station (Norway) project; it defines of future conditions by developing optimized adjustments and validates the possible scenarios to obtain a well-functioning and human-centered station. Examples of this action are the analysis of the pedestrian patterns for the detection of congestion hotspots or crowded spaces, visibility and safety issues and the vertical and horizontal connections. The project lays on the flexibility of the design choices for future expansions by using forecasted passenger data of the related functions in complex 3D environments (Figure 13).

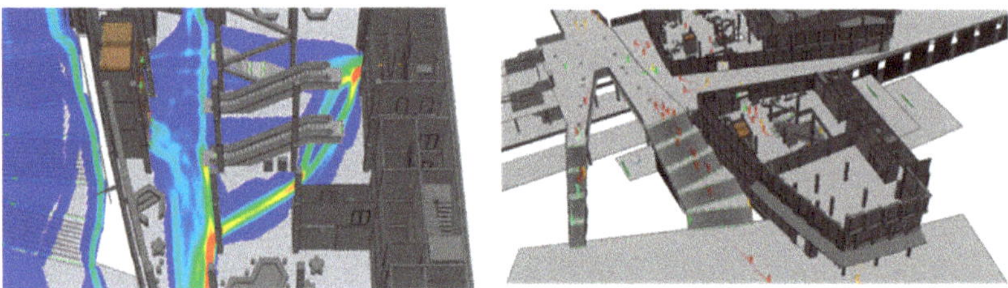

Figure 13. MassMotion applications in 3D complex environments.

6. Conclusions

As living laboratories, cities are facing tremendous complexities in accommodating a growing population and meeting the challenges of climate change and equal living for their citizens. Urban efficiency is being leveraged by uncertain environments with increasing transformations.

Being at the intersections of computer science, statistics and informatics, ML methods lead to more evidence-based solutions and decision-making processes along with the proposal of the dynamics of urban drives. They meet the challenge of 'big data' with a huge potential for smart and sustainable planning leading to resilient and inclusive urban configurations. From these perspectives, ML addresses socioeconomic issues including the challenges of inclusiveness, poverty, and environmental and sustainability issues. The smart city indicators related to digital integration and data management shaping a city more intelligently and autonomously could estimate and evaluate its evolutionary trends (e.g., land-use evolution and definition of estimated needs), anticipate phenomena (or crises) and regulate them accordingly to better direct the growth in a long-term horizon. ML provide new opportunities to better monitor, understand and predict the future and guarantee the wellbeing of future generations

In this digitalized era, and with the rapid growth of computational skills and advancements in artificial intelligence, ML uses in various applications are gaining a rising interest from scholars and practitioners and gaining popularity in many research fields. A particular lever of their implications is being developed in the framework of smart city development and urban design with the use of geospatial data in different aspects of the urban system consisting of multiple tangible (e.g., land use and coverage) and intangible aspects (e.g., social inequalities).

Reflecting the increasing interests in ML uses, several approaches have been proposed in the existing literature towards the direction of enhancing urban dynamics that go beyond the traditional techniques of urban modelling, which is an indispensable tool for planning decision support. A remarkable potential for addressing urban challenges is found in ML methods (e.g., land use/cover, energy efficiency, etc.) consisting of spatiotemporal analyses.

The key contribution of this paper has been to provide a critical angle on the ML taxonomy with respect to its use of the urban planning sector, the methods and tools for urban problems and associated challenges and future research directions. It discusses two representative examples of city applications developed by ARUP. In a scoping review, the authors discussed the insights from an urban planning view to identify the gap in specific applications concerning built and urban environments. They also provided an overview of the existing aspects of the field along with systematic reviews and a thorough bibliometric analysis through a database search (Scopus) to ensure the highest academic standards and the validity of the relevant outcomes, screening and review conduct in charting the main components of the topic.

The ML integration into urban planning strategies will meet the evolutionary challenges in its analysis, simulation and monitoring of the future urban form and its applications. However, there is still an uneven distribution in this area with limited studies that address the challenges and gains of its use for future urban developments. New methods are needed to link the research on ML and urban science with the use of big data and evidence-driven shifts in order to connect these analytic frameworks and support further synergies and in-depth explorations of the practical issues of city challenges, and to increase reproducibility with the construction of a common language and protocols.

Author Contributions: Conceptualization S.K. and C.S.I.; methodology, S.K.; investigation, S.K.; writing—original draft preparation, S.K.; writing—review and editing, S.K. and C.S.I.; supervision, C.S.I.; project administration, C.S.I. All authors have read and agreed to the published version of the manuscript.

Funding: This research received no external funding.

Institutional Review Board Statement: Not applicable.

Informed Consent Statement: Not applicable.

Data Availability Statement: Not applicable.

Conflicts of Interest: The authors declare no conflict of interest.

References

1.   Jordan, M.I.; Mitchell, T.M. Machine Learning: Trends. perspectives and prospects. *Science* **2015**, *349*, 255–260. [PubMed]
2.   Al-Garadi, M.A.; Mohamed, A.; Al-Ali, A.K.; Du, X.; Guizani, M. A survey of machine and deep learning methods for Internet of Things security. *IEEE Commun. Surv. Tutor.* **2020**, *22*, 1646–1685. [CrossRef]

3. Bhavsar, P.; Safro, I.; Bouaynaya, N.; Polikar, R.; Dera, D. Machine Learning in Transportation Data Analytics. In *Data Analytics for Intelligent Transportation Systems*; Elsevier: Amsterdam, The Netherlands, 2017; pp. 283–306.

4. Ross, T. The synthesis of intelligence-its implications. *Psychol. Rev.* **1938**, *45*, 185. [CrossRef]

5. Choung, Y.J.; Kim, J.M. Study of the relationship between urban expansion and PM10 concentration using multi-temporal spatial datasets and the machine learning technique: Case study fo Daegu. *South Korea. Appl. Sci.* **2019**, *9*, 1098. [CrossRef]

6. Fecht, D.; Beale, L.; Briggs, D. A GIS-based urban simulation model for environmental health analysis. *Environ. Model. Softw.* **2014**, *58*, 1–11. [CrossRef]

7. Samuel, A.L. Some studies in Machine Learning using the game of checkers. *IBM J. Res. Dev.* **1959**, *3*, 210–229. [CrossRef]

8. Li, X.; Cheng, S.; Lv, Z.; Song, H.; Jia, T.; Lu, N. Data analytics of urban fabric metrics for smart cities. *Futur. Gener. Comput. Syst.* **2020**, *107*, 871–882. [CrossRef]

9. Schwab, K. The Fourth Industrial Revolution, 1st ed. 2016. Available online: https://law.unimelb.edu.au/__data/assets/pdf_file/0005/3385454/Schwab-The_Fourth_Industrial_Revolution_Klaus_S.pdf (accessed on 12 November 2022).

10. Chaturvedi, V.; De Vries, W. Machine Learning Algorithms for Urban Land Use Planning. *A review. Urban Sci.* **2021**, *5*, 68. [CrossRef]

11. Shaikh Hameed, P.; Mohd Nor, N.; Nallagownden, P.; Elamvazuthi, I.; Ibrahim, T. A review on optimized control systems for building energy and comfort management of smart sustainable buildings. *Renew. Sustain. Energy Rev.* **2014**, *34*, 409–429. [CrossRef]

12. Neri, E.; Coppola, F.; Miele, V.; Bibbolino, C.; Grassi, R. Artificial Intelligence: Who is responsible for the diagnosis? *Radiol. Med.* **2020**, *125*, 517–521. [CrossRef]

13. Samardzic-Petrovic, M.; Kovacevic, M.; Bajat, B.; Dragicevic, S. Machine learning techniques for modelling short term land-use change. ISPRS Int. J. Geo-Inf. 2017, 6, 387.Batty, M. Big data and the city. *Built Environ.* **2016**, *42*, 321–337.

14. Hagenauer, J.; Omrani, H.; Helbich, M. Assessing the performance of 38 Machine Learning models: The case of land consumption rates in Bavaria. Germany. *Int. J. Geogr. Inf. Sci.* **2019**, *33*, 1399–1419. [CrossRef]

15. Robert, C. Machine Learning, a Probabilistic Perspective. *Chance* **2014**, *27*, 62–63. [CrossRef]

16. Gomez, J.A.; Patino, J.E.; Duque, J.C.; Passos, S. Spatiotemporal modeling of urban growth using machine learning. *Remote Sens.* **2020**, *12*, 109. [CrossRef]

17. Lim, T.S.; Loh, W.Y.; Shih, Y.S. A comparison of prediction accuracy. complexity. and training time of thirty-three old and new classification algorithms. *Mach. Learn.* **2000**, *40*, 203–228. [CrossRef]

18. Casali, Y.; Yonca Aydin, N.; Comes, T. Machine learning for spatial analyses in urban areas: A scoping review. *Sustain. Cities Soc.* **2022**, *85*, 104050.

19. Guigoz, Y.; Giuliani, G.; Nonguierma, A.; Lehmann, A.; Mlisa, A.; Ray, N. Spatial data infrastructures in Africa: A gap analysis. *J. Environ. Inform.* **2017**, *30*, 53–62. [CrossRef]

20. Leyk, S.; Gaughan, A.E.; Adamo, S.B.; Sherbinin, A.; Deborah, B.; Sergio, F.; Amy, R.; Stevens, F.R.; Brian, B.; Charlie, F. The spatial allocation of population: A review of large-scale gridded population data products and their fitness for use. *Earth Syst. Sci. Data* **2019**, *11*, 1385–1409. [CrossRef]

21. Varshney, H.; Khna, R.A.; Khan, U.; Verma, R. Approaches of Artificial Intelligence and Machine Learning in Smart Cities: A critical review. *IOP Conf. Ser. Mater. Sci. Eng.* **2021**, *1022*, 012019. [CrossRef]

22. Gao, J.; O'Neill, B.C. Data-driven spatial modeling of global long-term urban land development: The select model. *Environ. Model. Softw.* **2019**, *119*, 458–471. [CrossRef]

23. Murphy, K. Machine Learning: A Probabilistic Perspective. 2012. Available online: http://noiselab.ucsd.edu/ECE228/Murphy_Machine_Learning.pdf (accessed on 12 November 2022).

24. Fathi, S.; Srinivasan, R.; Fenner, A.; Fathi, S. Machine learning applications in urban building energy performance forecasting: A systematic review. *Renew. Sustain. Energy Rev.* **2020**, *133*, 110287. [CrossRef]

25. Chen, Z.; Deng, L.; Luo, Y.; Li, D.; Marcato, J.; Gonçalves, W.N.; Nurunnabi, A.; Li, J.; Wang, C.; Li, D. Road extraction in remote sensing data: A survey. *Int. J. Appl. Earth Obs. Geoinf.* **2022**, *112*, 102833. [CrossRef]

26. Poullis, C.; You, S. Delineation and geometric modeling of road networks. *ISPRS J. Photogramm. Remote Sens.* **2010**, *65*, 165–181. [CrossRef]

27. Wegner, J.D.; Montoya-Zegarra, J.A.; Schindler, K. Road networks as collections of minimum cost paths. *ISPRS J. Photogramm. Remote Sens.* **2015**, *108*, 128–137. [CrossRef]

28. Lu, P.; Du, K.; Yu, W.; Wang, R.; Deng, Y.; Balz, T. A new region growing-based method for road network extraction and its application on different resolution SAR images. *IEEE J. Sel. Top. Appl. Earth Obs. Remote Sens.* **2014**, *7*, 4772–4783. [CrossRef]

29. Perciano, T.; Tupin, F.; Hirata, R.; Cesar, R.M. A two-level markov random field for road network extraction and its application with optical. Sar and multitemporal data. *Int. J. Remote Sens.* **2016**, *37*, 3584–3610. [CrossRef]

30. Khesali, E.; Zoej Valadan, M.J.; Mokhtarzade, M.; Dehghani, M. Semi automatic road extraction by fusion of high resolution optical and radar images. *J. Indian Soc. Remote Sens.* **2016**, *44*, 21–29. [CrossRef]

31. Jochem, W.C.; Bird, T.J.; Tatem, A.J. Identifying residential neighbourhood types from settlement points in a machine learning approach. *Comput. Environ. Urban Syst.* **2018**, *69*, 104–113. [CrossRef]

32. Ma, J.; Cheng, J.C.; Jiang, F.; Chen, W.; Zhang, J. Analyzing driving factors of land values in urban scale based on big data and non-linear machine learning techniques. *Land Use Policy* **2020**, *94*, 104537. [CrossRef]

33. Novack, T.; Esch, T.; Kux, H.; Stilla, U. Machine learning comparison between worldview-2 and quickbird-2-simulated imagery regarding object-based urban land cover classification. *Remote Sens.* **2011**, *3*, 2263–2282. [CrossRef]
34. Shafizadeh-Moghadam, H.; Asghari, A.; Tayyebi, A.; Taleai, M. Coupling machine learning. tree-base and statistical models with cellular automata to simulate urban growth. *Comput. Environ. Urban Syst.* **2017**, *64*, 297–308. [CrossRef]
35. Koumetio Tekouabou, S.C.; Diop, E.B.; Azmi, R.; Jaligot, R.; Chenal, J. Reviewing the application of machine learning methods to model urban form indicators in planning decision support systems: Potential. issues and challenges. *J. King Saud Univ. Inf. Sci.* **2022**, *22*, 5943–5967. [CrossRef]
36. Hernandez Ruiz, I.E.; Shi, W.A. Random forests classification method for urban land-use mapping integrating spatial metrics and texture analysis. *Int. J. Remote Sens.* **2018**, *39*, 1175–1198. [CrossRef]
37. Wubie, A.M.; De Vries, W.T.; Alemie, B.K. A socio-spatial analysis of land-use dynamics and process of land intervention in the peri-urban areas of Bahir Dar City. *Land* **2020**, *9*, 445. [CrossRef]
38. Wu, E.; Silva, N. Artificial Intelligence solutions for urban land dynamics: A review. *J. Plan. Lit.* **2010**, *24*, 246–265.
39. Abdulijabbar, R.; Dia, H.; Liyanage, S.; Bagloee, S.A. Applications of Artificial Intelligence in transport: An overview. *Sustainability* **2019**, *11*, 189. [CrossRef]
40. Yigitcanlar, T.; Kankanamge, N.; Velia, K. How are smart city concepts and technologies perceived and utilized? A systematic Geo-Twitter analysis of smart cities in Australia. *J. Urban Technol.* **2021**, *28*, 135–154. [CrossRef]
41. Kamrowska-Zaluska, D. Impact of AI-based tool and urban big data analytics on the design and planning of cities. *Land* **2021**, *10*, 1209. [CrossRef]
42. Patel, A.; Crooks, A.; Koizumi, N. Spatial Agent-Based Modeling to Explore Slum Formation Dynamics in Ahmedabad, India. In *GeoComputational Analysis and Modeling of Regional Systems*; Springer: Berlin/Heidelberg, Germany, 2018; pp. 121–141.
43. Patt, T.R. Multiagent approach to temporal and punctual urban development in dynamic, informal contexts. *Int. J. Archit. Comput.* **2018**, *16*, 199–211.
44. Grekoussis, G.; Manetos, P.; Photis, Y.N. Modeling urban evolution using neural networks, fuzzy logic and GIS: The case of the Athens metropolitan area. *Cities* **2013**, *30*, 193–203. [CrossRef]
45. Hwang, S.; Lee, Z.; Kim, J. Real-time pedestrian flow analysis unsing networked sensors for a smart subway system. *Sustainability* **2019**, *11*, 6560. [CrossRef]
46. Ibrahim, M.R.; Haworth, J.; Cheng, T. URBAN-i: From urban scenes to mapping slums. transport modes and pedestrians in cities using deep learning and computer vision. *Environ. Plan. B Urban Anal. City Sci.* **2019**, *48*, 76–93. [CrossRef]
47. Soltani, A.; Karimzadeh, D. The spatiotemporal modeling of urban growth case study: Mahabad, Iran. *TeMA J. Land Use Mobil. Environ.* **2013**, *6*, 189–200.
48. Hao, J.; Zhu, J.; Zhong, R. The rise of Big Data on urban studies and planning practices in China: Review and open research issues. *J. Urban Manag.* **2015**, *4*, 92–124. [CrossRef]
49. Byon, Y.J.; Liang, S. Real-time transportation mode detection using smartphones and Artificial Neural Networks: Performance Comparisons between smartphones and conventional global positioning system sensors. *J. Intell. Transp. Syst. Technol. Plan. Oper.* **2014**, *18*, 264–272. [CrossRef]
50. Cheng, X.M.; Wang, J.Y.; Li, H.F.; Zhang, Y.; Wu, L.; Liu, Y. A method to evaluate task-specific importance of spatiotemporal units based on explainable artificial intelligence. *Int. J. Geogr. Inf. Sci.* **2020**, *35*, 2002–2025. [CrossRef]
51. Huang, J.; Obracht-Prondzynska, H.; Kamrowska-Zaluska, D. The image of the city on social media: A comparative study using 'Big Data' and 'Small Data' methods in the Tri-city region in Poland. *Landsc. Urban Plan.* **2021**, *20*, 103977. [CrossRef]
52. Quand, S.J.; Park, J.; Economou, A.; Lee, S. Artificial intelligence-aided design: Smart design for sustainable city development. *Environ. Plan. B* **2019**, *46*, 1581–1599.
53. Wang, W.J.; Wang, W.; Namgung, M. Linking people's perceptions and physical components of sidewalk environments-an application of rough sets theory. *Environ. Plan. B* **2010**, *37*, 234–247. [CrossRef]
54. Kwok, S.S.K.; Lee, E.W.M. A study of the importance of occupancy to building cooling load in prediction by intelligent approach. *Energy Convers. Manag.* **2011**, *52*, 2555–2564. [CrossRef]
55. Wenbin Tien, P.; Wei, S.; Darkwa, J.; Wood, C.; Calautit, J. Machine Learning and Deep Learning Methods for Enhancing Building Energy Efficiency and Indoor Environmental Quality—A Review. *Energy AI* **2022**, *10*, 100198. [CrossRef]
56. Schoenfeld, J. Using Machine Learning to Improve Building Energy Efficiency. 2022. Available online: https://www.buildingsiot.com/blog/using-machine-learning-to-improve-building-energy-efficiency-bd (accessed on 30 November 2022).
57. Amasyali, K.; El-Gohary, N.M. A review of data-driven building energy consumption prediction studies. *Renew. Sustain. Energy Rev.* **2018**, *81*, 1192–1205. [CrossRef]
58. Fan, C.; Wang, J.; Gang, W.; Li, S. Assessment of deep recurrent neural network-based strategies for short-term building energy predictions. *Appl. Energy* **2019**, *236*, 700–710. [CrossRef]
59. Ahmad, M.W.; Mourshed, M.; Mundow, D.; Sisinni, M.; Rezgui, Y. Building energy metering and environmental monitoring-A state of the art review and directions for future research. *Energy Build.* **2016**, *120*, 85–102. [CrossRef]
60. Chammas, M.; Makhoul, A.; Demerjian, J. An efficient data model for energy prediction using wireless sensors. *Comput. Electr. Eng.* **2019**, *76*, 249–257. [CrossRef]
61. Zhao, J.; Liu, X. A hybrid method of dynamic cooling and heating load forecasting for office buildings based on Artificial Intelligence and regression analysis. *Energy Build.* **2018**, *174*, 293–308. [CrossRef]

62. Cheng, C.C.; Lee, D. Artificial Intelligence-Assisted Heating Ventilation and Air Conditioning Control and the Unmet Demand for Sensors: Part I. Problem Forumation and the Hypothesis. *Sensors* **2019**, *19*, 1131. [CrossRef]

63. Ngaramble, J.; Yun, G.Y.; Santamouris, M. The use of Artificial Intelligence (AI) methods in the prediction of thermal comfort in buildings: Energy implications of AI-based thermal comfort controls. *Energy Build.* **2020**, *211*, 109807. [CrossRef]

64. Iddianozie, C.; Palmes, P. Towards smart sustainable cities: Addressing semantic heterogeneity in Building Management Systems using discriminative models. *Sustain. Cities Soc.* **2020**, *162*, 102367. [CrossRef]

65. Alanne, K.; Sierla, S. An overview of machine learning applications for smart buildings. *Sustain. Cities Soc.* **2022**, *76*, 103445. [CrossRef]

66. RICS. Artificial Intelligence: What It Means for the Built Environment? 2017. Available online: https://www.rics.org/uk/news-insight/research/insights/artificial-intelligence-what-it-means-for-the-built-environment/ (accessed on 27 November 2022).

67. Sacks, R.; Girolami, M.; Brilakis, I. Building Information Modelling, Artificial Intelligence and Construction Tech. *Dev. Built Environ.* **2020**, *4*, 100011. [CrossRef]

68. Gholizadeh, P.; Esmaeili, P.; Goorum, B. Diffusion of Building Information Modeling functions in the construction industy. *J. Manag. Eng.* **2018**, *34*, 04017060. [CrossRef]

69. Maher, M.L.; Fenves, S.J. HI-RISE—An Expert System for the Preliminary Structural Design of High Rise Buildings. In *Knowledge Engineering in Computer-Aided Design*; North-Holland: Amsterdam, The Netherlands, 1985.

70. Garrett, J.; Fenves, S.J. A knowledge-based standards processor for structural component design. *Eng. Comput.* **1987**, *2*, 219–238. [CrossRef]

71. Sacks, R.; Buyukozturk, O. Expert interactive design of R/C columns under biaxial bending. *J. Comput. Civ. Eng.* **1987**, *1*, 69–81. [CrossRef]

72. Al Qady, M.; Kandil, A. Concept relation extraction from construction documents using natural language processing. *J. Constr. Eng. Manag.* **2010**, *136*, 294–302. [CrossRef]

73. Brilakis, I.; Soibelman, L.; Shinagawa, Y. Material-based construction site image retrieval. *J. Comput. Civ. Eng.* **2005**, *19*, 341–355. [CrossRef]

74. Ballard, G. The Lean Project Delivery System: An update. *Lean Constr. J.* **2008**, 1–19.

75. Liu, M.; Hu, Y.M.; Li, C.L. Landscape metrics for three-dimensional urban building pattern recognition. *Appl. Geogr.* **2017**, *87*, 66–72. [CrossRef]

76. Suveg, I.; Vosselman, G. Automatic 3D building reconstruction. *Three-Dimens. Image Capture Appl. V* **2002**, *4661*, 59–69.

77. Alexander, C.; Smith-Voysey, S.; Jarvis, C.; Tansey, K. Integrating building footprints and LIDAR elevation data to classify roof structures and visualize buildings. *Comput. Environ. Urban Syst.* **2009**, *33*, 285–292. [CrossRef]

78. Awrangjeb, M.; Ravanbakhsh, M.; Fraser, C.S. Automatic detection of residential buildings using LIDAR and multispectral imagery. *ISPRS. J. Photogramm. Remote Sens.* **2010**, *65*, 457–467. [CrossRef]

79. Khosravi, I.; Momeni, M.; Rahnemoonfar, M. Performance evaluation of object-based and pixel-based building detection algorithms from very high spatial resolution imagery. *Photogramm. Eng. Remote Sens.* **2014**, *80*, 519–528. [CrossRef]

80. Mohan, R.; Nevatia, R. Using perceptual organization to extract 3-D structures. *IEEE Trans. Pattern Anal. Mach. Intell.* **1989**, *11*, 1121–1139. [CrossRef]

81. Weidner, U. Digital Surface Models for Building Extraction. In *Automatic Extraction of Man-Made Objects from Aerial and Space Images*; Springer: Berlin/Heidelberg, Germany, 1997.

82. Hashem, N.; Balakrishnan, P. Change analysis of land use/land cover and modelling urban growth in greated Doha, Qatar. *Ann. GIS* **2014**, *21*, 233–247. [CrossRef]

83. Heistermann, M.; Muller, C.; Ronneberger, K. Land in sight? Achievements, deficits and potentials of continental to global scale land-use modeling. *Agric. Ecosyst. Environ.* **2006**, *114*, 141–158. [CrossRef]

84. Kline, J.D.; Moses, A.; Lettman, G.J.; Azuma, D.L. Modeling forest and range land development in rural locations: With examples from Eastern Oregon. *Landsc. Urban Plan.* **2007**, *80*, 320–332. [CrossRef]

85. Schulp, C.J.; Naburrs, G.; Verburg, P.H. Future carbon sequestration in Europe-effects of land use change. *Agric. Ecosyst. Environ.* **2008**, *127*, 251–264. [CrossRef]

86. Clarke, K.C.; Gaydos, L.J. Loose-coupling a cellular automation model and GIS: Long-term urban growth prediction for San Francisco and Washington/Baltimore. *Int. J. Geogr. Inf. Sci.* **1998**, *12*, 699–714. [CrossRef]

87. Xia, C.; Gar-On Yeh, A.; Zhang, A. Analyzing spatial relationships between urban land use intensity and urban vitality at street block level: A case study of five Chinese megacities. *Landsc. Urban Plan.* **2020**, *193*, 103669. [CrossRef]

88. Liu, H.; Chen, H.; Hong, R.; Liu, H.; You, W. Mapping knowledge structure and research trends of emergency evacuation studies. *Saf. Sci.* **2020**, *121*, 348–361. [CrossRef]

89. White, R.; Engelen, G.; Ulijee, I. The use of constrained cellular automata for high-resolution modelling of urban land-use dynamics. *Environ. Plan. B Plan. Des.* **1997**, *24*, 323–343. [CrossRef]

90. Batty, M.; Couclelis, H.; Eichen, M. Urban systems as cellular automata. *Environ. Plan. B Plan. Des.* **1997**, *24*, 159–164. [CrossRef]

91. Liu, X.; Liang, X.; Li, X.; Xu, X.; Ou, J.; Chen, Y.; Li, S.; Wang, S.; Pei, F. A Future Land-Use model (FLUS) for simulating multiple land use scenarios by coupling human and natural effects. *Landsc. Urban Plan.* **2017**, *168*, 94–116. [CrossRef]

92. Rogan, J.; Franklin, J.; Stow, D.; Miller, J.; Woodcock, C.; Roberts, D. Mapping land-cover modifications over large areas: A comparison of machine learning algorithms. *Remote Sens. Environ.* **2008**, *112*, 2272–2283. [CrossRef]

93. Heung, B.; Chak Ho, H.; Zhang, J.; Knudgy, A.; Bulmer, C. An overview and comparison of machine learning techniques for classification purposes in digital soil mapping. *Geoderma* **2016**, *265*, 62–77. [CrossRef]
94. Omrani, H.; Parmentier, B.; Helbich, M.; Pijanowski, B. The land transformation model-cluster framework: Applying k-means and the Spark computing environment for large scale land change analytics. *Environ. Model. Softw.* **2019**, *111*, 182–191. [CrossRef]
95. Du, G.; Shin, K.J. A comparative approach to modelling multiple urban land use changes using tree-based methods and cellular automata: The case of Greater Tokyo Area. *Int. J. Geogr. Inf. Sci.* **2018**, *32*, 757–782. [CrossRef]
96. Hagenauer, J.; Helbich, M. Local modelling of land consumption in Germany with RegioClust. *Int. J. Appl. Earth Obs. Geoinf.* **2018**, *65*, 46–56. [CrossRef]
97. Veldkamp, A.; Lambin, E.F. Predicting land-use change. *Agric. Ecosyst. Environ.* **2001**, *85*, 1–6. [CrossRef]
98. Tayyebi, A.; Pijanowski, B.C. Modeling multiple land use changes using ANN. CART and MARS: Comparing tradeoffs in goodness of fit and explanatory power of data mining tools. *Int. J. Appl. Earth Obs. Geoinf.* **2014**, *28*, 102–116. [CrossRef]
99. Rienow, A.; Goetzke, R. Supporting SLEUTH-Enhancing a cellular automation with support vector machines for urban growth modeling. *Comput. Environ. Urban Syst.* **2015**, *49*, 66–81. [CrossRef]
100. Brown, D.; Verburg, P.H.; Pontius, R.; Lange, M.D. Opportunities to improve impact. integration and evaluation of land change models. *Curr. Opin. Environ. Sustain.* **2013**, *5*, 452–457. [CrossRef]
101. Song, J.; Kim, J.; Lee, J.K. NLP and deep learning-based analysis of building regulations to support automated rule checking system. In *Proceedings of the 35th International Symposium on Automation and Robotics in Construction (ISARC)*; IAARC Publications: Waterloo, ON, Canada, 2018.
102. Zhang, J.; El-Gohary, N. Integrating semantic NLP and logic reasoning into a unified system for fully-automated code checking. *Autom. Constr.* **2017**, *73*, 45–57. [CrossRef]
103. Witten, H.; Frank, E. Data Mining: Practical Machine Learning tools and techniques with java implementations. *ACM SIGMOD Rec.* **2002**, *31*, 76–77. [CrossRef]
104. Nagappan, S.D.; Mohd Daud, S. Machine Learning Predictors for Sustainable Urban Planning. *Int. J. Adv. Comput. Sci. Appl.* **2021**, *12*, 1–9. [CrossRef]
105. Shafizadeh-Moghadam, H.; Helbich, M. Spatiotemporal variability of urban growth factors: A global and local perspective on the megacity of Mumbai. *Int. J. Appl. Earth Obs. Geoinf.* **2015**, *35*, 187–198. [CrossRef]
106. Karimi, F.; Sultana, S.; Babakan Shirzadi, A.; Suthaharan, S. An enhanced support vector machine model for urban expansion prediction. *Comput. Environ. Urban Syst.* **2019**, *75*, 61–75. [CrossRef]
107. Mohamed, M.; Anders, A.; Schneider, C. Monitoring of changes in land use/land cover in Syria from 2010 to 2018 using multitemporal landsat imagery and GIS. *Land* **2020**, *9*, 226. [CrossRef]
108. Karakus, C.B. The impact of land use/land cover changes on land surface temperature in Sivas city center and its surroundings andd assessment of Urban Heat Island. *Asia Pacific Dev. J. Atmos. Sci.* **2019**, *55*, 69–684.
109. Deep, S.; Saklani, A. Urban sprawl modeling using cellular automata. *Egypt. J. Remote Sens. Sp. Sci.* **2014**, *17*, 179–187. [CrossRef]
110. Vaz, E.N.; Nijkamp, P.; Painho, M.; Caetano, M. A multi-scenario forecast of urban change: A study on urban growth in the algarve. *Landsc. Urban Plan.* **2012**, *104*, 201–211. [CrossRef]
111. Musa, S.I.; Hashim, M.; Reba, M.N.M. A review of geospatial-based urban growth models and modelling initiatives. *Geocarto Int.* **2016**, *32*, 813–833. [CrossRef]
112. Alsharif, A.A.A.; Pradhan, B. Urban sprawl analyis of Tripoli metropolitan city (Libya) using remote sensing data and mutlivariate logistic regression models. *J. Indian Soc. Remote Sens.* **2013**, *42*, 149–163. [CrossRef]
113. Tripathy, P.; Bandopadhyay, A.; Raman, R.; Singh, S.K. Urban growth modeling using logistic regression and geo-informatics: A case of Jaipur. India. *Int. J. Sci. Technol.* **2018**, *13*, 47–62.
114. Motieyan, H.; Mesgari, M.S. An agent-based modeling approach for sustainable urban planning from land use and public transit perspectives. *Cities* **2018**, *81*, 91–100. [CrossRef]
115. Shirzadi Babakan, A.; Taleai, M. Impacts of transport development on residence choice of renter households: An agent-based evaluation. *Habitat Int.* **2015**, *49*, 275–285. [CrossRef]
116. Zhang, Q.; Vatsavai, R.R.; Shashidharan, A.; Van Berkel, D. Agent-based urban growth modeling framework on apache spark. In *Proceedings of the 5th ACM SIGSPATIAL Internationl Workshop on Analytics for Big Geospatial Data, BigSpatial*; Association for Computing Machinery: New York, NY, USA, 2016.
117. Jin, H.; Mountrakis, G. Integration of urban growth modelling products with image-based urban change analysis. *Int. J. Remote Sens.* **2013**, *66*, 127–137. [CrossRef]
118. Samardzic-Petrovic, M.; Dragicevic, S.; Kovacevic, M.; Bajat, B. Modeling urban land use changes using support vector machines. *Transcactions GIS* **2016**, *20*, 718–734. [CrossRef]
119. Deng, Y.; Srinivasan, S. Urban land use change and regional access: A case study in Beijing. China. *Habitat Int.* **2016**, *51*, 103–113. [CrossRef]
120. Samara, F.; Tampekis, S.; Sakellariou, S.; Christopoulou, O. Sustainable indicators for land use planning evaluation: The case of a small greek island. In Proceedings of the 4 th International Conference on Environmental Management, Engineering, Planning and Economics (CEMEPE), Mykonos, Greece, 24–28 June 2013.
121. Rahmati, O.; Golkarian, A.; Biggs, T.; Keesstra, S.; Mohammadi, F.; Daliakopoulos, I.N. Land subsidence hazard modeling: Machine learning to identify predictors and the role of human activities. *J. Environ. Manag.* **2019**, *236*, 466–480. [CrossRef]

122. Tarabon, S.; Bergès, L.; Dutoit, T.; Isselin-Nondedeu, F. Environmental impact assessment of development projects improved by merging species distribution and habitat connectivity modelling. *J. Environ. Manag.* **2019**, *241*, 439–449. [CrossRef] [PubMed]

123. Lin, J.; Li, H.; He, X.; Zhuang, Y.; Liang, Y.; Lu, S. Estimating potential illegal and development in conservation areas based on a presence-only model. *J. Environ. Manag.* **2022**, *321*, 115994. [CrossRef] [PubMed]

124. Li, W.; Guo, Q. A maximum entropy approach to one-class classification of remote sensing imagery. *Int. J. Remote Sens.* **2010**, *31*, 2227–2235. [CrossRef]

125. Townsend Peterson, A.; Papes, M.; Eaton, M. Transferability and model evaluation in ecological niche modeling: A comparison of GARP and Maxent. *Ecography* **2007**, *30*, 550–560. [CrossRef]

126. Cavendish, W. Artificial Intelligene and Machine Learning. Human Creativity Augmented by the Immense Power of Machines. 2022. Available online: https://www.arup.com/services/digital/artificial-intelligence-and-machine-learning (accessed on 24 November 2022).

127. Sagris, T.; Zhao, M. Shanghai's Urban Drainage Masterplan—A Vision for 2030. 2022. Available online: https://www.ciwem.org/the-environment/shanghai\T1\textquoterights-urban-drainage-masterplan (accessed on 24 November 2022).

128. Rivers, E.; Jaynes, C.; Kimball, A.; Morrow, E. Using case study data to validate 3D Agent-based pedestrian simulation tool for building egress modeling. *Transp. Res. Procedia* **2014**, *2*, 123–131. [CrossRef]

129. Montjoy, V. Powerful Crowd Simulation Software for Human-Centered Design. 2022. Available online: https://www.archdaily.com/990775/powerful-crowd-simulation-software-for-human-centered-design (accessed on 30 November 2022).

 *land*

Article

Uncovering Network Heterogeneity of China's Three Major Urban Agglomerations from Hybrid Space Perspective-Based on TikTok Check-In Records

Bowen Xiang [1], Rushuang Chen [2] and Gaofeng Xu [3,*]

[1] School of Urban Design, Wuhan University, Wuhan 430072, China
[2] China Southwest Architectural Design and Research Institute Corp. Ltd., Chengdu 610041, China
[3] School of Architecture and Design, Beijing Jiaotong University, Beijing 100044, China
* Correspondence: gfxu@bjtu.edu.cn

Abstract: Urban agglomeration is an essential spatial support for the urbanization strategies of emerging economies, including China, especially in the era of mediatization. From a hybrid space perspective, this paper invites TikTok cross-city check-in records to empirically investigate the vertical and flattened distribution characteristics of check-in networks of China's three major urban agglomerations by the hierarchical property, community scale, and node centrality. The result shows that (1) average check-in flow in the Yangtze River Delta, Beijing-Tianjin-Hebei, and Pearl River Delta network decreases in descending order, forming a Z-shaped, single-point radial, and N-shaped structure, respectively. (2) All three urban agglomerations exhibit a nexus community structure with the regional high-flow cities as the core and the surrounding cities as the coordinator. (3) Geographically proximate or recreation-resource cities have a high degree of hybrid spatial accessibility, highlighting their nexus role. Finally, the article further discusses the flattened evolutionary structure of the check-in network and proposes policy recommendations for optimizing check-in networks at both the digital and geospatial levels. The study gains from the lack of network relationship perspective in the study of location-based social media and provides a novel research method and theoretical support for urban agglomeration integration in the context of urban mediatization.

Keywords: urban network; hybrid space; TikTok; three major urban agglomerations of China

Citation: Xiang, B.; Chen, R.; Xu, G. Uncovering Network Heterogeneity of China's Three Major Urban Agglomerations from Hybrid Space Perspective-Based on TikTok Check-In Records. *Land* **2023**, *12*, 134. https://doi.org/10.3390/land12010134

Academic Editor: Chuanrong Zhang

Received: 21 November 2022
Revised: 21 December 2022
Accepted: 27 December 2022
Published: 31 December 2022

1. Introduction

In the context of rapid globalization, urban agglomerations, as an advanced form of regional urbanization, formulate multiple cities into a mega-city system with continuous spatial patterns and close functional connections. Promoting the development of urban agglomeration has been considered an essential part of urbanization strategies in China and even in emerging economies worldwide. Meanwhile, with the innovation of communication and information technology, the dominant urban network is no longer "local space" but "flow space" [1]. The interaction between local space and flow space gradually transforms the traditional hierarchical structure into a networked one [2], which further brings about changes in spatial form, structure, and function of cities and regions [3], and the resulting network organization eventually becomes an essential structural element of the economic and social system. With the signing of the United States-Mexico-Canada Agreement and the official implementation of the Horizon 2020 plan, economic and scientific cooperation between different countries and cities has been promoted, further strengthening urban network development [4,5]. Similarly, China has also implemented regional integration and spatial network development policies in major urban agglomerations such as the Yangtze River Delta Integration and the Pearl River Delta Integration [6,7]. These phenomena reflect the importance of strengthening urban networks for spatial optimization and the high-quality development of urban agglomerations. Therefore, identifying the flow network characteristics of urban agglomerations is vital for optimizing the regional spatial structure and promoting regional collaborative development.

Castell [8], Hall [9], and Taylor [10] have laid the theoretical foundations for the study of urban agglomeration networks. Related research has focused on transport, economic, innovation, and tourism networks. Road [11,12], rail [13–15], and airflow [16,17] data were used to characterize transport networks. Corporate headquarters branch [18,19], listed companies' off-site investment data [4], and energy consumption [20] data were used to characterize economic networks. Academic papers [21,22], invention patents [23], logistics, and transportation [24] were used to represent the innovation network structure. Questionnaires [25], online travelogue texts [26], online travel booking data [27,28], and taxi tracks [29] were used to characterize the tourism flow network. Multiple factor flows are used to synthetically describe the structure of urban networks within urban agglomerations and provinces [30–33]. These studies explore the overall topological features and spatial structure characteristics of urban networks and identify the characteristics of urban networks, such as scale-free, small-world, hierarchical hierarchy, and spatial agglomeration [34,35]. Chinese research mainly focuses on the major urban agglomerations, such as the Yangtze River Delta, the Pearl River Delta, and the Beijing-Tianjin-Hebei region. It is found that the structural characteristics of different types of flow networks exhibit different spatial patterns. Still, the three major urban agglomerations' structures show a shift from a hierarchical system to a flat network and present a multi-core network shape [36,37]. Some scholars have also suggested after comparative analysis that the Yangtze River Delta cities have the most robust horizontal connections and the strongest integration, the Pearl River Delta is the second, and Beijing-Tianjin-Hebei is the weakest [38]. In conclusion, the above studies have explored the urban network structure characteristics of urban agglomerations from a multidimensional perspective. As SMPs are increasingly integrated into residents' daily lives and influence their travel patterns, it is necessary to dissect the mobility patterns of media users in urban agglomerations and thus examine the impact of social media platforms on regional integration.

The theory of hybrid space offers a new perspective on the urban agglomeration network. With the increasing popularity of information and communication technologies, location-based social media platforms (SMPs) have become one of the most common virtual spaces in everyday life, linked to physical space through various geo-tagged and real-time logging data, delimiting neospatiality with its logic and structure [39]. Souza introduces the concept of hybrid space at the beginning of the 21st century and pointed out three main characteristics of hybrid space as mobile and social space, namely the blurring of the physical-digital spatial boundary, the physical carrying-in of the static-mobile interface, and the reconfiguration of urban space [40]. Similarly, Soja proposes a triadic dialectic of 'history-society-space' and constructs the third space theory [41]. He points out that the third space is a hybrid space that transcends physical space (the first space) and imaginary space (the second space) and is composed of sensory experiences, intuitive experiences, and abstract symbols. He asserts that the third space is characterized by complete openness, reconstruction, and the transcendence of relations of production and space [42]. The theory of hybrid space and third space lays the foundation for the spatial epistemology of media and communication geography. Along with the multi-functional development of short video applications, they can satisfy not only the essential functions of browsing, entertainment, and recording daily life, but also multiple functions such as socializing with fans, professional learning, and live shopping. As a result, short video applications such as Tik Tok, Auto Quicker, and Xiaohongshu are rapidly overtaking Weibo, WeChat, and traditional news media in terms of downloads and views. The third space it represents breaks the long-standing binary separation between immaterial media texts and material geographical landscapes, integrating into the space of everyday life and completing the reproduction of spatial relations.

Hybrid spaces influence the spatial dynamics of cities and the correlation between multiple geospatial units by shaping "communication networks". Social annotative, or "check-in" behavior, is a typical spatial, social practice shaped by such correlation. Check-in users obtain urban spatial information in digital space and then take videos in geospatial space and upload them to digital space, thus forming a set of check-in behaviors. In this process, check-in users, on the one hand, descend from digital space to geographic space, driving the infiltration and interaction between digital space and geographic space; on the other hand, they move from one space to another, strengthening the geographic interaction between different urban spaces, and even between different cities. Previous studies have used the number of geo-tagged images, records, comments, and check-ins to capture people's activities in physical space. Paldino et al. analyzed the number of geo-tagged Flickr images [43], and Sulis et al. used spatial information recorded by Twitter to characterize the spatial distribution of Londoners' activities [44]. Several studies in China have also used geolocation tags [45,46] on social networking sites, the inter-city Baidu index [47,48], Baidu Post Bar [49,50], and Douban [51] to characterize the urban network patterns [52] constructed by information flow. Studies on check-in behavior generally regard it as a representation of spatial vitality, focusing on identifying the spatial characteristics of user activity [53,54] and followership [55], but neglecting the impact of SMPs on geospatial interactions. SMPs change the popularity of geographic space but also the mobility of people between different geographic spatial units, shaping the spatial interaction pattern between cities. With location-based social media embedded in daily life, social platforms such as TikTok and Instagram have gradually replaced television advertisements as a channel for people to obtain spatial information about cities. Social-targeted behaviors such as visiting and check-in have become increasingly popular, intensifying the shaping of inter-city connectedness. Therefore, we consider the hybrid space a valuable addition to describing the urban network. It is necessary to analyze how check-in behavior shapes the interactions between cities to respond to the increasing mediatization of cities.

Given this, the current paper adopts a hybrid space perspective to construct urban association networks in three major urban agglomerations using cross-city check-in data from Tik Tok and uses social network analysis to analyze the hierarchical attributes, community scale, and node centrality of cross-city check-in networks, based on which, it attempts to summarize the spatial organizational patterns of check-in networks.

The significance of this paper is mainly reflected in the following aspects. This paper constructs a framework for analyzing urban networks based on a hybrid spatial perspective and conducts a comparative study with three major urban agglomerations as a case study, which provides a new view and method for the analysis of urban networks and urban agglomeration integrations, and provides theoretical support for promoting the synergistic regional development of urban agglomerations. In addition, this paper introduces Jitterbug cross-city punch card data to characterize inter-city association patterns. It verifies the method validity, which gains from the lack of spatial interaction in geographic annotation behavior research, and provides new data and methods for media geography research.

The remainder of the paper is structured as follows: Section 2 introduces the study area of this paper, the data sources, and the methods used in this paper. Section 3 shows the results of these methods. In Section 4, we have a further discussion of these research results. Section 5 is the conclusion of the paper.

2. Data and Methods

2.1. Study Area

The area of this paper is the three major urban agglomerations in China, i.e., the Yangtze River Delta (YRD), the Pearl River Delta (PRD), and the Beijing-Tianjin-Hebei (BTH) (Figure 1). Among the 19 urban agglomerations in China, the YRD, PRD, and BTH are the three most economically active urban agglomerations, with high shares of tertiary industries, penetration rates of geographic media facilities, and increased numbers of media users. According to the "Statistics Yearbook of 2021 China's Top nineteen Urban Agglomerations (Giant Engine Urban Institute)", the YRD, PRD, and BTH urban agglomerations rank among the top three in terms of TikTok online prosperity. This shows that the three major urban agglomerations are more mature in terms of hardware and software for location-based social media, which is a typical model for examining the movement of people between cities in a hybrid space.

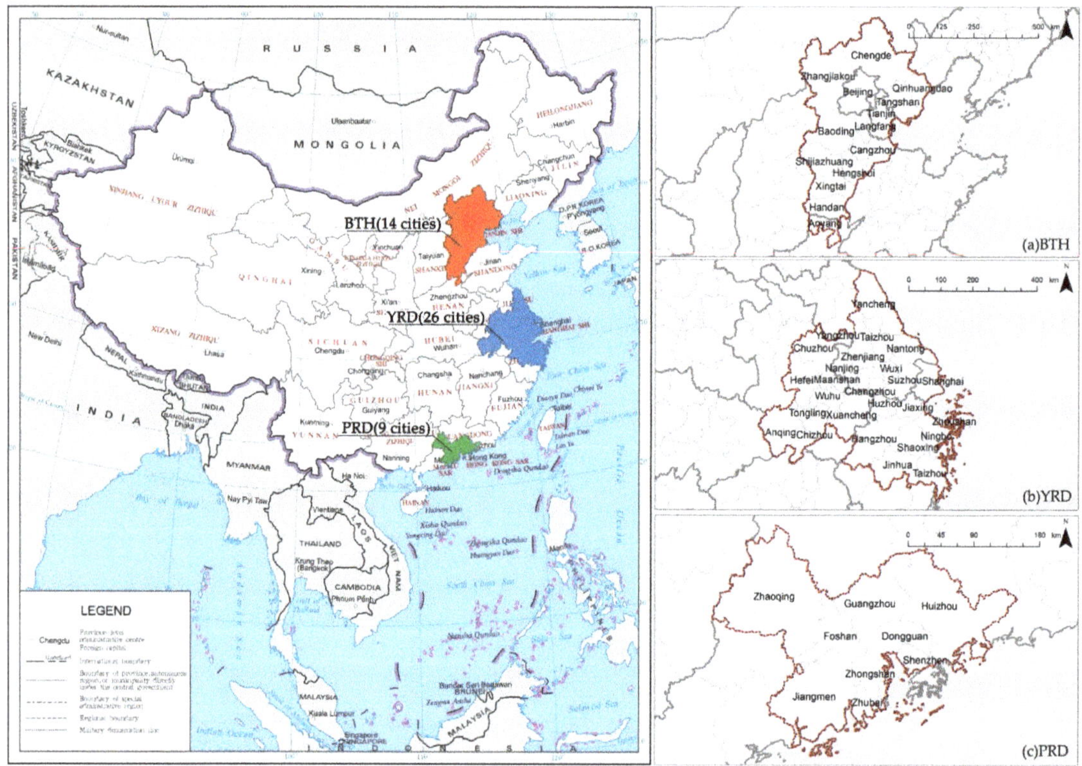

Figure 1. Location of three major urban agglomerations of China.

2.2. Data Sources

This paper chooses the TikTok short video platform (www.douyin.com accessed on 1 August 2022) as the location-based social media check-in data source. TikTok is currently one of China's most popular platforms for producing and disseminating short videos. As of June 2022, the number of active users of TikTok was 697.93 million, which ranked first in the sector. With the increase in mobile phone penetration, short videos have become one of the critical digital scenarios for users to access information, and the public has accepted TikTok. To strengthen offline-online interaction and promote the physical tourism industry, the TikTok platform has launched a series of online and offline check-in activities over the past few years, attracting a large number of users to spontaneously share and spread the word, leading to the creation of urban online scenes and boosting the "check-in economy" with recreational activities as the primary purpose. Based on this, this paper writes a crawler by Python and obtained 263,791 check-in records within the three major urban agglomerations from 1 August to 7 August 2022 (Table 1). We pay particular attention to only the check-in locations and origin city attached to check-in users rather than the videos themselves.

Table 1. Cross-city check-in data samples.

Oid	Date	User_City	Check-In_City
1	2022/8/1	Zhaoqing	Zhuhai
2	2022/8/1	Foshan	Zhaoqing
3	2022/8/1	Jiangmen	Guangzhou
4	2022/8/1	Guangzhou	Shenzhen
5	2022/8/1	Shenzhen	Guangzhou
6	2022/8/1	Dongguan	Huizhou
7	2022/8/1	Foshan	Guangzhou
8	2022/8/1	Guangzhou	Foshan
9	2022/8/1	Zhongshan	Huizhou
10	2022/8/1	Guangzhou	Foshan

The process of check-in behavior is usually as follows: after being attracted to a scenario on an Internet platform, media users tend to travel to the physical space recorded in the digital space. Then the users usually record the interaction between humans and the physical space along with the geographic location in a short video and upload the SMPs again, thus completing the "check-in" of a geospatial space (Figure 2). Therefore, the check-in behavior is an offline representation of virtual space, which connects the virtual and physical spaces and enhances the geographical interaction between different cities. During this process, users wander through virtual and physical spaces and create more and more artworks online, eventually improving the vitality of urban spaces and influencing the correlation between multiple geospatial units.

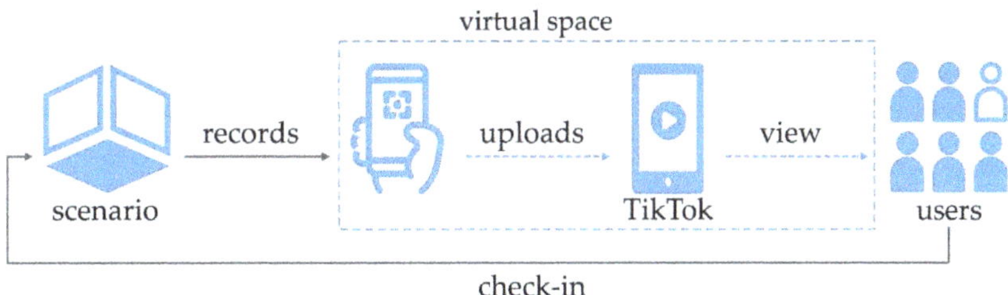

Figure 2. Diagram of the check-in process.

2.3. Methods

The technical route of this study is shown in Figure 2 below. According to Figure 3, our work was divided into three parts: data collection, check-in network modeling, and network characteristics evaluation.

2.3.1. Modeling the check-in network

Drawing on the current research on travel flows based on travelogue data, we propose the following methods to model a check-in network:

(1) Address resolution. Based on the check-in data, we locate the city from where the check-in users originate $(City_{origin})$ and the city to which the check-in location belongs $(City_{target})$. City names are geocoded through the AMap Web API (https://restapi.amap.com/v3/geocode/geo?parameters, access date: 21 September 2022) to get latitude and longitude coordinates.

(2) Data filtering. The data from the same city as $City_{origin}$ and $City_{target}$ are eliminated, and the 23,301 records from different cities are retained as cross-city check-in data.

(3)   Network modeling. We aggregate cross-city check-in data according to city units and transform it into an OD matrix. Point O is the city from which the check-in users originated ($City_{origin}$), point D is the city to which the check-in location belongs ($City_{target}$).

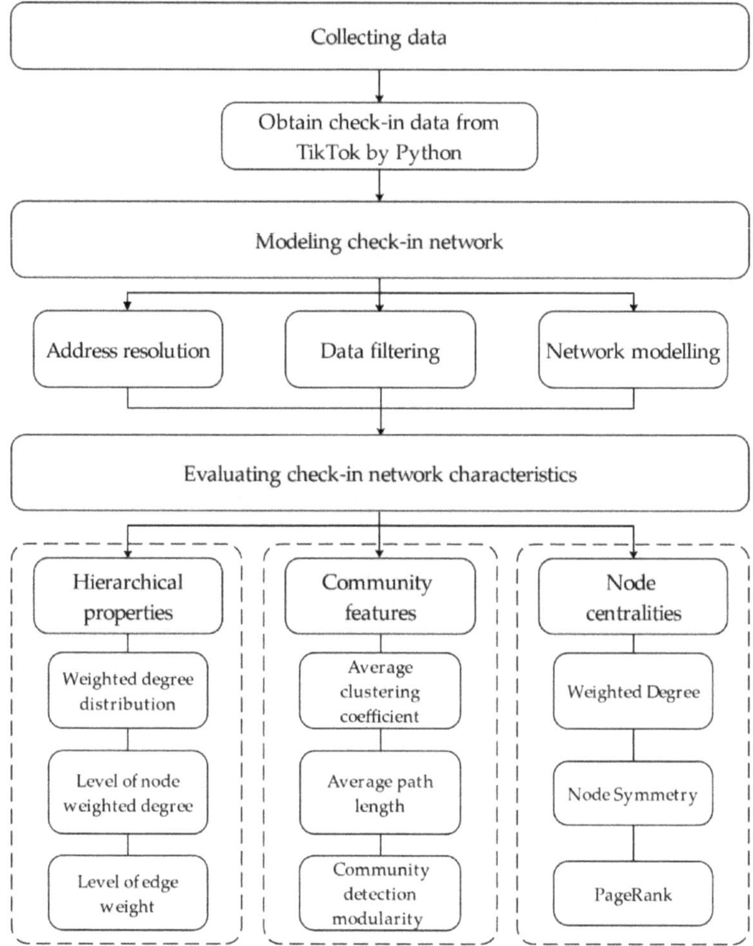

Figure 3. The technical route diagram.

Finally, the matrix was fed into the Gephi software to generate a graphical network of check-in flows. The network type is a directed, weighted network. The network nodes are the municipalities within the study area. The network edge weights are the check-in flows delivered from one city to another, characterized by the sum of the above check-in frequencies.

2.3.2. Evaluating the Characteristics of the Check-In Network

This paper mainly adopts the social network analysis (SNA) method to evaluate the check-in network characteristics. SNA method is a quantitative analysis method developed on the mathematical method and graph theory, which conceptualizes each subject in the social relationship into independent points, converts various relationships between subjects into lines, and analyzes the laws and characteristics of social structure through different quantitative data of nodes and networks [56]. This method has been widely used in the urban network, urban cluster structure, and population mobility. The check-in network studied in this paper is a mobile network of cross-city check-in holders, which belongs to one of the types of population mobility networks, and what it characterizes is the spatial interaction between cities in a diverse spatial perspective and explores the structure of urban clusters, which applies to the network research paradigm.

Integrating a hybrid space perspective with existing spatial network research, this paper selects indicators related to the social network analysis method to examine the check-in networks of the three major urban agglomerations in terms of hierarchical property, community range, and node centrality, respectively.

The traditional vertical town system has been impacted in the information age and evolved into a flat structure. In a hybrid spatial perspective, the network of punching streams is influenced by both digital and geographical space; the check-in behavior is more likely to be embedded in the short-distance recreation function, which is more susceptible to the geographical distance factor. Therefore, it needs to be further examined whether the urban spatial network constructed by the check-in flow is a vertical-distributed structure or has shifted to a flattened one. This paper examines the vertical and flattened distribution of the check-in network by analyzing hierarchical and community characteristics, respectively. In addition, based on the overall network characteristics portrayed above, this study conducts individual network characteristics through node characteristics analysis.

1.   Hierarchical property

The weighted degree is a fundamental indicator of complex networks. The weighted degree in a check-in network indicates the total number of check-in flows generated in the city. The weighted degree distribution refers to the probability of the weighted degree of the network nodes. In this paper, we analyze the hierarchical properties among the nodes by examining the scale-free property of the weighted degree distribution. The scale-free property means that most nodes in a complex network have minimal weighting, but conversely, a few nodes have a tremendous amount of weighting. Existing research on urban networks has found that innovation, trade, enterprise, and tourism flow networks are scale-free. Still, it remains to be examined whether check-in networks have this property. In this paper, we use a power function in the logarithmic form to fit the scale-free property of check-in networks. The algorithm is as follows:

$$K_h = P\left(K_h^*\right)^a \tag{1}$$

$$\ln K_h = \ln P + a \ln K_h^* \tag{2}$$

where K_h denotes the weighted degree of node h; K_h^* denotes the ranking of the weighted degree values of node h; P is a constant; and a denotes the slope of the weighted degree distribution curve. The larger the value of a, the more pronounced the network hierarchy is.

Further, the natural discontinuity method is used to classify the node weighting degree and edge weights. A spatial network map is drawn based on ArcGIS to analyze inter-city check-in flows' spatial vertical distribution characteristics.

In addition, we used the average weighted degree, the average degree, and the number of nodes to examine the size of the network. The average weighted degree is the average sum of the weighted degrees of the entire network and characterizes the average amount of punching traffic formed by each node. The average degree is the average of the whole network of degrees, representing the average number of cities connected per city.

2.   Community scale

A community is a structural unit within a network, with relatively dense connections between nodes within a community and sparse connections between communities, creating a parallel rather than vertical structure. Analyzing the community structure of the check-in network identifies the well-connected urban assemblages in a diverse spatial perspective. It reveals the degree of flattening of the check-in network structure. The modularity algorithm is commonly used to classify communities, which is an efficient and accurate method for medium-sized networks but fails to consider weighted information. This paper uses a weighted modularity algorithm for community segmentation of check-in networks. Q-value is a metric to evaluate the results of community segmentation. A higher Q-value means the more significant the module segmentation feature. This means the more obvious division between communities and the more flattening of the check-in network. A value greater than 0.3 is generally considered to be a significant degree of network modularity. The algorithm is as follows.

$$Q = \frac{1}{2m} \sum_{i,j} \left[w_{ij} - \frac{k_i}{k_j} \right] \delta(c_i, c_i) \tag{3}$$

where Q is the module degree value, w_{ij} is the edge weight between city i and j; k_i and k_j are the degree values of city i and j in the unweighted network; c_i and c_j are the communities into which city i and j are divided; $m = \frac{1}{2} \sum_{i,j} w_{ij}$ is the sum of all weights in the network.

In addition, this paper examines the small-world phenomenon of the check-in network through the average clustering coefficient and the average path length. The network structure of sparse random long connections accompanied by rich partial connections revealed by the small-world phenomenon is essentially an interpretation of community structure. Established research has commonly examined whether the small-world properties of real networks are significant by comparing them with stochastic models. Specifically, a network is said to have a small-world phenomenon if its average clustering coefficient is much larger than a random network. In contrast, its average path length is comparable.

3. Node centrality

This paper applies weighted degrees to examine the intensity of check-in flows. It assesses whether cities prefer to export or receive check-in flows in a cross-city network by comparing the weighted indegree with the weighted-out degree. The activities carried out by the cross-city check-in flows regulate income distribution by tourism consumption. The difference is that, in a hybrid space perspective, the outward flow of check-in from the city reflects the flow of economic factors. It can also be interpreted as a flow of media resources. As a participant in mobile social media, the act of check-in across cities can be interpreted as carrying media resources into another city and sharing media resources through the act of check-in. Thus, the weighted out-degree is a measure of a city's capability to export check-in users or media resources, while the weighted in-degree is the one to attract check-in users or provide check-in users places to the outside world.

Based on existing research, Node Symmetry is applied to reflect the inflow and outflow of individual nodes.

$$NSI_i = \frac{S_i^{in} - S_i^{out}}{S_i^{in} + S_i^{out}} \tag{4}$$

S_i^{in} denotes the weighted in-degree of node i and S_i^{out} denotes the weighted out-degree of node i. If NSI_i is greater than 0, it means that the city is an input-flow city; if NSI_i is equal to 0, it means that the city is a balanced-flow city; if NSI_i is less than 0, it means that the city is an output-flow city.

Compared to the city delivering the check-in flow, the city receiving the check-in flow is where the check-in behavior occurs. In the node evaluation of the urban network, the weighted in-degree can be used to assess the visibility of a city in digital space as it more intuitively characterizes the frequency of completion of check-in behavior. However, the weighted in degree only takes into account the total volume aggregated by a city, but not the number of cities towards that city, thus losing the overall network perspective in terms of examining the importance of city nodes. This feature has been confirmed in numerous network studies and is also reflected in check-in networks. In detail, a node may gather a large amount of check-in flow that originates from one city while not connected to other cities. It is structurally at the edge of the network. Such nodes do not significantly shape the network's structure, and their importance is relatively limited.

The PageRank algorithm avoids the isolated perspective of the weighted in degree described above and examines the node's importance in link quantity and quality. It is, therefore, widely used for identifying core nodes in directed networks such as virtual communities, academic collaboration networks, and social networks. The PageRank algorithm, proposed by Google, is an algorithm for ranking the importance of web pages. The core idea is that the extent of a page on the World Wide Web depends on the number and volume of the other pages pointing to it and that pages pointed to by multiple high-importance pages will also have high priority. The algorithm measures the extent of a node by its PageRank value (PR). The formula is as follows:

$$PR_i = \sum_{j \in B_i} \frac{PR_j}{N_j} \tag{5}$$

where i and j denote nodes, PR_i and PR_j denote their PR values, B_i denotes the set of nodes pointing to node i, and N_j denotes the number of nodes pointed to by node j. PageRank defines a random wander model, a first-order Markov chain, on a directed graph that describes the behavior of random wanderers visiting individual nodes at random along the directed graph. Through iteration, a stable PR is eventually computed for all nodes in the network. Based on this, the PageRank algorithm can be understood as modeling the probability of a user's attention reaching each web page on the Internet.

This algorithmic mechanism for modeling the flow of attention has theoretical applicability to the analysis of check-in networks. In a hybrid space perspective, digital space overlaps with geographical space. The media user's attention first flows in the digital space, then descends to the geographic space through the check-in behavior to transform into a check-in flow. As the check-in behavior is completed, it is uploaded to the digital space again to enhance the attention of the check-in place. Accordingly, the PageRank algorithm can be applied to the check-in network to simulate the probability of a media user arriving in each city in an urban agglomeration and completing a check-in behavior. The higher the PR of a city, the greater the mixed spatial accessibility.

3. Results

3.1. Hierarchical Attributes

The hierarchical characteristics of the three city clusters are prominent. In terms of statistical indicators, the highest average weighted degree of the check-in network of the three urban agglomerations is in the YRD, with BTH and the PRD in decreasing order, while the most apparent vertical hierarchical feature of the network is in BTH, with the PRD and YRD in decreasing order. Spatially, the check-in network of YRD shows a Z-shaped skeleton with Suzhou, Shanghai, and Hangzhou as the core. The PRD check-in network shows an N-shaped structure with Guangzhou and Shenzhen as the core. The BTH check-in network shows a Beijing single-point radiation-type core skeleton. The specific results are as follows.

First, the highest average weighted degree of the check-in network of the three urban agglomerations is the YRD, with BTH and the PRD in decreasing order (Table 2). Each YRD city is connected to an average of 10.077 cities in the check-in network, and the average check-in flow of each city is 433, which is much higher than that of BTH (293.357) and PRD (264.556). The weighted degree distributions of the three networks all conform to the power-law distribution (R2 > 0.8), indicating that they are scale-free networks. It also illustrates the fact that a small number of cities create large-scale check-in flow, while the majority of cities create only a minimal amount. In addition, the distribution fit coefficients of check-in networks of YRD, PRD, and BTH are 0.91903, 0.96918, and 1.02552, respectively, indicating that the vertical hierarchy of the network is most clearly characterized by BTH, with the PRD and YRD diminishing in that order (Figure 4).

Table 2. Topological eigenvalues of check-in network of three major urban agglomerations.

Eigenvalues	YRD	PRD	BTH
Number of nodes	26	9	14
Average degree	10.077	5.111	6
Average weighted degree	433	264.556	293.357
Fit coefficient (a)	-0.91903	-0.96918	-1.02552

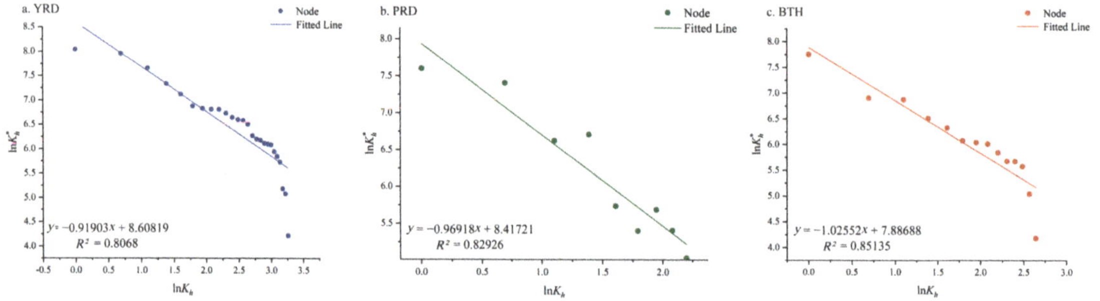

Figure 4. Fitting results of the weighted degree distribution of check flow network of three major urban agglomerations.

Further, we divide the nodes and edges into five levels according to the weighted degree and edge weight by the natural breakpoint method and draw the networks' topological map and spatial distribution map based on Gephi and ArcGIS, respectively (Figure 5).

In the YRD check-in network, the first level nodes include Shanghai (3112), Suzhou (2859), and Hangzhou (2121), and the first level edges include Shanghai-Hangzhou (403) and Shanghai-Suzhou (388), which forms the open triangle pattern. At the second level, Hangzhou connects to Huzhou, Shaoxing, and Jiaxing, strengthening the internal check-in connection of Zhejiang Province cities. The third level emerges with cities south of the Yangtze River in Jiangsu Province, such as Nanjing (1535), Yancheng (973), Changzhou (840), and Nantong (735), as well as Zhejiang Province cities, such as Ningbo (911) and Jinhua (771), generating check-in flow among cities within each province. The fourth level mainly includes check-in connections among existing nodes. At the same time, Anhui Province cities such as Hefei and Wuhu also emerge and form less intense check-in connections with Suzhou Province cities such as Nanjing, Chuzhou, and Suzhou. Chizhou and Tongling appear in the fifth level, complementing all YRD cities.

In the PRD check-in network, Guangzhou-Foshan forms the first level with an edge weight of 799. At the second level, Shenzhen, as the core, connects to Dongguan and Huizhou with edge weights of 545 and 481, respectively. At the third level, Guangzhou forms the two-way links with Shenzhen-Dongguan and Foshan. Huizhou also establishes connections with Guangzhou and Dongguan, strengthening the relationship between the central and eastern cities. At the fourth and fifth levels, Zhuhai, Jiangmen, and Zhongshan emerge, yielding a relatively lower check-in connection.

In the BTH check-in network, Beijing, as the core, connects to Langfang and Baoding with edge weights of 417 and 282, respectively, forming the first and second levels. At the third level, Beijing complements the two-way links with Langfang and Baoding on the one hand. It connects to Shijiazhuang, Tianjin, Chengde, Handan, and Zhangjiakou on the other hand. At the fourth level, Beijing complements the two-way connection with Qinhuangdao, Cangzhou, and Xingtai. At the fourth and fifth levels, relationships are formed mainly among nodes outside Beijing.

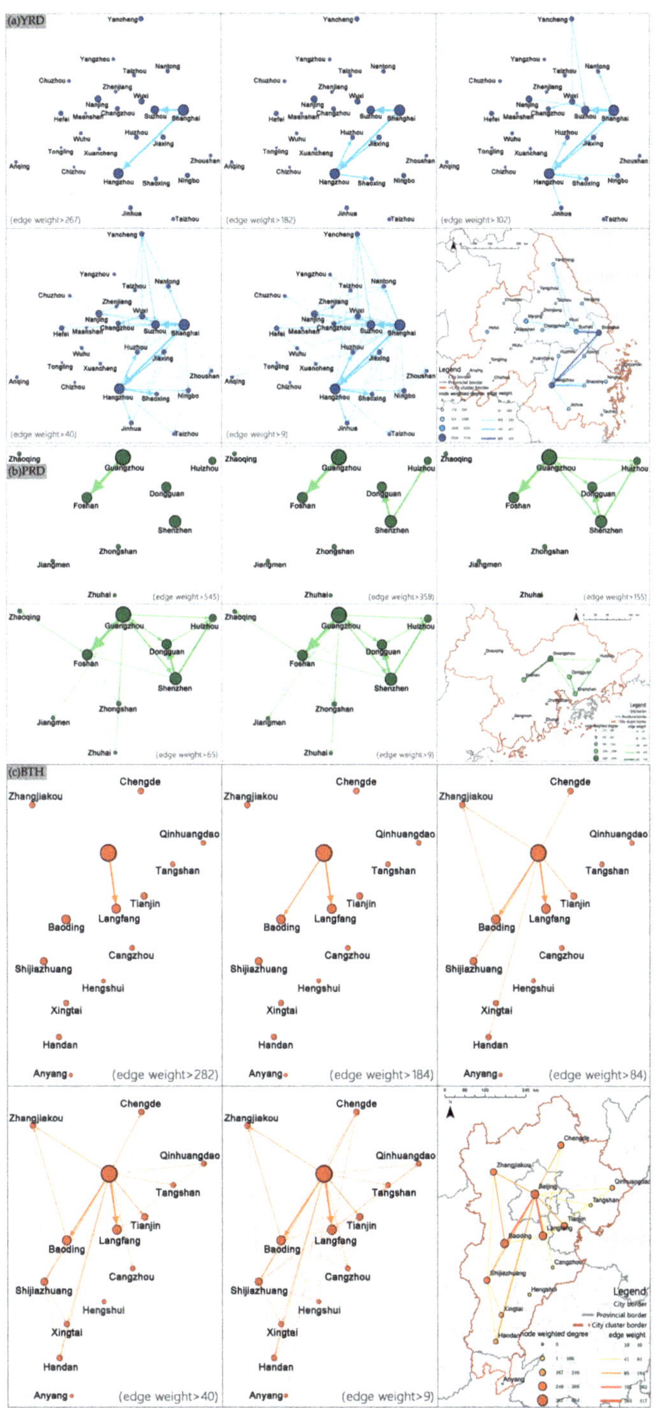

Figure 5. Topology and spatial distribution of the check-in networks of three major urban agglomerations (**a**), YRD; (**b**), PRD; (**c**), BTH.

In general, the check-in network of YRD is spatially centered on Suzhou-Shanghai-Hangzhou, forming a Z-shaped spatial structure. The PRD network forms the N-shaped spatial structure, with Guangzhou and Shenzhen as the dual cores. The BTH check-in network forms a single-point radial spatial structure with Beijing as the core.

3.2. Communities Scale

The three check-in networks show obvious small-world characteristics, but the flattening characteristics are immature, and the community division needs to be further clarified. The community division of the three major urban agglomerations shows a spatial structure with the regional high check-in flow cities as the core and the neighboring cities as the coordinator. The specific results are as follows.

First, a small-world network has a similar average shortest path and a more significant clustering coefficient when compared with a random network of the same size [57]. The check-in networks of three urban agglomerations show small-world characteristics (Table 3). Specifically, the average clustering coefficients of the YRD, BTH, and PRD networks are 0.658, 0.626, and 0.597, respectively, which are larger than the average clustering coefficients of the random networks. The average path lengths of the YRD, BTH, and PRD networks are 1.6, 1.556, and 1.361, which are smaller than those of the random networks. It means that the networks are characterized by high aggregation and high topological accessibility, which further confirms the small-world characteristics of the three check-in networks. Among the three agglomerations, the average clustering coefficient of the check-in network of YRD is the largest, indicating it has the most significant small-world characteristics and the highest degree of flatness.

Table 3. Small-world of check flow network of three major urban agglomerations.

Index	YRD	BTH	PRD
Average clustering coefficient	0.658 (0.377) [1]	0.626 (0.435)	0.597 (0.586)
Average path length	1.6 (1.623)	1.556 (1.571)	1.361 (1.375)

[1] The eigenvalues of the actual network and the eigenvalues of the zero model are shown in parentheses. The zero model is a random network with the same number of nodes and edges as the actual network, computed by Gephi.

Second, we used the weighted modularity community detection algorithm for the three check-in networks. The urban communities in each network were obtained, as shown in Figure 6, and the community attributes were shown in Table 4. The modularity of the three networks is below 0.3, indicating that the communities are not clearly divided, which further reflects that none have significant flat distribution characteristics.

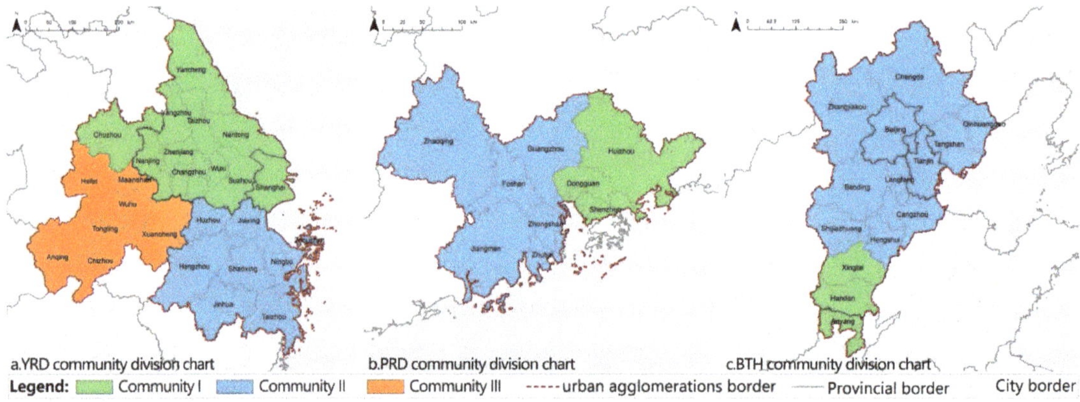

Figure 6. Community distribution of three major urban agglomerations.

Table 4. Community statistics of three major urban agglomerations.

Urban Agglomeration	Community Number	Number of Nodes	Density	Flow	Flow Ratio	Core City
YRD	1	11	0.82	4519	39.96%	Shanghai, Suzhou, Nanjing
YRD	2	8	0.75	2688	23.77%	Hangzhou
YRD	3	7	0.355	522	4.62%	Hefei
PRD	1	6	0.835	2569	39.95%	Guangzhou
PRD	2	3	1	1728	26.87%	Shenzhou
BTH	1	11	0.59	3421	83.01%	Beijing
BTH	2	3	0.665	125	3.03%	— —

The modularity of the YRD network is the highest (0.278). It emerges three major communities based on provincial boundaries. Specifically, there are 11 cities in community 1, with nearly 40% of the YRD check-in network. In this community, Shanghai is the core and connects cities in Jiangsu Province. In addition, Chuzhou in Anhui Province is also integrated into this community by emerging a close connection with Nanjing. In community 2, there are 8 cities and 23.77% check-in flow. Hangzhou is the center, connecting cities in Zhejiang Province. Community 3 has a shallow check-in flow (4.62%). As the center of the community, Hefei connects cities in Anhui Province.

The PRD check-in network has a low modularity (0.195) and forms two communities on the east and west sides of the Pearl River Estuary. Community 1 emerges the triangle structure of Shenzhen-Dongguan-Huizhou with 26.87% of the PRD check-in network. Other cities from the PRD form community 2 with 39.95% check-in flow. In this community, Guangzhou-Foshan-Zhaoqing is the core triangle, connecting Zhuhai, Zhongshan, and Jiangmen.

The BTH check-in network has the lowest modularity (0.058), forming two significantly unbalanced communities. Beijing, as the core, coordinates and organizes the surrounding cities, creating community 1 with high check-in flow (83.01%). Anyang, Xingtai, and Handan are not included in community 1 because they are at the periphery of the urban agglomeration but form community 2 with a low check-in flow (3.03%).

3.3. Node Centrality

Through the node centrality analysis, we found that megacities such as Beijing, Shanghai, Guangzhou, and Shenzhen perform as an outward export type, sending many media resources outward and promoting the integration of urban agglomerations. Cities with geographical proximity to the core nodes or specific recreational resources, such as Dongguan, Foshan, and Chengde, have a stronger weighted indegree and present inward aggregation type. These cities have hybrid spatial accessibility. The specific results are as follows.

4. Node weighted degree and NSI

This paper assesses a city's ability to generate check-in flow by node weighted degree and evaluates whether a city is an inward aggregator or outward exporter by nodal symmetry (NSI) (Table 5). Shanghai and Hangzhou are the two centers of check-in flow generated in YRD with a weighted degree of 2208 and 1588, respectively. The difference is that Shanghai shows apparent spillover, with the weighted out-degree (2208) being much higher than the weighted in-degree (904), with an NSI of −0.419, while Hangzhou is relatively balanced (−0.111). The check-in flow of Suzhou, Nanjing, and Wuxi are above 1000, with Nanjing showing some spillover (−0.165) and the other two cities showing a not-so-subtle aggregation phenomenon (0.05). Yancheng, Hefei, Ningbo, and Jiaxing all have a weighted degree above the average value (870). In contrast, the weighted degree of other cities is relatively low, most of which show strong aggregation characteristics, especially Zhoushan (0.785), Huzhou (0.575), Nantong (0.376), and Zhenjiang (0.359). These cities have created many internet-famous spots with their high-quality tourism resources and become important nodes for gathering check-in flow.

Table 5. Node attributes of check-in network of three major urban agglomerations.

YRD [1]					PRD					
City	WID [2]	WOD [2]	WD [2]	NSI	PR	City	WID	WOD	WD	NSI	PR
Shanghai	904	2208	3112	−0.42	0.091	Guangzhou	1243	2003	3246	−0.234	0.214
Hangzhou	1271	1588	2859	−0.11	0.109	Foshan	1254	754	2008	0.249	0.158
Suzhou	1116	1005	2121	0.052	0.082	Dongguan	1059	821	1880	0.126	0.14
Naning	641	894	1535	−0.17	0.058	Shenzhen	777	1649	2426	−0.359	0.136
Wuxi	654	585	1239	0.056	0.054	Huizhou	974	310	1284	0.517	0.112
Yancheng	389	584	973	−0.2	0.032	Zhongshan	445	222	667	0.334	0.085
Hefei	430	497	927	−0.07	0.036	Jiangmen	233	223	456	0.021	0.056
Ningbo	510	401	911	0.12	0.056	Zhuhai	221	154	375	0.178	0.049
Jiaxing	501	405	906	0.106	0.035	Zhaoqing	225	295	520	−0.134	0.046
Changzhou	539	301	840	0.283	0.044	Average value	715	715	1429	0.077	0.111
Jinhua	408	363	771	0.058	0.037	**BTH**					
Nantong	506	229	735	0.377	0.035	City	WID	WOD	WD	NSI	PR
Huzhou	570	154	724	0.575	0.041	Beijing	664	1668	2332	−0.43	0.223
Shaoxing	473	197	670	0.412	0.042	Langfang	539	464	1003	0.074	0.081
Taizhou-J [3]	246	284	530	−0.07	0.024	Baoding	474	499	973	−0.025	0.072
Taizhou-Z [3]	291	204	495	0.176	0.029	Shijiazhuang	318	358	676	−0.059	0.063
Zhoushan	431	52	483	0.785	0.045	Tianjin	339	224	563	0.204	0.059
Wuhu	225	229	454	−0.01	0.022	Zhangjiakou	386	52	438	0.762	0.065
Chuzhou	223	224	447	−0	0.02	Chengde	346	77	423	0.635	0.155
Yangzhou	172	269	441	−0.22	0.016	Handan	202	209	411	−0.017	0.042
Anqing	161	222	383	−0.16	0.017	Xingtai	206	142	348	0.183	0.039
Xuancheng	190	158	348	0.092	0.018	Tangshan	166	128	294	0.129	0.051
Zhenjiang	210	99	309	0.359	0.021	Cangzhou	141	152	293	−0.037	0.029
Maanshan	77	101	178	−0.14	0.013	Qinhuangdao	248	18	266	0.864	0.081
Chizhou	104	57	161	0.292	0.012	Hengshui	92	64	156	0.179	0.022
Tongling	68	0	68	1	0.01	Anyang	0	66	66	−1	0.011
Average value	435	435	870	0.122	0.038	Average value	294	294	589	0.105	0.071

[1] The weighted degree refers to the total weighted degree, which is the sum of the Weighted in-degree and Weighted out-degree. [2] WID refers to Weighted in-degree, WOD refers to Weighted out-degree, WD refers to Weighted degree. [3] Two cities in the YRD are called Taizhou. In order to distinguish, Taizhou in Zhejiang Province is named Taizhou-Z, and Taizhou in Jiangsu Province is named Taizhou-J in this paper.

In PRD, Guangzhou (3246) and Shenzhen (2426) have the highest weighted degree and both show strong spillover characteristics (−0.234, −0.359), indicating that a large number of check-in flows are delivered to other cities in PRD from these two cities. Due to the geographical proximity to Guangzhou and Shenzhen, Foshan and Dongguan have a high check-in flow, which is 2008 and 1880, respectively, and show strong aggregation characteristics (0.249, 0.127). The weighted degrees of other cities are below the average value (1429), among which Huizhou, Zhongshan, and Zhuhai show strong aggregation characteristics, especially Huizhou with weighted indegree and NSI as high as 974 and 0.517, respectively. These cities are the strongest aggregation in PRD, reflecting tourist cities' ability to gather social media resources from outside.

The only core node in BTH is Beijing, with a weighted degree of 2332. It far exceeds those of Langfang (1003), Baoding (973), and Shijiazhuang (676). In addition, Beijing also shows significant spillover characteristics (−0.431), while the check-in flow of Langfang and Baoding is relatively balanced, with no obvious spillover or aggregation characteristics. The other cities are all weighted below the average value (588). It is worth noting that the other cities, although all weighted below the mean (588), generally show a strong aggregation. In particular, Zhangjiakou (0.762), Chengde (0.636), and Qinhuangdao (0.865), although not sending strong outward check-in flows (<450), attract a large number of media users originating from Beijing through their positioning as suburban Beijing tourist cities.

5.   PageRank

PageRank (PR) is applied to simulate the probability of media users arriving and completing the check-in behavior in each city and further examine the hybrid space accessibility of each city. In YRD, the PR of Hangzhou is the highest (0.109), indicating that media users in YRD have the highest probability of arriving in Hangzhou for check-in activities. Suzhou, Nanjing, Ningbo, Wuxi, Zhoushan, Changzhou, Shaoxing, and Huzhou follow with Hangzhou, with PR above the average (0.047), among which Zhoushan, Changzhou, Shaoxing, and Huzhou are all weighted below the average. It indicates that although these cities do not generate a very high check-in flow, they are the media resource input for many cities with a large amount of check-in flow.

In the PRD, Guangzhou has an absolute advantage in PR (0.214), gathering a wide range of check-in flow from various cities. Foshan (0.159) and Dongguan (0.140) have a high PR by gathering check-in flow from Guangzhou and Shenzhen, reflecting the co-location effect of Guangzhou-Foshan and Shenzhen-Dongguan. On the contrary, although Shenzhen's weighted degree is significant, it only has a small amount of check-in flow from big weighted degree cities due to its outward-oriented export characteristics. It leads to Shenzhen's lower PR (0.137) than Foshan and Dongguan's.

With the highest PR of 0.224, Beijing is the primary node for gathering check-in flow from other cities in the BTH region. Although Chengde and Qinhuangdao have a lower weighted degree, they attract check-in flow from several high check-in flow nodes through their high-quality tourism resources. Their PRs are second only to that of Beijing, with 0.156 and 0.081, respectively. In contrast, although Langfang and Baoding both have higher weighted degrees, they are less connected to other nodes as most of the check-in flow originates from Beijing only. They are at the edge of the network structure and therefore do not lead in PR. The other cities have lower PR than the average (0.071) and thus lower accessibility in the hybrid space due to the long geographical distance from Beijing and the lack of recreational resources to attract mobile media.

4. Discussion

4.1. Key Findings and Significance of This Study

As mobile Internet devices represented by cell phones are increasingly integrated into people's daily lives, social media platforms, such as TikTok and Facebook, have become virtual places to experience, shape, and communicate city imagery. It leads to an increase in the popularity of geo-tagging and the dissolution of the boundaries between urban geospatial and digital spaces, resulting in hybrid spaces. Existing studies have focused on the impact of SMPs on urban spatial dynamics while neglecting the ability to influence cross-city connections at a more macroscopic scale. On the other hand, the integration of urban agglomerations is the current theme of regional research. In the face of the increasing trend of urban mediatization, it is necessary to examine the impact of geo-tagged behavior on regional integration. We invite the TikTok data to conduct the check-in networks of YRD, PRD, and BTH. The structural features of the check-in networks are examined in terms of hierarchical attributes, community scale, and node centrality. The study yields some interesting findings:

The first important finding of this paper is that YRD, PRD, and BTH respectively exhibit Z-shaped, N-shaped, and single-point radial spatial distribution as well as the vertical hierarchical characteristics of check-in networks. This spatial distribution is similar to the urban network structure of the three major urban agglomerations in terms of service [36] and finance [37]. In terms of urban system structure, the YRD has the most robust flattening characteristics, followed by the PRD and BTH. This is consistent with the results of comparative studies on integrating the three major urban agglomerations [51,58]. The innovative finding of this paper is that the flatness is weaker, and the vertical distribution feature is stronger in the check-in network compared to the demographic migration network, such as the tourism network [38]. This may be because check-in behavior is more in line with the characteristics of short-distance leisure behavior, and the distance friction effect of the check-in network is more significant. The behavior of check-in users is motivated by recording geospatial experiences to build virtual personality and media image. In this quality, the urban spatial experience is as important as arriving at the destination and recording electronically. Therefore, geographic distance becomes an essential influencing factor. This is a powerful response to the question of the death of geography. The second important finding of this paper is that the megacities such as Beijing, Shanghai, Guangzhou, and Shenzhen perform as an outward export type, sending many media resources outward and promoting the integration of urban agglomerations. Cities with geographical proximity to the core nodes, such as Dongguan, Foshan, and Chengde, have a stronger weighted indegree and present inward aggregation type. This is somewhat inconsistent with the role exhibited by mega-cities in existing tourism networks [38,59]. According to the established theories, the hub role and agglomeration effect of core cities in the network are the fundamental driving force for their growth into mega-cities, which attract more flows than those sent outward [60,61]. While the innovative finding of this paper is that cities with high check-in flow tend to send outward check-in flow more than gathering check-in flow. Guangzhou, Shenzhen, Shanghai, and Beijing all demonstrate the diffusion effect. This is a manifestation of shared media resources. When high-ranking cities send check-in flow to low-ranking cities, they also send media resources. With the geo-tagging behavior of check-in in high-ranking cities, low-ranking cities will further expand their visibility in digital space. It is helpful to promote the balanced development of cities in urban agglomerations.

In addition, this paper also found that cities with high-quality tourism resources can break the geographical proximity effect to a certain extent and be fed into the check-in flow by multiple cities, thus becoming essential nodes in the check-in network. This feature can be seen when comparing the node characteristics of Chengde and Langfang in the BTH check-in network. Langfang gathers a large number of check-in flow from Beijing through its proximity to Beijing but is not strongly connected to other cities except Beijing. In contrast, although Chengde is relatively far from other cities, it plays a pivotal role in the network structure by attracting check-in flow from many cities through its high-quality tourism resources. It is essential to notice that Zhuhai, a famous tourist city in PRD, is recognized as a core node in established tourism flow network studies but is at the edge of the network in this study. This is partly due to its distance from Guangzhou and Shenzhen. On the other hand, it is because of the significant decline in tourism activities in Macau due to the COVID-19 epidemic, and as the spatial hinterland of Macau tourism, Zhuhai's tourism industry is also more significantly affected, especially the number of cross-city types of tourism activities is sharply reduced.

4.2. Spatial Organizational Pattern of Three Major Urban Agglomerations

The spatial organization patterns of the check-in networks of three urban agglomerations are plotted (Figure 7) to analyze whether the spatial structures of the urban agglomerations maintain a vertical distribution structure or have shifted to a flattened distribution from a hybrid space perspective. In general, the check-in networks of the three urban agglomerations still maintain a vertical structure with a strong hierarchy but also show a tendency to evolve into a flattened structure. The YRD urban agglomeration has the strongest characteristics of a flat structure, the BTH urban agglomeration has the most significant vertical structure, and the PRD is in the middle of the two.

The YRD presents a composite spatial organization model with multi-level cores. As the first spillover core of YRD, Shanghai spreads the check-in flow to all cities in the YRD in a hierarchical manner, forming a composite spatial organization model of one main and many vice. Hangzhou, Suzhou, Nanjing, and Hefei are the major cities that carry the check-in flow from Shanghai, playing the role of the regional core hub of Zhejiang Province, Suzhou Province, and Anhui Province. Among them, Hangzhou, as the capital of Zhejiang Province, promotes the descending of the check-in flow inside the province, gathering the check-in flow and then passing it to normal node cities such as Huzhou, Jinhua, Shaoxing, and Jiaxing. Because the check-in flow among general nodes is small, Zhejiang Province constitutes a spatial organization pattern of monocentric diffusion. In contrast, Jiangsu Province presents a polycentric network structure since cities in the province are closely connected and have a balanced check-in flow intensity. In Anhui Province, Hefei is the core city, but it has a low card flow with the neighboring cities, and the peripheral nodes are less connected, which constitutes a monocentric discrete spatial organization pattern.

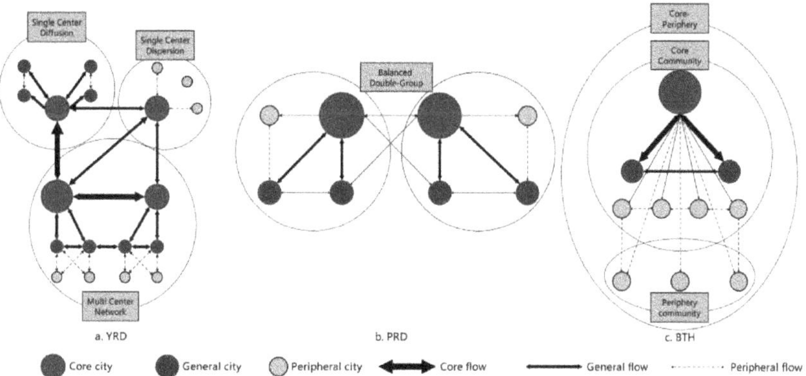

Figure 7. Spatial organizational patterns of three major urban agglomerations.

The PRD presents a balanced double-group model with a double-core structure. Guangzhou and Shenzhen, as the dual cores, coordinate the surrounding cities to form a relatively balanced double cluster model, namely the Pearl River west coast cluster with Guangzhou as the core and the East Coast Cluster with Shenzhen as the core. Foshan, Dongguan, and Huizhou carry the main check-in flow from the core cities because of their geographical proximity, while other cities such as Zhaoqing, Jiangmen, and Zhongshan are at the edge of the network due to their distance from the core nodes.

The BTH presents a core-periphery model with a single-center radial structure. Beijing, as the core of BTH, connects with other cities, which presents the single-center radial structure. The intensity of check-in flow decreases with the geographical distance from Beijing, forming the core and the periphery communities.

5. Conclusions

The widespread application of SMPs has changed the mobility between cities in hybrid spaces. This paper proposes a method and analysis system for the construction of urban punching flow networks from the perspective of hybrid space and conducts an empirical study on three major urban clusters in China using Jitterbug cross-city punching data. The hierarchical attributes, community scope, and node centrality are analyzed, and the vertical and flat distribution characteristics are examined. The results are as follows.

(1) The highest average weighted degree of the check-in network of the three urban agglomerations is the YRD, with BTH and the PRD in decreasing order. The most apparent vertical hierarchical feature of the network is BTH, with the PRD and YRD in decreasing order. In terms of space, the YRD check-in network presents a Z-shaped skeleton with Suzhou-Shanghai-Hangzhou as the core. The PRD check-in network presents an N-shaped structure with Guangzhou and Shenzhen as the dual-core. The BTH check-in network presents a Beijing single-point radial core skeleton.

(2) The three check-in networks show prominent small-world characteristics, but the community division needs to be further clarified and the flattening characteristics are still immature. The community division of the three major urban agglomerations shows a spatial structure with the regional high check-in flow cities as the core and the neighboring cities as the coordinator. Among them, the YRD forms three communities based on the

provincial boundary effect, the PRD forms two communities on the east and west sides of the Pearl River Estuary, and BTH creates a core community and a peripheral community.

(3) Due to the enormous population scale, SMPs penetration rate, and many internet celebrity spaces, megacities such as Beijing, Shanghai, Guangzhou, and Shenzhen are the core nodes of each check-in network. Generally, they perform as an outward export type, sending many media resources outward and promoting the integration of urban agglomerations. Cities with geographical proximity to the core nodes or specific recreational resources, such as Dongguan, Foshan, and Chengde, have a stronger weighted indegree and present inward aggregation type. These cities have a solid hybrid spatial accessibility and act as network hubs to shape the formation of check-in networks by gathering media resource inputs from multiple cities.

The primary significance of this paper is as follows. (1) We introduce the hybrid space perspective to study urban agglomeration integration and respond to the increasing trend of mediatization. (2) We introduce the cross-city check-in data of TikTok and conduct a modeling method and framework for the check-in network, which provides new data and methods for inter-city association pattern research and communication geography. (3) The structure of check-in networks in three major Chinese urban agglomerations is studied in comparison, providing theoretical support for the integration of urban agglomerations.

In addition, suggestions can be made to optimize the check-in network and enhance the integration pattern of urban agglomerations at digital and geospatial levels. (1) On the one hand, by actively releasing short videos on the theme of cultural tourism, tourism resource-based cities can portray cities' leisure and cultural labels to enhance the visibility and attractiveness of cities in the digital space. On the other hand, through short video content or short video recommendation mechanism, cities can strengthen the virtual connection of specific city combinations in the digital space and deepen the intention of co-location, thus promoting media users to travel between cities. (2) In the geographic space, on the one hand, the attractiveness of cities to media users is enhanced by creating high-quality recreational spaces, and the conversion mechanism of online enthusiasm-offline vitality is strengthened. On the other hand, the transportation infrastructure is optimized to improve inter-city accessibility, thus reducing the frictional effect of geographical distance and promoting the offline mobility of media users across cities.

This study also has some limitations that are worth exploring further. First, in terms of data, although TikTok is the SMPs with the largest share of users in China, there is other software such as RED and Kuaishou. The number of users in the software is also large, and there are differences in user characteristics. For example, female users dominate RED, and users in small cities and rural areas dominate Kuaishou. Therefore, using only a single software may miss certain media users, resulting in inaccurate study results. We will combine data from multiple SMPs for future analysis. Second, in terms of methods, the node centralities metrics used in this paper need to fully reveal the importance of each node in the check-in flow network. In the node centrality analysis, the metrics used in this paper mainly examine the importance of cities in terms of check-in flow. However, examining the nodes' characteristics from the topological structure features is also essential. Other node centralities metrics in SNA can be used in the future to fully reveal the functions played by cities in the check-in network. For example, intermediary centrality can be used to analyze the hub role of nodes in the network, and proximity centrality can be used to analyze the topological accessibility of nodes in the network. Third, it should be clarified that check-in activity represents only one type of spatial activity and is more inclined to describe leisure and recreational activities. It cannot fully characterize the spatial activities of urban agglomerations. This type of data can be combined with other types of activity data for further study. In addition, this study has only described and summarized the network characteristics of check-in flow. It is hoped that methods such as ERGM can be introduced in future studies to analyze the mechanism further. In addition, this paper selects a specific time cross-section, which can be extended to multiple time cross-sections.

Author Contributions: Conceptualization, B.X. and G.X.; methodology, B.X. and R.C.; software, B.X. and R.C.; validation, G.X., B.X. and R.C.; formal analysis, B.X. and G.X.; resources, G.X.; data curation, B.X.; writing—original draft preparation, G.X., B.X. and R.C.; writing—review & editing, G.X., B.X. and R.C.; visualization, R.C.; supervision, G.X.; project administration, G.X.; funding acquisition, G.X. All authors have read and agreed to the published version of the manuscript.

Funding: This research was funded by Talent Fund of Beijing Jiaotong University, grant number 2021RCW122, China Postdoctoral Science Foundation, grant number 2022M720393, National Natural Science Youth Program, grant number 52108040, and the Beijing Social Science Foundation, grant number 22GLC071.

Institutional Review Board Statement: Not applicable.

Informed Consent Statement: Not applicable.

Data Availability Statement: The datasets used and/or analyzed during the current study are available from the corresponding author upon reasonable request.

Acknowledgments: The authors would like to thank all the anonymous reviewers and editors who contributed their time and knowledge to this study. The authors also thank Zhao, M. who provided insightful opinion the precious support for the previous survey.

Conflicts of Interest: The authors declare no conflict of interest.

References

1. Castells, M. *The Informational City: Information Technology, Economic Restructuring, and the Urban-Regional Process*; Blackwell Oxford: Oxford, UK, 1989.
2. Castells, M. *The Rise of the Network Society*; Wiley: Hoboken, NJ, USA, 2009.
3. Derudder, B. Mapping Global Urban Networks: A Decade of Empirical World Cities Research. *Geogr. Compass* **2008**, *2*, 559–574. [CrossRef]
4. Zhang, Y.; Wang, T.; Supriyadi, A.; Zhang, K.; Tang, Z. Evolution and Optimization of Urban Network Spatial Structure: A Case Study of Financial Enterprise Network in Yangtze River Delta, China. *ISPRS Int. J. Geo-Inf.* **2020**, *9*, 611. [CrossRef]
5. Balland, P.-A.; Boschma, R.; Ravet, J. Network dynamics in collaborative research in the EU, 2003–2017. *Eur. Plan. Stud.* **2019**, *27*, 1811–1837. [CrossRef]
6. Li, X.D. Spatial structure of the Yangtze river delta city network based on the pattern of listed companies network. *Prog. Geogr.* **2017**, *33*, 1587–1600. [CrossRef]
7. Yeh, A.G.; Yang, F.F.; Wang, J. Producer service linkages and city connectivity in the mega-city region of China: A case study of the Pearl River Delta. *Urban Stud.* **2015**, *52*, 2458–2482. [CrossRef]
8. Castells, M. Informationalism, networks, and the network society: A theoretical blueprint. *Netw. Soc. Cross-Cult. Persxpect.* **2004**, 3–45. [CrossRef]
9. Hall, P.G.; Pain, K. *The Polycentric Metropolis: Learning from Mega-City Regions in Europe*; Routledge: Oxford, UK, 2006.
10. Taylor, P.J.; Hoyler, M.; Verbruggen, R. External Urban Relational Process: Introducing Central Flow Theory to Complement Central Place Theory. *Urban Stud.* **2010**, *47*, 2803–2818. [CrossRef]
11. Wang, Q.; Cheng, Y. Characteristics and Performance of City Network from the Perspective of High-way Freight—The Case of Three Major Urban Agglomerations in China. *Urban Plan. Forum.* **2020**, *10*, 32–39. [CrossRef]
12. Chen, W.; Liu, W.; Ke, W.; Wang, N. Understanding spatial structures and organizational patterns of city networks in China: A highway passenger flow perspective. *J. Geogr. Sci.* **2018**, *28*, 477–494. [CrossRef]
13. Huang, Y.; Lu, S.; Yang, X.; Zhao, Z. Exploring Railway Network Dynamics in China from 2008 to 2017. *ISPRS Int. J. Geo-Inform.* **2018**, *7*, 320. [CrossRef]
14. Xu, W.; Zhou, J.; Qiu, G. China's high-speed rail network construction and planning over time: A network analysis. *J. Transp. Geogr.* **2018**, *70*, 40–54. [CrossRef]
15. Wang, Y.; Niu, X.; Song, X. Spatial Organizational Characteristics of the Yangtze River Delta Urban Agglomeration Based on Intercity Trips. *City Plan. Rev.* **2021**, *45*, 43–53. [CrossRef]
16. Teixeira, S.H.D.O.; Catelan, M.J. New Articulations of the Brazilian Cities Network: An Analysis of the Heterarchies by the Airflow System. *Soc. Nat.* **2019**, *31*, e42622. [CrossRef]
17. Zhao, Y.; Zhang, G.; Zhao, H. Spatial Network Structures of Urban Agglomeration Based on the Improved Gravity Model: A Case Study in China's Two Urban Agglomerations. *Complexity* **2021**, *2021*, 6651444. [CrossRef]
18. Zhao, M.; Liu, X.; Derudder, B.; Zhong, Y.; Shen, W. Mapping producer services networks in mainland Chinese cities. *Urban Stud.* **2015**, *52*, 3018–3034. [CrossRef]
19. Zhao, M.; Derudder, B.; Huang, J. Examining the transition processes in the Pearl River Delta polycentric mega-city region through the lens of corporate networks. *Cities* **2017**, *60*, 147–155. [CrossRef]

20. Wang, Y.; Yin, S.; Fang, X.; Chen, W. Interaction of economic agglomeration, energy conservation and emission reduction: Evidence from three major urban agglomerations in China. *Energy* **2022**, *241*, 122519. [CrossRef]
21. Cao, Z.; Dai, L.; Wu, K.; Peng, Z. Structural Features and Driving Factors of the Evolution of the Global Interurban Knowledge Collaboration Network. *Geogr. Res.* **2022**, *41*, 1072–1091. [CrossRef]
22. Read, R. Knowledge counts: Influential actors in the education for all global monitoring report knowledge network. *Int. J. Educ. Dev.* **2019**, *64*, 96–105. [CrossRef]
23. Tang, C.; Dou, J. Exploring the Polycentric Structure and Driving Mechanism of Urban Regions from the Perspective of Innovation Network. *Front. Phys.* **2022**, *10*, 855380. [CrossRef]
24. Liu, L.; Luo, J.; Xiao, X.; Hu, B.; Qi, S.; Lin, H.; Zu, X. Spatio-Temporal Evolution of Urban Innovation Networks: A Case Study of the Urban Agglomeration in the Middle Reaches of the Yangtze River, China. *Land* **2022**, *11*, 597. [CrossRef]
25. Yan, S.; Jin, C. Characteristics of Spatial Network Structure of Tourist Flow in Urban Area of Luoyang. *Sci. Geogr. Sin.* **2019**, *39*, 1602–1611. [CrossRef]
26. Chen, H.; Wang, M.; Zheng, S. Research on the Spatial Network Effect of Urban Tourism Flows from Shanghai Disneyland. *Sustainability* **2022**, *14*, 13973. [CrossRef]
27. Wei, T. Application of GIS in Spatial Characteristics of Tourist Flow Based on Online Booking Data: A Case Study of Yangtze River Delta. *Iran. J. Sci. Technol. Trans. Civ. Eng.* **2022**, *22*, 1–11. [CrossRef]
28. Seok, H.; Barnett, G.A.; Nam, Y. A social network analysis of international tourism flow. *Qual. Quant.* **2021**, *55*, 419–439. [CrossRef]
29. He, B.; Liu, K.; Xue, Z.; Liu, J.; Yuan, D.; Yin, J.; Wu, G. Spatial and Temporal Characteristics of Urban Tourism Travel by Taxi—A Case Study of Shenzhen. *ISPRS Int. J. Geo-Inf.* **2021**, *10*, 445. [CrossRef]
30. Gan, C.; Voda, M.; Wang, K.; Chen, L.; Ye, J. Spatial network structure of the tourism economy in urban agglomeration: A social network analysis. *J. Hosp. Tour. Manag.* **2021**, *47*, 124–133. [CrossRef]
31. Lin, Q.; Xiang, M.; Zhang, L.; Yao, J.; Wei, C.; Ye, S.; Shao, H. Research on Urban Spatial Connection and Network Structure of Urban Agglomeration in Yangtze River Delta—Based on the Perspective of Information Flow. *Int. J. Environ. Res. Public Health* **2021**, *18*, 10288. [CrossRef]
32. Chu, N.; Wu, X.; Zhang, P.; Zhang, P. Urban Spatial Network Characteristics from the Perspectives of Reality and Virtual Flow in Northeast China. *Econ. Geogr.* **2022**, *42*, 66–74. [CrossRef]
33. An, D.; Hu, Y.; Wan, Y. Urban Network Association and Spillover Effects of Economic growth in China: A Study Based on Big Data and Network Analysis. *Geogr. Res.* **2022**, *41*, 2465–2481. [CrossRef]
34. Duan, D.; Du, D.; Chen, Y.; Zhai, Q. Spatial-temporal complexity and growth mechanism of city innovation network in china. *Sci. Geogr. Sin.* **2018**, *38*, 1759–1768. [CrossRef]
35. Zhou, C.; Zeng, G.; Cao, X. Chinese inter-city innovation networks structure and city innovation capability. *Geogr. Res.* **2017**, *36*, 1297–1308. [CrossRef]
36. Chen, H.; Wu, S. Comparison of the Development Level and Structural Characteristics of Urban Networks in the three Metropolitan Areas: An Empirical Study Based on Six Major Segments of the Producer Service Industry. *Econ. Geogr.* **2020**, *40*, 110–118. [CrossRef]
37. Ren, H.; Ye, M.; Yu, Y. Spatial Structure and Evolution Characteristics of Financial Network in Three Major Urban Agglomerations of China: A Case Study of Beijing-Tianjin-Hebei, Yangtze River Delta and Pearl River Delta. *Econ. Geogr.* **2021**, *41*, 63–73. [CrossRef]
38. Fang, Y.; Su, X.; Huang, Z.; Guo, B. Structural Characteristics and Resilience Evaluation of Tourism Flow Networks in Five Major Urban Agglomerations in Coastal China: From the Perspective of Evolutionary Resilience. *Econ. Geogr.* **2022**, *42*, 203–211. [CrossRef]
39. Ash, J.; Kitchin, R.; Leszczynski, A. Digital turn, digital geographies? *Prog. Hum. Geogr.* **2018**, *42*, 25–43. [CrossRef]
40. Silva, A.D.S.E. From Cyber to Hybrid: Mobile Technologies as Interfaces of Hybrid Spaces. *Space Cult.* **2006**, *9*, 261–278. [CrossRef]
41. Soja, E.W. Thirdspace: Journeys to Los Angeles and other Real-and-Imagined Places. *Cap. Cl.* **1998**, *22*, 137–139. [CrossRef]
42. Wang, W.; Zhang, M. Geomedia and thirdspace: The progress of research of geographies of media and communication in the West. *Prog. Geogr.* **2022**, *41*, 1082–1096. [CrossRef]
43. Paldino, S.; Bojic, I.; Sobolevsky, S.; Ratti, C.; González, M.C. Urban magnetism through the lens of geo-tagged photography. *EPJ Data Sci.* **2015**, *4*, 5. [CrossRef]
44. Sulis, P.; Manley, E.; Zhong, C.; Batty, M. Using mobility data as proxy for measuring urban vitality. *J. Spat. Inf. Sci.* **2018**, *16*, 137–162. [CrossRef]
45. Long, Y.; Huang, C. Does block size matter? The impact of urban design on economic vitality for Chinese cities. *Environ. Plan. B Urban Anal. City Sci.* **2019**, *46*, 406–422. [CrossRef]
46. Zhang, W.; Chong, Z.; Li, X.; Nie, G. Spatial patterns and determinant factors of population flow networks in China: Analysis on Tencent Location Big Data. *Cities* **2020**, *99*, 102640. [CrossRef]
47. Jiang, H.; Luo, S.; Qin, J.; Liu, R.; Yi, D.; Liu, Y.; Zhang, J. Exploring the Inter-Monthly Dynamic Patterns of Chinese Urban Spatial Interaction Networks Based on Baidu Migration Data. *ISPRS Int. J. Geo-Inf.* **2022**, *11*, 486. [CrossRef]
48. Liu, Y.; Liao, W. Spatial Characteristics of the Tourism Flows in China: A Study Based on the Baidu Index. *ISPRS Int. J. Geo-Inf.* **2021**, *10*, 378. [CrossRef]

49. Deng, C.; Song, X.; Xie, B.; Li, M.; Zhong, X. City Network Link Analysis of Urban Agglomeration in the Middle Yangtze River Basin Based on the Baidu Post Bar Data. *Geogr. Res.* **2018**, *37*, 1181–1192. [CrossRef]
50. Li, X.; Liu, H.; Tian, S.; Gong, Y. Network structure and influencing factors of urban human habitat activities in the three provinces of Northeast China: Based on Baidu Post Bar data. *Prog. Geogr.* **2019**, *38*, 1726–1734. [CrossRef]
51. Li, Z.; Zhao, M. City Networks in Cyberspace: Using Douban-Event to Measure the Cross-City Activities in Urban Agglomeration of China. *Hum. Geogr.* **2016**, *31*, 102–108. [CrossRef]
52. Wang, P.; Liu, K.; Wang, D.; Fu, Y. Measuring Urban Vibrancy of Residential Communities Using Big Crowdsourced Geotagged Data. *Front. Big Data* **2021**, 34. [CrossRef]
53. Zhao, M.; Xu, G.; De Jong, M.; Li, X.; Zhang, P. Examining the Density and Diversity of Human Activity in the Built Environment: The Case of the Pearl River Delta, China. *Sustainability* **2020**, *12*, 3700. [CrossRef]
54. Zhang, Y.; Li, Y.; Zhang, E.; Long, Y. Revealing virtual visiting preference: Differentiating virtual and physical space with massive TikTok records in Beijing. *Cities* **2022**, *130*, 103983. [CrossRef]
55. Ding, Z.; Ma, F.; Zhang, G. Spatial Differences and Influencing Factors of Urban Network Attention by Douyin Fans in China. *Geogr. Res.* **2022**, *41*, 2548–2567.
56. Peng, H.; Lu, L.; Lu, X.; Ling, S.; Li, Z.; Deng, H. The network structure of cross-border tourism flow based on the social network method:A case of Lugu Lake Region. *Sci. Geogr. Sin.* **2014**, *34*, 1041–1050. [CrossRef]
57. Wang, J.; Mo, H.; Wang, F.; Jin, F. Exploring the network structure and nodal centrality of China's air transport network: A complex network approach. *J. Transp. Geogr.* **2011**, *19*, 712–721. [CrossRef]
58. Duan, D.; Chen, Y.; Du, D. Regional Integration Process of China's Three Major Urban Agglomerations from the Perspective of Technology Transfer. *Sci. Geogr. Sin.* **2019**, *39*, 1581–1591. [CrossRef]
59. Lin, Z.; Chen, Y.; Liu, X.; Ma, Y. Spatio-temporal pattern and influencing factors of cooperation network of China's inbound tourism cities. *Acta Geogr. Sin.* **2022**, *77*, 2034–2049.
60. Forstall, R.L.; Greene, R.P.; Pick, J.B. Which Are the Largest? Why Lists of Major Urban Areas Vary so Greatly. *Tijdschr. Voor Econ. En Soc. Geogr.* **2009**, *100*, 277–297. [CrossRef]
61. Fang, C.; Yu, D. Urban agglomeration: An evolving concept of an emerging phenomenon. *Landsc. Urban Plan.* **2017**, *162*, 126–136. [CrossRef]

 *land*

Article

A Machine Learning Framework for Assessing Urban Growth of Cities and Suitability Analysis

Anne A. Gharaibeh [1], Mohammad A. Jaradat [2,3,*] and Lamees M. Kanaan [1]

[1] Department of City Planning and Design, College of Architecture and Design, Jordan University of Science and Technology, Irbid 22110, Jordan

[2] Department of Mechanical Engineering, College of Engineering, American University of Sharjah, Sharjah 26666, United Arab Emirates

[3] Department of Mechanical Engineering, College of Engineering, Jordan University of Science and Technology, Irbid 22110, Jordan

* Correspondence: mjaradat@aus.edu

Abstract: Rural–urban immigration, regional wars, refugees, and natural disasters all bring to prominence the importance of studying urban growth. Increased urban growth rates are becoming a global phenomenon creating stress on agricultural land, spreading pollution, accelerating global warming, and increasing water run-off, which adds exponentially to pressure on natural resources and impacts climate change. Based on the integration of machine learning (ML) and geographic information system (GIS), we employed a framework to delineate future urban boundaries for future expansion and urban agglomerations. We developed it based on a Time Delay Neural Network (TDNN) that depends on equal time intervals of urban growth. Such an approach is used for the first time in urban growth as a predictive tool and is coupled with Land Suitability Analysis, which incorporates both qualitative and quantitative data to propose evaluated urban growth in the Greater Irbid Municipality, Jordan. The results show the recommended future spatial expansion and proposed results for the year 2025. The results show that urban growth is more prevalent in the eastern, northern, and southern areas and less in the west. The urban growth boundary map illustrates that the continuation of urban growth in these areas will slowly further encroach upon and diminish agricultural land. By means of suitability analysis, the results showed that 51% of the region is unsuitable for growth, 43% is moderately suitable and only 6% is suitable for growth. Based on TDNN methodology, which is an ML framework that is dependent on the growth of urban boundaries, we can track and predict the trend of urban spatial expansion and thus develop policies for protecting ecological and agricultural lands and optimizing and directing urban growth.

Keywords: machine learning; Artificial Neural Network (ANN); GIS; urban growth; land suitability analysis; Time Delay Neural Network (TDNN)

Citation: Gharaibeh, A.A.; Jaradat, M.A.; Kanaan, L.M. A Machine Learning Framework for Assessing Urban Growth of Cities and Suitability Analysis. *Land* **2023**, *12*, 214. https://doi.org/10.3390/land12010214

Academic Editor: Chuanrong Zhang

Received: 11 December 2022
Revised: 4 January 2023
Accepted: 6 January 2023
Published: 9 January 2023

1. Introduction

The urbanization process means transforming rural society into an urban one accompanied by changes in the landscape. However, urban expansion denotes transforming vacant land or natural environment to constructed urban fabrics including residential, industrial and infrastructure development [1].

In the year 1800 CE, the global urban population was about 3%; however, in the 1950s, this had increased to 30%, while studies in the 2000s indicated that more than 47% of the world's population was living in urban areas [2]. Based on both World Bank statistics for 2015 and the United Nations 2014 revision, the urban population now constitutes about 53.857% of the global population. In the study area of Jordan, the urban population reached 83% in 2014, which is an alarming percentage [3]. Estimates for 2050 indicate a further 12% increase in urban population globally, with Jordan specifically facing a 6% increase. More developed regions are no exception, with the urban population expected to reach 89% by 2050 in Jordan [3].

Understanding the urbanization process is no easy task, since it has evolved over the years as a result of a complex network of changes in human behavior or land use policy in addition to societal pressures and activities in cities making difficult any measured urban development, such as the effects of ethnicity, religion, culture, and lifestyle on spatial growth in urban areas [4].

Some countries were aware of the urban growth early in the 20th century. For example, the UK introduced laws that, if implemented, would ensure a greenbelt policy is followed to control sprawling [5]. More studies in America and China referred to it as an urban development boundary (UDB) providing guidelines for the decision-makers to control and plan urban boundaries [6]. The urban growth boundary (UGB) became the focus of many studies especially the ones using artificial intelligence [7]. Some studies refer the expansion to the growth in economic urban activities, while others consider it the main result of population growth [8]. Regardless of the reasons, the negative consequence is the encroachment of agricultural and ecological land usage, culminating in urban sprawl [9]. The outstanding growth in relation to the demand for expansion made it necessary to plan for future growth, and models such as Future Land Use Model (FLUS), Markov, Patch-generating Land Use Simulation (PLUS) and Artificial Neural Network (ANN) are employed to predict Land Use Land Cover (LULC) change [7,9–11]. Some Remote Sensing (RS) research offers essential information on LULC change in connection to Land Surface Temperature (LST) that is valuable for predicting changes that may impact climate change and assist policymakers in developing effective land resource management plans [12,13].

Over the years, urban growth models have proven to be effective in describing and predicting urban development, providing sufficient information to help planners make informed decisions about urban planning. An Artificial Neural Network (ANN) is used to find the urban growth boundary and is increasingly used in many fields because of its powerful attributes and software flexibility, especially since urban growth is highly non-linear through time [7,14].

Tayyebi et al. [7] used ANN for a comprehensive study of Tehran, Iran, to predict growth boundaries to limit urban expansion, upgrade urban services, ensure landscape maintenance, and aid environmental protection. Planners employed neural networks, remote sensing systems models, and geographic information systems to predict future urban growth boundaries for 2012 given a set of variables such as roads, slopes, green spaces, service stations, elevation, aspect, and built areas.

Another study in Iran [15] utilized ANN to study changes in land use in previous years in the city of Kermanshah and to predict future changes. In addition to data for the past 19 years, satellite images were used for each of the years 1987, 2000, and 2006. ANN and the Markov model were used to predict land use for the next 19 years, from 2006 to 2025. Others developed a methodology for delineating an urban development boundary based on the Minimum Cumulative Resistance (MCR) model and CA-Markov model [16,17]. Aithani et al. [18] generated urban growth zonation maps using feedforward ANN for Dehradun city. However, Al-Kheder in [19] utilized a fuzzy logic-based intelligent system to model urban growth using satellite images.

Suitability analysis can be defined as a model used to select a suitable spatial site to perform a particular function; it is one of the most important functions of geographic information systems (GIS), assisting in the choice of the site by applying specific parameters in making the selection, which describes the study area landmarks (terrain, road network, etc.), thus contributing the data required for spatial site suitability analysis. Some studies coupled GIS with genetic algorithms to optimize specific land uses based on demand and allocation criteria [20].

The present research focuses on suitability analysis to investigate and determine the optimum sites for urban growth of the city of Irbid, following input of the required data. GIS tools can identify and calculate the weights of the urban growth factors based on their importance, wherein lies the main challenge to achieving the appropriate analysis, in determining the relative weights.

In the available literature, studies have made use of the suitability analysis tool to determine sites best suited for population expansion and urban growth [21,22]; for determining the most suitable areas for rangelands [23]; and for determining the suitability trends for settlement [24]. In all of these studies, researchers used a wide range of factors depending on the nature of the study area and the elements influencing growth [19,25,26]. The determination of urban growth-dependent factors is based on choosing the most suitable direction, including physical factors (slope, elevation), environmental and topographical factors (such as agricultural land, valleys), accessibility factors (distance from main streets), as well as consideration of economic and social factors.

The study by Berry et al. [27] used suitability analysis to focus on increasing sea levels as a result of global climate change. Suitability analysis was based on the map overlay, with the integration of sea level rise expectations based on several factors: elevation, slope, distance to coast, rock type, land cover, and sea level rise.

Raddad in [28] conducted a study that employed suitability analysis to evaluate the most feasible places for development in the southeast Jerusalem region. Built-up areas, geopolitical categorization, agricultural land, the separation wall, settlement areas, highways, terrain, heritage, and water sensitivity zones were identified as the primary determinants influencing urban growth. He utilized Arc GIS processing modules to generate final suitability maps based on these variables. This study, however, reveals an anomaly in the spatial dimension of population distribution in Irbid, where expansion is taking place at the cost of agricultural land.

One of the main causes of this fast-growing urban expansion is the increasing number of refugees that immigrate to Jordan from neighboring countries suffering war and turmoil. They are mostly concentrated in GIM due to its location in the north of Jordan. Jordan refugee statistics for 2021 were 3,047,612.00 granted asylum, making up a percentage of 33.6% of the local residents [29]. Another cause of urban expansion is the local authority's decision to subdivide agricultural lands in the 2000s in Irbid and change the use to residential and services to absorb the increased number of refugees and locals returning home following the Iraq war. Another recent wave of refugees followed the Syrian war from 2011 onward. This has caused far more expansion than is needed in the next 50 years. This could not have happened without the Cities, Villages, and Buildings Planning Law No. 79 of 1966, which established zoning plans (Al-Tantheem) by the Ministry of Municipalities and Towns. This usually refers to zones on a municipal zoning map that are changed on a regular basis. The Al-Taqseem Law of 1968 established subdivision plans (Al-Taqseem), which are implemented per basin. Unfortunately, zoning is not a requirement for permission, and Al-Taqseem can be used in both zoning-enforced and unzoned regions [30]. Therefore, choosing Irbid to implement this AI methodology will be a beneficial task to develop and provide service to this municipality in particular.

The primary objective of this study is to increase and broaden our understanding of spatial urban growth extent. It focuses on anticipating urban expansion so that strategies for land preservation and land use modification may be developed. The framework integrates machine learning (ML), geographic information systems (GIS), and image analysis for this purpose. The framework suggested is based on a Time Delay Neural Network (TDNN) that is dependent on equal time intervals of urban growth. In addition, it takes into account characteristics that influence land use change, such as soil fertility, the location of town and city centers, built-up areas, streams, and slopes. The ML-based prediction model is paired with Land Suitability Analysis, which includes both qualitative and quantitative data, in order to suggest analyzed urban expansion in the Greater Irbid Municipality, Jordan.

This paper is organized as follows: Section 2 covers the method, Section 3 illustrates and explains the prediction results and the suitability analysis, Sections 4 and 5 present the discussion and conclusions.

2. Methodology

Artificial neural networks (ANN) resemble the brain in two ways. 1—The network acquires knowledge from its surroundings via a learning process. 2—The strength of interneuron connections, termed "synaptic weights", is employed to store gained information [31]. ANN is composed of several layers, an input layer, one or more hidden layers and an output layer [7]. The training is achieved by exposing the network to examples of similar problems, and the network adapts itself (learns). After sufficient training, the neural network can model the problem data to the solutions, and it is then able to offer a solution to the problem [32]. During the training, the network predicts output and compares it with the correct available answer; if there is an output error, it works to modify the weights (w) of the links of each layer of the network and reprocesses the output. Time Delay Neural Network (TDNN) is among the ANNs. The TDNN consists of the lapped time delay with focused memory structure in the input layer of the network [33], where equal time intervals may be used as input data. Up to now, this network has not been used in urban growth forecasting, and it is worth investigating, especially where the data corresponded with equal intervals.

TDNN is a non-linear predictor that can train networks faster and easier with the least prediction errors [34]. This is completed based on the tapped delay line with a focused memory structure in the input layer of the network [33]. The more training it receives, the more accurate it can be.

As previously mentioned, this research aims to predict the boundary of the city of Irbid in 2025 by using TDNN and determining the best places for future urban growth of the city of Irbid based on selected criteria, namely: slope, soil fertility, streams, built-up area, and distance from the city center. The (TDNN), like other neural networks, consists of three layers: an input layer, an output layer and middle hidden layers. It relies on data input for specified time periods. In this research, the time period is optimized to be 10 years between the first and the second input; the Matlab software toolbox is utilized to implement and train the TDNN for prediction.

To predict the extent of urban growth in 2025 using (TDNN), training data were extracted from maps obtained from the Irbid municipality, as shown in Figure 1. The base of the analysis was Irbid Tal at the city center, which represents the kernel for the historical urban growth of the city. The growth maps from the center in each phase were used as input layers in the TDNN. Maps for the suitability analysis layers were created in GIS and other related software. The use of the TDNN allowed for working on a two-dimensional time series: one dimension is the time, while the other is the angle. In other words, the network would build its prediction based on an awareness of the growth radiuses (ρ) of all the angles (θ) rather than one angle only starting from the city center. The growth radius is transformed into the percentage of growth by dividing the new radius over the older one each time as follows:

$$G_r = \frac{\rho_c}{\rho_p} \tag{1}$$

where G_r is the growth percentage, ρ_c is the current radius, and ρ_p is the previous radius. In this series, each radius is normalized relative to the previous one, where if we had the value 2 for example, this meant that the current radius was twice the previous one. The values ranged between 1 and 6.9. The average in most years was about 1.98. The TDNN used in this work was a two-step time delay, which meant that the network takes the last two consequent radiuses and predicted the third one, as shown in Figure 2.

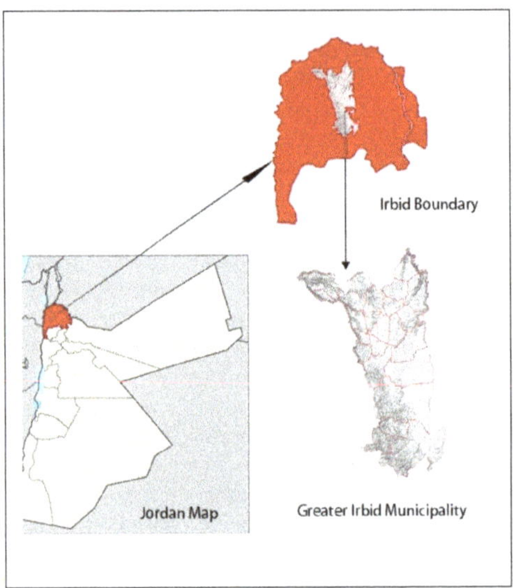

Figure 1. Study area. Map of Greater Irbid Municipality and City of Irbid boundaries (GIM, 2013) [35].

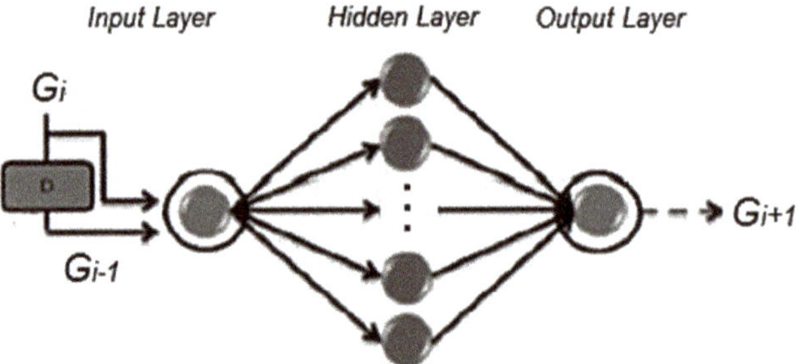

Figure 2. General structure for the TDNN [36].

To achieve the objectives of the TDNN network in providing the optimized input/output relationship for the predicted growth percentage G_{i+1}, the output layer collects the weighted inputs from the last hidden layer, and each hidden layer collects the weighted output from the previous layer until the first hidden layer, which collects the weighted inputs to the network as in Equations (2) and (3) [36].

$$G_{i+1} = F_o(\sum_0^h w_{ho} A_h + b_o) \tag{2}$$

$$A_h = F_h(w_{1h}G_{i-1} + w_{jh}G_i + b_h) \tag{3}$$

where G_{i+1} is the growth percentage predicted by the network, F_o is the activation function of the neuron in the output layer, F_h is the activation function of the neuron in the hidden layer, w is the weight for the link to be optimized during training between the neurons, and b_o and b_h are the neuron bias for the output and hidden layers, respectively. A_h is the collected output from the hidden layer, h is the number of neurons in the hidden layer, and the number of inputs of the network is j. To reach the optimal weight during the training process, the backpropagation training approach is utilized, which is considered among the most popular approaches for multilayer NN weight optimization [36]. The objective function is to minimize the Mean Square Error (*MSE*) [37] between the terms of the actual

2D input series of the training data (*Ga*) and the equivalent estimation of the network (*G_e*) as in Equations (4) and (5) [36,38].

$$MSE = \frac{1}{m} \sum_{x=1}^{m} e^2(x) \tag{4}$$

$$e = G_a - G_e \tag{5}$$

where *e* is the error, and *m* is the length of the series for each angle in the training data.

Following the usage of neural networks in the prediction of the boundary of the city for 2025 (Figure 3), suitability analysis was performed to determine the most suitable areas for urban growth based on several criteria (topographical maps, soil fertility maps, distance from the city center, built-up areas, and maps indicating streams). Maps are created using GIS software (Figure 4).

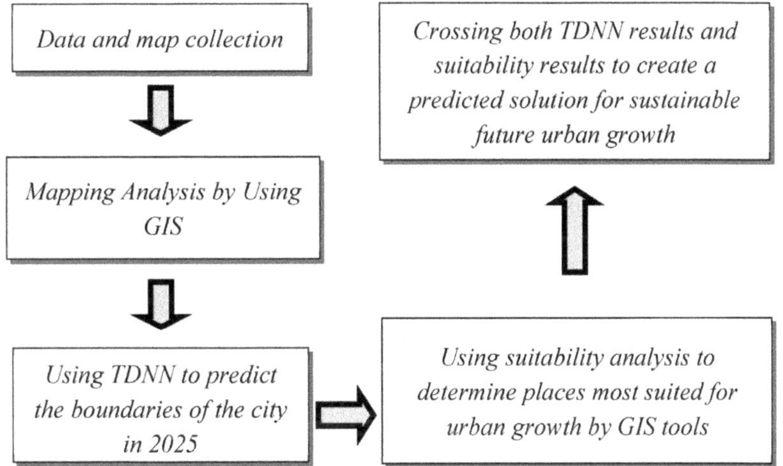

Figure 3. General structure of the proposed framework based on Time Delay Neural Network (TDNN).

Streams are important determinants that must be considered in order to preserve ecological corridors [28]. Distance from the major urban center and town agglomeration; closeness to services; level of city compactness; and physical limits to urban sprawl were all mentioned by [39,40]. Due to the importance of these factors, towns with a population of more than 11,000 people, based on Jordan's Department of Statistics report in 2015, were adopted as towns with urban centers (Table 1).

Table 1. Towns for which urban centers have been adopted in the suitability analysis [41].

Towns	Population Numbers	Towns	Population Numbers
Alhusun	*37,141*	*Beit ras*	*18,019*
Alsareeh	*19,227*	*Hawara*	*12,801*
Iydun	*18,592*	*Bushra*	*11,377*

Additionally, slope is one of the most important factors affecting planning and the direction of urbanization. Steep slopes are considered unsuitable for urban expansion because planning and construction are very expensive in these areas. Therefore, areas with a gradient of up to about 10% are suitable for residential development [28,42,43]. The increase in the urban built-up area makes it imperative to include this as a factor to encourage urban growth in vacant land [24]. Areas rich with fertile soil and agricultural land are worth protection and preservation [24]. After considering the previous literature and the considered variable weights, this research classified the variables and discussed each one based on the requirement of the place as well (Table 2).

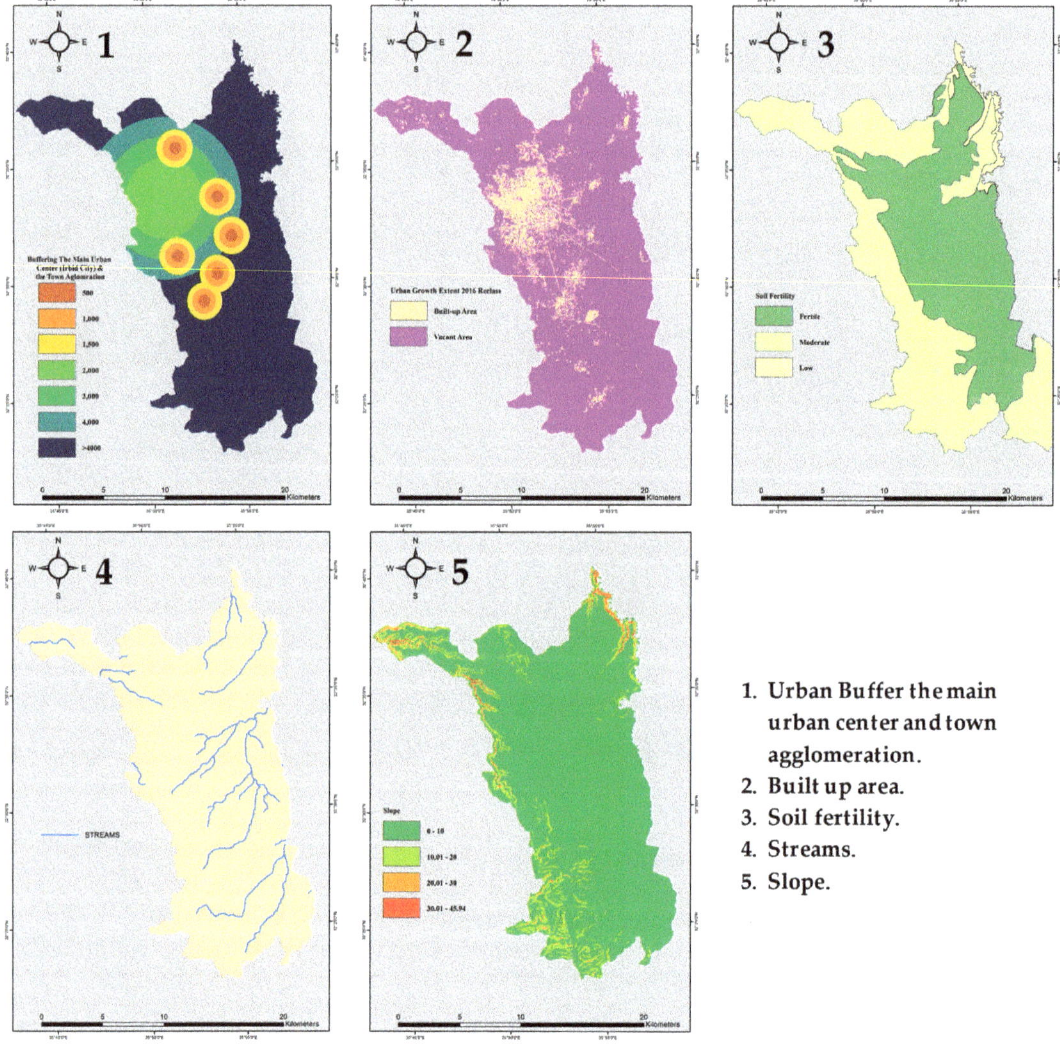

Figure 4. Variables affecting urban expansion in Irbid.

Table 2. Criteria classes and weighting.

Criteria	Classes	References	Weight	Weight Range	Note
Slope (degree)	0–10%	[24]	0.22	0.195	The slope factor in this research was given a weight based on the rate of the weight for the same factor in similar research. So, the weight = 0.19
	10.1–20%	[44] [43]	0.3		
	20.1–30% >30%	[28]	0.22 0.04		

Table 2. *Cont.*

Criteria	Classes	References	Weight	Weight Range	Note
Distance from the main urban center and town agglomeration	Main urban center 0–2000 m 2000–3000 3000–4000 Town center 0–500 m 500–1000 m 1000–1500 m	[39] [40]	0.22 0.26	0.24	Due to the population density of many towns in the study area, the urban center of the towns with a population of more than 11,000 was taken into consideration, including Huwwara, Al-Sareeh, Bushra, Idun, Beit-ras, and Al-Huson. To keep the towns and cities compact and prevent urban sprawl, the weight = 0.25
Streams (*m*) Buffering the water course with 40 m on either side of the center line.	0–40 m >40	[28] [45]	0.04 0.11	0.07	The study area is interspersed with many streams that have negative effects, especially flooding in the winter, so the weight = 0.11 based on [45].
Soil fertility	Fertile Moderate Low	[40] [22]	0.09 0.21	0.15	This factor was given the highest weight since the study area is rich in fertile soil and agricultural land and is being engulfed by urban growth. This is one of the most important factors that should control the future growth process so the weight = 0.3
Built-up area	Built-up Vacant land	[24] [45] [43]	0.13 0.15 0.12	0.133	The built-up area factor gained a weight of 0.15 based on [45]. The aim was to move away from the built-up area as it constituted an obstacle to growth

3. Results

3.1. Phases of Urban Expansion

The map of urban expansion of the Greater Irbid Municipality (GIM) is analyzed by separating each phase of the expansion using GIS tools (Figure 5 [46]). Earlier stages (1924–2001) were excerpted from GIM. Earth Explorer is the source for checking the later stages for specific years. We adopted and digitized the years 2005, 2010, and 2015.

The growth radius (ρ) was collected from the city center (Tal Irbid) to the boundary of each stage at a succession of equal angles (θ). This research experimented with a succession of 5 degree angles constituting a sum of 72 wedges, as shown in Figure 6.

The TDNN is adopted for the prediction process in the following years: (1955, 1965, 1975, 1985, 1995, 2005, and 2015). However, as a result of emigration in 1967, the population density increased, and urban expansion increased rapidly. There was a large difference between urban expansion in 1965 and 1975, which created illogical TDNN results. To eliminate the sudden jump, another experiment was performed by taking the years (1975, 1985, 1995, 2005 and 2015) as shown in Figure 7. Before the prediction process, the network was trained on the years 1975, 1985, 1995, and 2005 to predict 2015. However, in order for the data to be suitable for prediction, the growth radius was transformed into the percentage of growth by dividing the new radius over the older one each time, as in Equation (1). The average in most years was about 1.98. However, noticeably, in 1995–2005, larger values were registered, which was possibly because of the increased immigration to Jordan in this period due to the regional turmoil and instabilities in neighboring countries.

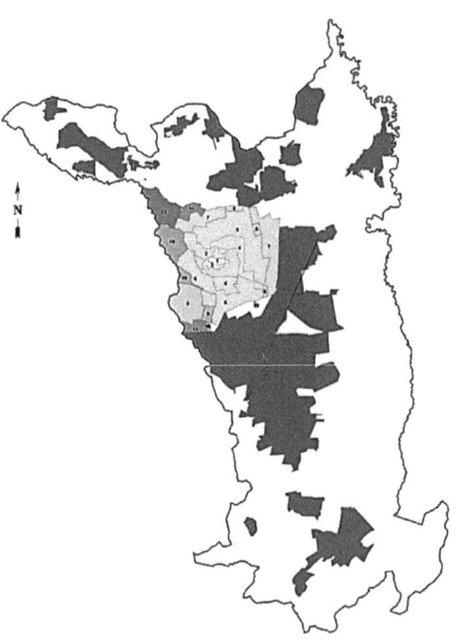

Phase	Year	Area (m^2)	Phase	Year	Area (m^2)	Phase	Year	Area (m^2)
1	1924	285.739 K	6	1970	658.860 K	11	1994	1480.731 K
2	1953	596.792 K	7	1978	6134.087 K	12	2000	2443.510 K
3	1955	2770.525 K	8	1985	2999.913 K	13	2001	82312.254 K
4	1960	3286.844 K	9	1986	474.657 K			
5	1967	11396.903 K	10	1990	2820.436 K			

Figure 5. Growth stages of greater Irbid municipality (GIM, 2005) [46].

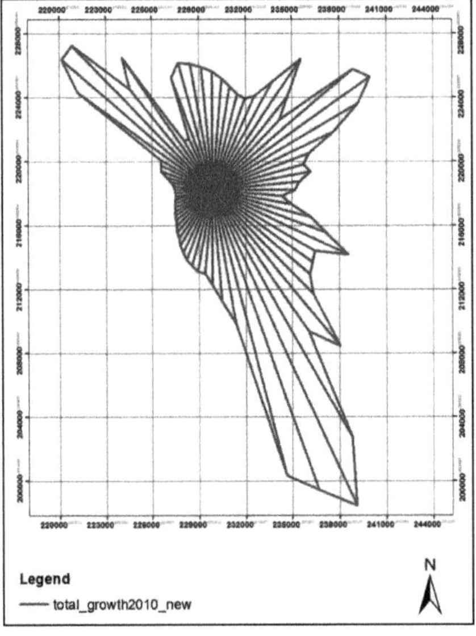

Figure 6. The succession of growth radiuses at θ = 5-degree angle for the 2010 stage using GIS.

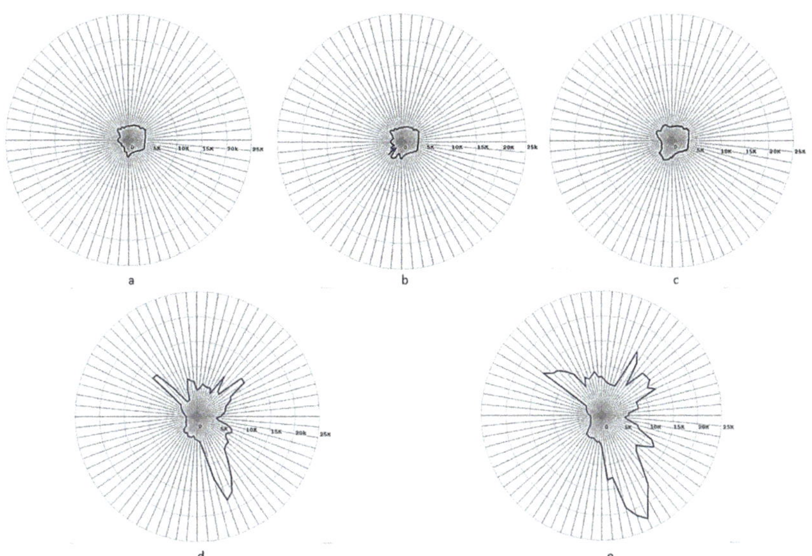

Figure 7. The succession of growth radiuses at θ = 5-degree angles for the years: 1975, 1985, 1995, 2005, and 2015, (**a–e**) respectively.

The general shape of the neural network and its components is shown in Figure 8; the number of inputs is 72 and the number of outputs is 72, which is equivalent to (360/5 = 72). Several iterations were made to reach the optimum network with the least error value starting from different random initial weights; 70% of the data was used for training. Finally, the network that had the least error consisted of two hidden layers; the first hidden layer consisted of five neurons and the second consisted of 71 neurons with the minimum MSE as shown in Figure 9.

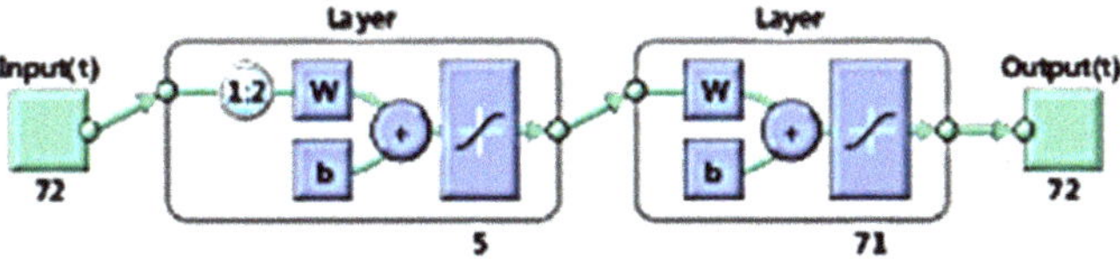

Figure 8. The general structure of TDNN.

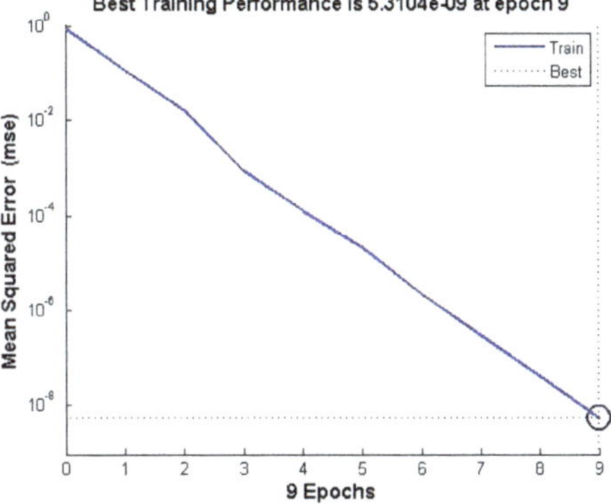

Figure 9. The mean square error of the network during the training epochs.

Another measure of how well the neural network has fit the data was the regression plot as shown in Figure 10. The regression plot illustrated the plot of the expected output based on associated target values. If the network had learned to fit the data well, the linear fit to this output–target relationship should closely intersect with the bottom-left and top-right corners of the plot [47].

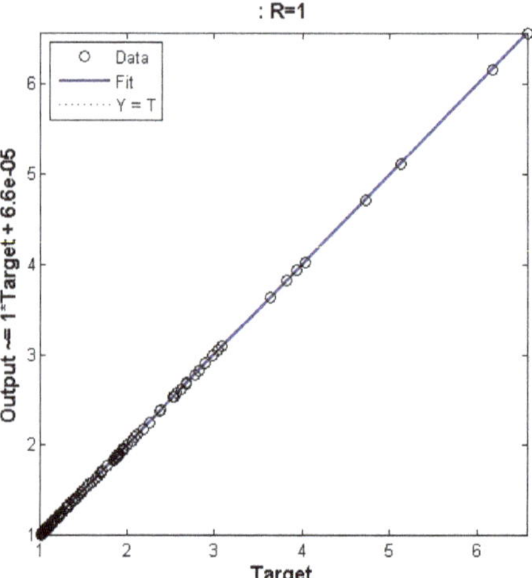

Figure 10. The regression plot measure of the suitability of TDNN.

Figure 11 illustrates a third measure of how well the neural network has fit data; the error histogram plot showed how the error sizes were distributed. Typically, most errors were near zero, with very few errors far from that. The value at the bottom of the blue rectangle is the error value. The value on the vertical axis is the number of times this error value appeared among the data.

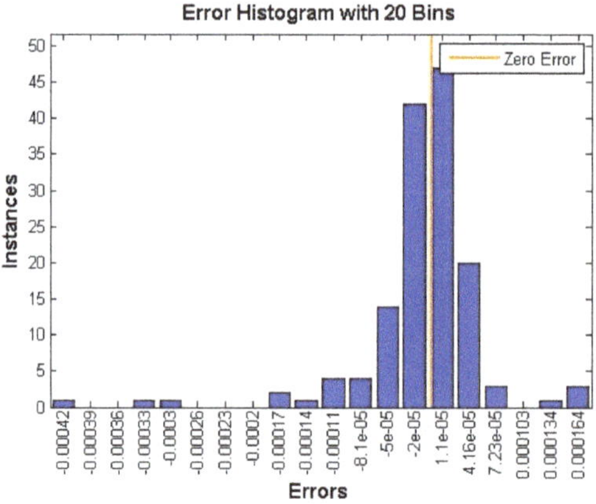

Figure 11. The error histogram measure of the suitability of the TDNN for data.

For presentation purposes, GIS was used to illustrate the existing and predicted maps. Each radius was drawn with an angle above the layer of 2015, as shown in Figure 12. As it is noticed, the results indicated that urban expansion was more prevalent in the eastern, northern, and southern areas and less in the west due to the presence of some valleys in this area.

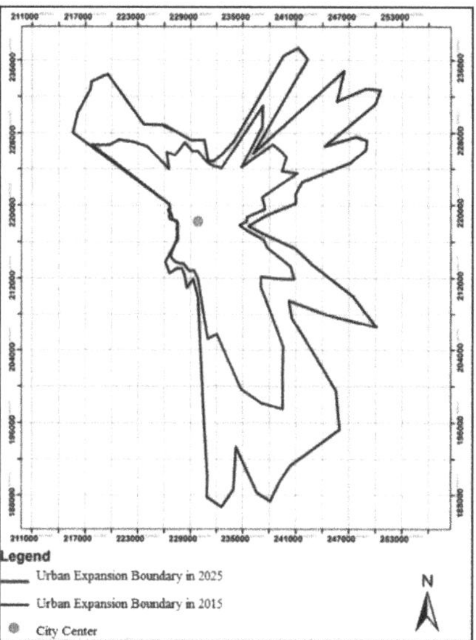

Figure 12. Prediction map for 2025 resulting from the use of TDNN.

Additionally, the urban growth boundary map illustrated that the continuation of urban growth in these areas will further impinge upon and diminish agricultural land, especially in the southern and eastern regions [48,49]. Unless prevented by urban development policies with firm and timely interventions of the concerned authorities, urban growth will continue to invade agricultural land.

3.2. Suitability Analysis

After obtaining the prediction map, the suitability analysis phase begins. Each variable is classified in a separate map (urban buffer, built-up area, soil fertility, streams, and slope) (Figure 13). The maps were reclassified and given weights depending on their importance and impact and depending on previous literature (Table 3). A multi-criteria analysis was made to create suitability maps by combining factors and weighting them. The resulting map was classified into three classes, namely: low suitability, moderate suitability, and high suitability for urban expansion (Figure 14).

Table 3. Criteria reclassification and weighting.

Criteria	Classes	Reclassified	Weight
Slope (degree)	0–10%	4 (the best slope for growth)	0.19
	10.1–20%	3	
	20.1–30%	2	
	More than 30%	0 (building restricted and challenging)	
Distance from the main urban center and town agglomeration	Main urban center		0.25
	0–2000 m	4 (the best for growth)	
	2000–3000	3	
	3000–4000	1	
	>4000	0	
	Town center		
	0–500 m	4	
	500–1000	3	
	1000–1500	1	
	>1500	0	

Table 3. *Cont.*

Criteria	Classes	Reclassified	Weight
Streams (m) Buffering the water course with 40 m on either side of the center line.	0–40 m More than 40	0 4	0.11
Soil fertility	Fertile Moderate Low	0 2 4	0.30
Built-up area	Built up Vacant land	0 4	0.15

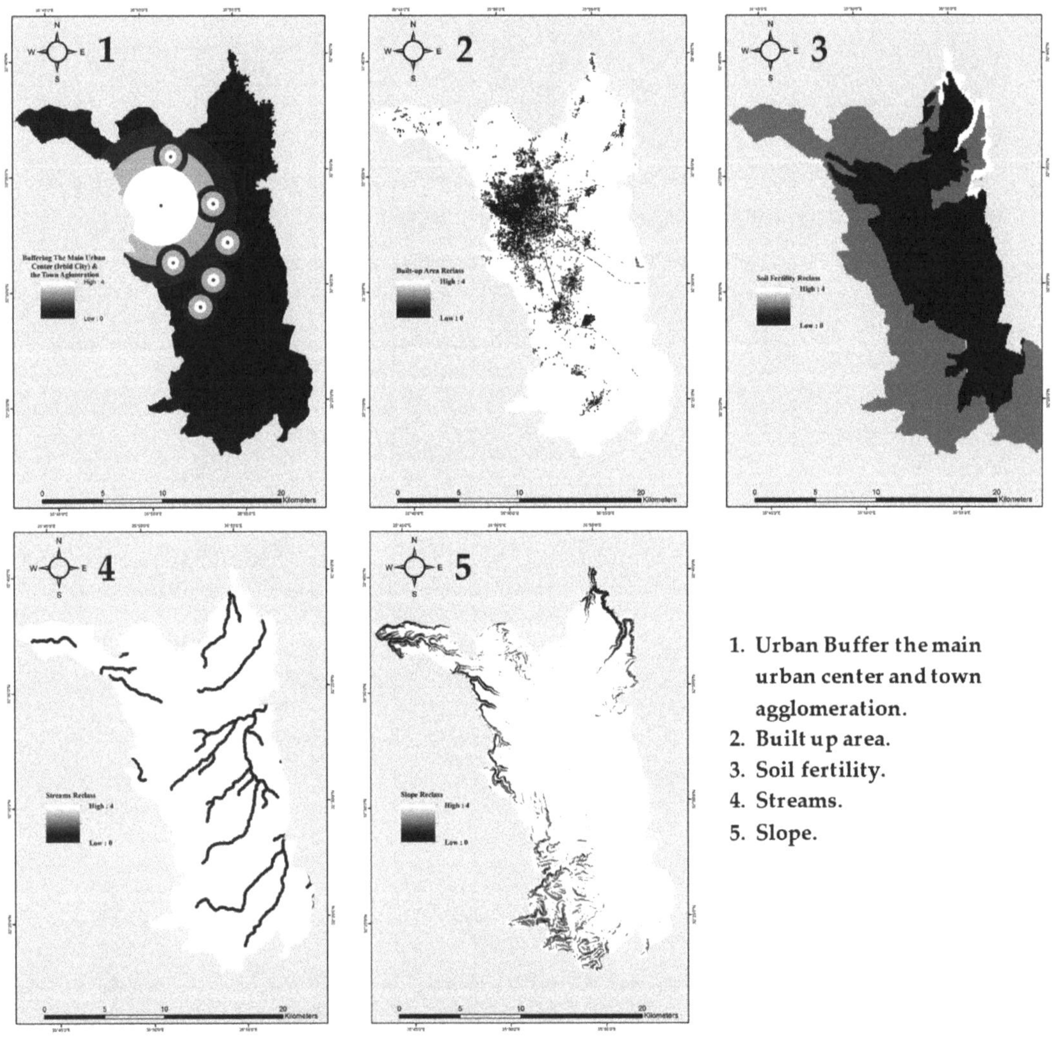

Figure 13. The factors after reclassified.

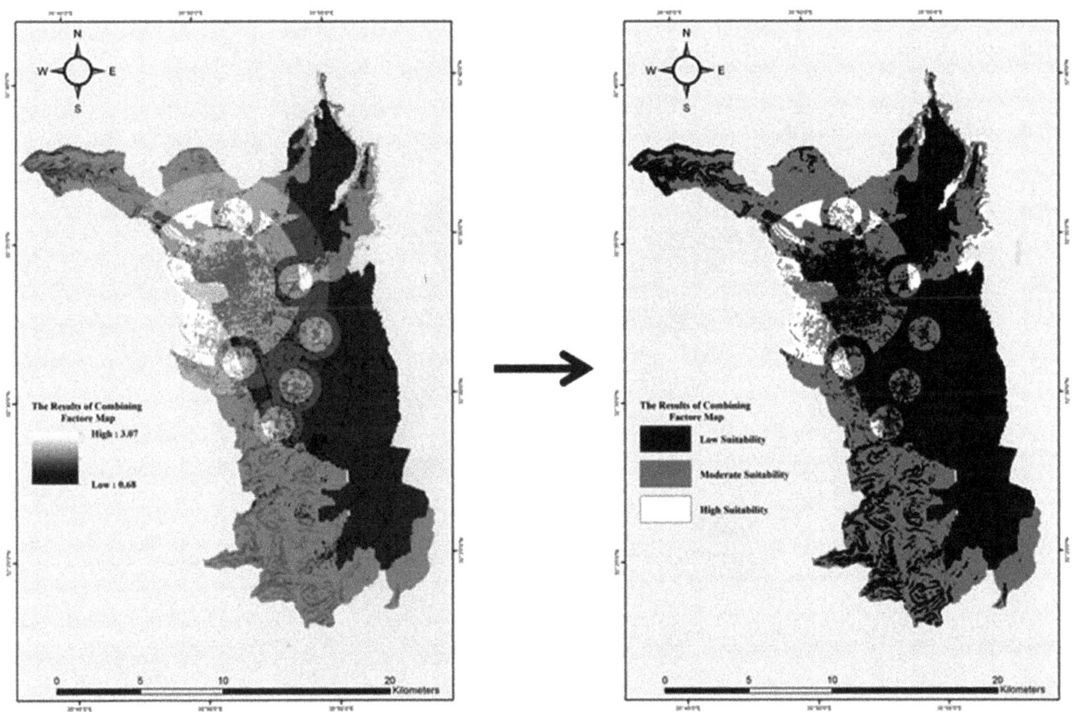

The result of combining factors map   The result of suitability analysis

Figure 14. The result of combining factors and suitability analysis.

3.3. Juxtaposing the Suitability Analysis and the TDNN Map

Factors with higher weights (white color) were a major influence to make the region suitable for growth. The results indicated that many of the areas suitable for urban growth are located around urban centers (Figure 14). Therefore, we recommend that the city grows vertically and fill the vacant city parcels with more development. This will create a more compact urban fabric while preserving agricultural lands. Most of the areas located between 52 degrees and 120 degrees (2 and 1) and 282 degrees and 300 degrees (4 and 3) had a medium suitability (Figure 15). They grow from the city center toward the north or south in general. The eastern areas are more suited for agricultural land because of the soil fertility and their nature as plains. This answers the research question concerning the preferred direction of urban growth and land suitability for urban growth.

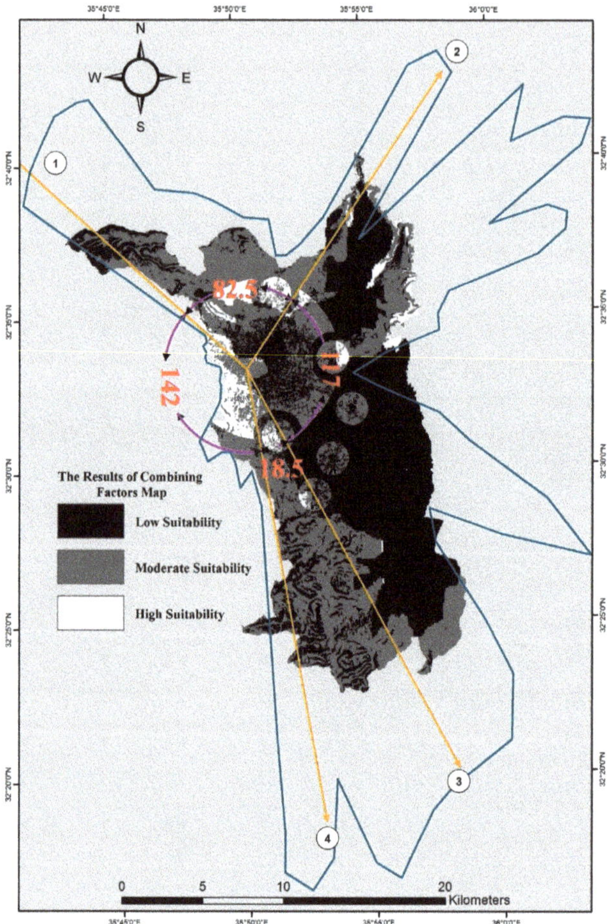

Figure 15. The result of suitability analysis overly and TDNN prediction for 2025. Line color representation: blue is the expected urban growth extent, yellow are the lines delimiting the preferred angle of growth between the arrows 1&2 and 3&4. Delimited angles between 2&3 and 1&4 define the unwanted growth direction avoiding fertile soils (117°) and rugged terrains (142°) respectively.

4. Discussion

If we want to be governed by this predictive study and maintain more resilient urban growth, we must create policies to control the urban edge, the permissible areas of growth, and the tolerable densities. We must think of re-zoning the city and nominate the types of buildings that suit the multi-cultural melting pot we are experiencing in this part of the world. As it seems, there will be more people to settle in secured urban areas, and we must be prepared to handle forecasts. We must think of our future and long-term interests in growing our foods in the places most suited for that and in the places that have the ability to grow abundant amounts of basic foods to create food security. With the United Nation's sustainable development goals (SDGs), this issue has an enormous importance for the coming generations: in particular, Zero Hunger (SDG 2), Sustainable Cities and Communities (SDG 11), Climate Action (SDG 13), and Life on Land (SDG 15) [50].

If we continue conducting business as usual, we will end up with overly expanded metropolitan areas, unaffordable infrastructures, the loss of agricultural lands, and most probably, loss of identity. There is an opportunity to develop a better organized spatial strategy that considers both the scope and capacity of urban growth as well as the need for urban green spaces to accommodate expanding populations. This is an opportunity to preserve ecological corridors, renovate heritage buildings, and organize land uses.

It is in our interest to encourage people to live in their home towns and prevent rural urban migration, which is another cause of this outstanding urban growth. Since most people tend to have daily commutes to obtain services at their urban centers, it is wise to serve them in their hometowns and prevent excessive commuting and the tendency to immigrate to urban areas to fulfill consumption aspirations. A study concentrating on the rural–urban relationship in the north of Jordan revealed that the commutes were not based on obtained jobs at the urban center but rather in search of attractive and necessary services [51]. Therefore, most of this sprawling is not essential. If we are discussing policies, we also have to consider providing services especially entertainment and shopping in addition to job opportunities at key rural locations to prevent the rural–urban migration and the fast-growing urban areas.

This study demonstrated how the TDNN predictive tool, coupled with suitability analysis, can predict and weigh urban growth. This is similar to other methods but with more precise outputs as a result of the self-learning tool and the suitability analysis. The machine learning tools allow self-correction and adaptation to change. The more it learns, the more precise the prediction. When such tools are accessible to decision makers, they can modify and study growth and manage city requirements with great precision. Giving building permits, preparing infrastructures, proposing services, and many other land uses will be made easier for them. With such predictive tools that are able to learn and evolve every day, we can also plan for resource management to lead a more resilient growth.

5. Conclusions

The purpose of this research is to predict the boundary of urban growth in GIM in 2025 by using ANN as a model, specifically TDNN to support the planning process and help decision-makers see the future status of the city. This was expected to aid them in the future expansion of the city and prepare for greater sustainability in the foreseeable future. The research is based on the analysis of the urban expansion map throughout history, using the resulting data from the analysis as input for TDNN, and by employing suitability analysis of the resulting expected growth. Results indicated that urban growth will be significantly southward, with little northward and eastward, and very limited westward. The results showed that 51% of the region is unsuitable for growth, 43% is moderately suitable and only 6% is suitable for growth. The ML model is very useful in determining the future of the city based on learning from previous spatial urban growth.

We note from the analysis that the eastern and southeastern areas are generally unsuitable for urban growth due to the fertility of the land there and the presence of agricultural land in abundance in addition to the increased distance from the city center. The area in the proximity to the city center was suitable for growth despite the presence of a built-up area. Suitability was strongly affected by other factors, such as the soil, which despite being fertile has not so far been used as agricultural land because of urban growth.

Since the more suitable lands for urban expansion are the lands within the urban areas and in close proximity to their centers, the results highlight the importance of densification in this case. Densification will save agricultural land and create more sustainable and manageable urban growth. Smaller towns also act as sub-centers that will help with densification and more controllable urban growth. Encouraging growth on the hillier sides of the city toward the southwest is another foreseen alternative. The more challenging slopes will provide more bare land than agricultural lands with moderate suitability for urban growth. This will result in a more linear city footprint but a more resilient existence in this part of the country. The waves of refugees in this part of the world keep coming to Jordan as things escalate within the Middle East region. Enforcing policies to direct growth appears to be critical for the city in order to control its excessive urban growth activities and create a more resilient future with a served city limit and reduced sprawl. Based on the demonstrated results, this work can be further extended in the future and applied to other growing cities.

Author Contributions: The work was developed and implemented as part of the thesis work by L.M.K. and supervised by the project advisors; A.A.G. and M.A.J. All authors have read and agreed to the published version of the manuscript.

Funding: This research received no external funding.

Data Availability Statement: Not applicable.

Conflicts of Interest: The authors declare no conflict of interest.

References

1. Shenghe, L.; Sylvia, P. *Spatial Patterns and Dynamic Mechanisms of Urban Land Use Growth in China: Case Study in Beijing and Shanghai*; IIASA Interim Report; IIASA: Laxenburg, Austria, 2002.
2. United Nations. *Department of Economic and Social Affairs. Population Division Population Distribution, Urbanization, Internal Migration and Development: An International Perspective*; United Nations: New York, NY, USA, 2011.
3. United Nations. *World Urbanization Prospects, United Nations, 2014 revision, Department of Economic and Social Affairs*; United Nations: New York, NY, USA, 2015.
4. Herold, M.; Goldstein, N.C.; Clarke, K.C. The spatiotemporal form of urban growth: Measurement, analysis and modeling. *Remote Sens. Environ.* **2003**, *86*, 286–302. [CrossRef]
5. Han, A.T.; Go, M.H. Explaining the national variation of land use: A cross-national analysis of greenbelt policy in five countries. *Land Use Policy* **2019**, *81*, 644–656. [CrossRef]
6. Zheng, B.; Liu, G.; Wang, H.; Cheng, Y.; Lu, Z.; Liu, H.; Zhu, X.; Wang, M.; Yi, L. Study on the Delimitation of the Urban Development Boundary in a Special Economic Zone: A Case Study of the Central Urban Area of Doumen in Zhuhai, China. *Sustainability* **2018**, *10*, 756. [CrossRef]
7. Tayyebi, A.; Pijanowski, B.; Tayyebi, A. An urban growth boundary model using neural networks, GIS and radial parameterization: An application to Tehran, Iran. *Landsc. Urban Plan.* **2011**, *100*, 35–44. [CrossRef]
8. Gao, L.; Tao, F.; Liu, R.; Wang, Z.; Leng, H.; Zhou, T. Multi-scenario simulation and ecological risk analysis of land use based on the PLUS model: A case study of Nanjing. *Sustain. Cities Soc.* **2022**, *85*, 104055. [CrossRef]
9. Feng, D.; Bao, W.; Fu, M.; Zhang, M.; Sun, Y. Current and Future Land Use Characters of a National Central City in Eco-Fragile Region—A Case Study in Xi'an City Based on FLUS Model. *Land* **2021**, *10*, 286. [CrossRef]
10. Xu, T.; Zhou, D.; Li, Y. Integrating ANNs and Cellular Automata–Markov Chain to Simulate Urban Expansion with Annual Land Use Data. *Land* **2022**, *11*, 1074. [CrossRef]
11. Cağlıyan, A.; Dağlı, D. Monitoring Land Use Land Cover Changes and Modelling of Urban Growth Using a Future Land Use Simulation Model (FLUS) in Diyarbakır, Turkey. *Sustainability* **2022**, *14*, 9180. [CrossRef]
12. Hussain, S.; Mubeen, M.; Karuppannan, S. Land use and land cover (LULC) change analysis using TM, ETM+ and OLI Landsat images in district of Okara, Punjab, Pakistan. *Phys. Chem. Earth Parts A/B/C* **2022**, *126*, 103117. [CrossRef]
13. Hussain, S.; Mubeen, M.; Ahmad, A.; Majeed, H.; Qaisrani, S.A.; Hammad, H.M.; Amjad, M.; Ahmad, I.; Fahad, S.; Ahmad, N.; et al. Assessment of land use/land cover changes and its effect on land surface temperature using remote sensing techniques in Southern Punjab, Pakistan. *Environ. Sci. Pollut. Res.* 2022, pp. 1–17. [CrossRef]
14. Triantakonstantis, D.; Mountrakis, G. Urban Growth Prediction: A Review of Computational Models and Human Perceptions. *J. Geogr. Inf. Syst.* **2012**, *4*, 26323. [CrossRef]
15. Razavi, B.S. Predicting the Trend of Land Use Changes Using Artificial Neural Network and Markov Chain Model (Case Study: Kermanshah City). *Res. J. Environ. Earth Sci.* **2014**, *6*, 215–226. [CrossRef]
16. Yi, S.; Zhou, Y.; Li, Q. A New Perspective for Urban Development Boundary Delineation Based on the MCR Model and CA-Markov Model. *Land* **2022**, *11*, 401. [CrossRef]
17. Gharaibeh, A.; Shaamala, A.; Obeidat, R.; Al-Kofahi, S. Improving land-use change modeling by integrating ANN with Cellular Automata-Markov Chain model. *Heliyon* **2020**, *6*, e05092. [CrossRef] [PubMed]
18. Maithani, S.; Arora, M.K.; Jain, R.K. An artificial neural network based approach for urban growth zonation in Dehradun city, India. *Geocarto Int.* **2010**, *25*, 663–681. [CrossRef]
19. Al-Kheder, S.; Wang, J.; Shan, J. Fuzzy inference guided cellular automata urban-growth modelling using multi-temporal satellite images. *Int. J. Geogr. Inf. Sci.* **2008**, *22*, 1271–1293. [CrossRef]
20. Gharaibeh, A.A.; Ali, M.H.; Abo-Hammour, Z.S.; Al Saaideh, M. Improving Genetic Algorithms for Optimal Land-Use Allocation. *J. Urban Plan. Dev.* **2021**, *147*, 04021049. [CrossRef]
21. Jiao, S.; Gao, Q.; Wei, C.-Y.; Liu, B.; Li, X.-D.; Zeng, G.-M.; Yuan, Z.-X.; Liang, J. Ecological suitability evaluation for urban growth boundary in red soil hilly areas based on fuzzy theory. *J. Cent. South Univ.* **2012**, *19*, 1364–1369. [CrossRef]

22. Mohammed, K.S.; Elhadary, Y.A.E.; Samat, N. Identifying Potential Areas for Future Urban Development Using Gis-Based Multi Criteria Evaluation Technique. In *SHS Web of Conferences*; EDP Sciences: Les Ulis, France, 2016; p. 03001. [CrossRef]

23. Jafari, S.; Zaredar, N. Land Suitability Analysis using Multi Attribute Decision Making Approach. *Int. J. Environ. Sci. Dev.* **2010**, *1*, 441–445. [CrossRef]

24. Omar, N.Q.; Raheem, A.M. Determining the suitability trends for settlement based on multi criteria in Kirkuk, Iraq. *Open Geospat. Data Softw. Stand.* **2016**, *1*, 10. [CrossRef]

25. Bagheri, M.; Sulaiman, W.N.A.; Vaghefi, N. Application of geographic information system technique and analytical hierarchy process model for land-use suitability analysis on coastal area. *J. Coast. Conserv.* **2013**, *17*, 1–10. [CrossRef]

26. Kumar, M.; Shaikh, V.R. Site suitability analysis for urban development using GIS based multicriteria evaluation technique. *J. Indian Soc. Remote Sens.* **2013**, *41*, 417–424. [CrossRef]

27. Berry, M.; BenDor, T.K. Integrating sea level rise into development suitability analysis. *Comput. Environ. Urban Syst.* **2015**, *51*, 13–24. [CrossRef]

28. Raddad, S. Integrated GIS and multi criteria evaluation approach for suitability analysis of urban expansion in south eastern Jerusalem region–Palestine. *Geogr. Inf. Syst.* **2016**, *5*, 24–31.

29. Macrotrends. Jordan Refugee Statistics 1960–2022. 2022. Available online: https://www.macrotrends.net/countries/JOR/jordan/refugee-statistics (accessed on 1 December 2022).

30. UN-Habitat's Urban Practices Branch, Planning, Finance and Economy Section. Urban Planning & Infrastructure in Migration Contexts, Irbid Spatial Profile, Jordan. A report in Collaboration of UN-Habitat, Greater Irbid Municipality, and Swiss State Secretariat for Economic Affairs (SECO). 2022. Available online: https://unhabitat.org/sites/default/files/2022/04/220411-final_irbid_profile.pdf (accessed on 1 December 2022).

31. Jensen, R.R.; Binford, M.W. Measurement and comparison of Leaf Area Index estimators derived from satellite remote sensing techniques. *Int. J. Remote Sens.* **2004**, *25*, 4251–4265. [CrossRef]

32. Zhou, L. Integrating Artificial Neural Networks, Image Analysis and GIS for Urban Spatial Growth Characterization. Ph.D Thesis, College of Social Sciences and Public Policy, Florida State University, Tallahassee, FL, USA, 2012; pp. 1–154.

33. Hagan, M.; Demuth, H. *Neural Network Design*; PWS Publishing: Boston, MA, USA, 1996.

34. Winograd, T. Shifting viewpoints: Artificial intelligence and human–computer interaction. *Artif. Intell.* **2006**, *170*, 1256–1258. [CrossRef]

35. Al-Kofahi, S.D.; Gharaibeh, A.A.; Bsoul, E.Y.; Othman, Y.A.; Hilaire, R.S. Investigating domestic gardens' densities, spatial distribution and types among city districts. *Urban Ecosyst.* **2019**, *22*, 567–581. [CrossRef]

36. Saadeddin, K.; Abdel-Hafez, M.F.; Jaradat, M.A.; Jarrah, M.A. Optimization of Intelligent Approach for Low-Cost INS/GPS Navigation System. *J. Intell. Robot. Syst.* **2014**, *73*, 325–348. [CrossRef]

37. Negnevitsky, M. *Artificial Intelligence*, 1st ed.; Harlow Addison-Wesley: Boston, MA, USA, 2011.

38. Jaradat, M.A.K.; Abdel-Hafez, M.F. Non-Linear Autoregressive Delay-Dependent INS/GPS Navigation System Using Neural Networks. *IEEE Sens. J.* **2016**, *17*, 1105–1115. [CrossRef]

39. Liu, Y.; Hu, Y.; Long, S.; Liu, L.; Liu, X. Analysis of the Effectiveness of Urban Land-Use-Change Models Based on the Measurement of Spatio-Temporal, Dynamic Urban Growth: A Cellular Automata Case Study. *Sustainability* **2017**, *9*, 796. [CrossRef]

40. Alanbari, M.A.; Al-Ansari, N.; Jasim, H.K. GIS and Multicriteria Decision Analysis for Landfill Site Selection in Al-Hashimyah Qadaa. *Nat. Sci.* **2014**, *6*, 282–304. [CrossRef]

41. Department of Statistics (DOS). Main Results of the General Census of Population and Housing 2015. Published in 2016. 2016. Available online: http://dosweb.dos.gov.jo/products/census_mainresults2015/ (accessed on 1 December 2022).

42. Aliani, H.; Ghanbari Motlagh, M.; Danesh, G.; Aliani, H. Land suitability analysis for urban development using TOPSIS, WLC and ANP techniques (Eastern cities of Gilan-Iran). *Arab. J. Geosci.* **2021**, *14*, 1276. [CrossRef]

43. Satir, O. Mapping the Land-Use Suitability for Urban Sprawl Using Remote Sensing and GIS Under Different Scenarios. Master's Thesis, Department of Landscape Architecture, Faculty of Agriculture, Yuzuncu Yil University, Van, Turkey, 2016.

44. Ahmad, F.; Goparaju, L.; Qayum, A. Agroforestry suitability analysis based upon nutrient availability mapping: A GIS based suitability mapping. *AIMS Agric. Food* **2017**, *2*, 201–220. [CrossRef]

45. Kaoje, I.U.; Dankani, I.M.; Ishiaku, I. Site Suitability Analysis for Municipal Solid Waste Disposal in Birnin Kebbi, Nigeria. *IOSR J. Humanit. Soc. Sci.* **2016**, *21*, 1–10. [CrossRef]

46. *Growth Stages of Greater Irbid Municipality*; Greater Irbid Municipality (GIM): Irbid, Jordan, 2005.

47. Beale, M.; Martin, T.H.; Howard, B.D. *MATLAB Neural Network Toolbox User's Guide (version R2017b)*; The MathWorks, Inc.: Natick, MA, USA, 2017; p. 158.

48. Gharaibeh, A.A.; Shaamala, A.H.; Ali, M.H. Multi-Criteria Evaluation for Sustainable Urban Growth in An-Nuayyimah, Jordan; Post War Study. *Procedia Manuf.* **2020**, *44*, 156–163. [CrossRef]

49. Gharaibeh, A.A.; Jaradat, R.A.; Okour, Y.F.; Al-Rawabdeh, A.M. Landscape Perception and Landscape Change for the City of Irbid, Jordan. *J. Archit. Plan.* **2017**, *29*, 89–104.

50. UN General Assembly, Transforming Our World: The 2030 Agenda for Sustainable Development, 21 October 2015. A/RES/70/1. Available online: https://www.refworld.org/docid/57b6e3e44.html (accessed on 1 January 2023).
51. Gharaibeh, A.A.; Alhamad, M.N.; Al-Hassan, D.A.; Abumustafa, N.I. The impact of the spatial configuration of socioeconomic services on rural–urban dependencies in Northern Jordan. *GeoJournal* **2022**, *87*, 4475–4490. [CrossRef] [PubMed]

 *land*

Article

Quantitative Morphometric 3D Terrain Analysis of Japan Using Scripts of GMT and R

Polina Lemenkova * and Olivier Debeir

Laboratory of Image Synthesis and Analysis (LISA), École Polytechnique de Bruxelles (Brussels Faculty of Engineering), Université Libre de Bruxelles (ULB), Building L, Campus du Solbosch, ULB—LISA CP165/57, Avenue F. D. Roosevelt 50, 1050 Brussels, Belgium
* Correspondence: polina.lemenkova@ulb.be; Tel.: +32-471-86-04-59

Abstract: In this paper, we describe two related scripting methods of cartographic data processing and visualization that provide 2D and 3D mapping of Japan with different algorithm complexity. The first algorithm utilizes Generic Mapping Toolset (GMT), which is known as an advanced console-based program for spatial data processing. The modules of GMT combine the functionality of scripting with the aspects of geoinformatics, which is especially effective for the rapid analysis of large geospatial datasets, multi-format data processing, and mapping in 2D and 3D modes. The second algorithm presents the use of the R programming language for cartographic visualization and spatial analysis. This R method utilizes the packages 'tmap', 'raster', 'maps', and 'mapdata' to model the morphometric elements of the Japanese archipelago, such as slope, aspect, hillshade and elevation. The general purpose graphical package 'ggplot2' of R was used for mapping the prefectures of Japan. The two scripting approaches demonstrated an established correspondence between the programming languages and cartography determined with the use of scripts for data processing. They outperform several well-known and state-of-the-art GIS methods for mapping due to their high automation of data processing. Cartography has largely reflected recent advances in data science, the rapid development of scripting languages, and transfer in the approaches of data processing. This extends to the shift from the traditional GIS to programming languages. As a response to these new challenges, we demonstrated in this paper the advantages of using scripts in mapping, which consist of repeatability and the flexible applicability of scripts in similar works.

Keywords: terrain modelling; script; geoscience; R language; generic mapping tools; computer science; data visualization; 3D modelling; cartography

PACS: 91.10.Da; 95.75.Rs; 95.75.-z; 92.70.Iv; 92.70.Pq; 93.30.-w; 93.30.Db; 92.40.Pb

MSC: 97Pxx; 97P40; 97P50; 97M50; 97P10; 91D20; 68P05; 92F05

JEL Classification: C60; C61; Q50; Q51; Q55; Q56

Citation: Lemenkova, P.; Debeir, O. Quantitative Morphometric 3D Terrain Analysis of Japan Using Scripts of GMT and R. *Land* **2023**, *12*, 261. https://doi.org/10.3390/land12010261

Academic Editor: Chuanrong Zhang

Received: 25 December 2022
Revised: 11 January 2023
Accepted: 13 January 2023
Published: 16 January 2023

1. Introduction

Cartographic visualization is an important component of many Earth science applications. In numeric land modelling, the discriminative power of 2D and 3D mapping is a key factor in the topographic analysis of land features since it provides the most direct and quickest way to evaluate geospatial information. During the last four decades, a variety of Geographic Information Systems (GIS) have been developed, with commercial ArcGIS software certainly being the most widely used. Land surface modelling in GIS shows that qualitative data analysis, cartographic visualization and the interpretation of the topographic features visualized on maps activate the evaluation of spatial heterogeneity and the variability of the objects and environmental processes on the Earth.

The performance of various GIS is, in many cases, tailored to their specific tasks, among which vector and raster spatial data processing is arguably the most prominent and important functionality of these software. For instance, some are better suited for image processing, such as Erdas Imagine [1], Idrisi GIS [2,3], Integrated Land and Water Information System (ILWIS) GIS [4], and ENvironment for Visualizing Images (ENVI) GIS, while others are best at vector data analysis and visualization, e.g., ArcGIS [5–9]. Most of these GIS are based on a standard interface with a restricted functionality that requires the manual processing of data, although recently machine learning techniques have been applied to spatial data modelling [10] and image analysis [11,12]. However, while these methods drive the geospatial analysis to gather explicit information on the terrain structure and morphometric variations, the core question emerges as the cartographic approach underlying the computational complexity of geoinformation processing. This raises the multi-disciplinary goal of succeeding in improving the cartographic workflow over the performance of the conventional algorithms. For instance, the use of scripts, besides the pre-defined algorithms in GIS, can result in an enlarged cartographic workflow functionality for data being modelled.

GIS-based mapping is a tedious and time-consuming process for cartographic performance as it may involve a large number of separated tasks and operations that are normally made using different commands in the menu toolbar. For this reason, the workflow is split into various steps of data processing. Better still, however, is using the full functionality of the programming and machine learning, applications of scripting languages for plugins, and auxiliary tools that ensure data processing using scripts [13]. Spatial analysis can benefit from the automation of cartographic processes because scripting enables the repeatability of the process. The advanced modelling enables the performance of a more comprehensive analysis of various factors affecting land surfaces and processes.

On the other hand, the importance of machine learning for topographic and geomorphological mapping is well known since it is less error-prone and time-consuming compared to the traditional state-of-the-art GIS. At the very least, this can be a use for scripting for data visualization, such as in Generic Mapping Tools (GMT) [14,15]. Despite certain difficulties in mastering the program, such as its high learning curve, console-based non-visual operation mode, and complex steps of fine-tuning map elements, GMT nevertheless provides a much more powerful functionality of cartographic workflow and increased flexibility in data processing. Furthermore, it yields deep insights into how to process geospatial information in relation to the individual traits of datasets, including the transformation of coordinate systems and georeferencing, extracting attributes and labelling, processing binary formats for gridded datasets, modelling elevation data for morphometric analysis, reading interleaved data formats in one script, and many more.

In this paper, we extend this idea of using scripts to the morphometric mapping of Japan by incorporating several libraries of R and modules of GMT for cartographic data processing. We propose a scripting framework that, by facilitating the mapping process using powerful methods of programming, can plot 2D and 3D maps more effectively than standard techniques of GIS. We devise a systemic way to use and process geoinformation derived from raster grids for visualising morphometric elements and parameters of terrain and use them to develop a series of scripts that could be extended to other regions to improve cartographic performance in 2D and 3D modelling.

The accurate visualization of complex terrain models can lead to the increased precision and efficacy of maps. Cartographic layout is usually seen as the final stage of spatial analysis summarising and presenting the results in a graphical form. That said, scripting by R land GMT largely facilitates visualization since similar parts of scripts may be reused. The contribution of the paper is twofold. First, we demonstrated that script-based visualization offers the increased precision of the morphological analysis through automation of data processing. Second, a marked improvement in the development of cartographic methods is observed, since using R and GMT in a mapping workflow paved the way for the machine learning methods in geospatial data analysis.

1.1. Background and Motivation

Topography is an important characteristic of the land surface, which is reflected in the landscape and associated with numerous environmental factors. Specifically, the geomorphic and topographic features control and affect vegetation coverage [16,17], habitats and landforms [18,19], hydrology, soil distribution, and local micro-climate settings that are dependent on the topographic exposure, slope steepness, and curvature of the relief. Spatial characteristics that can be retrieved from the topographic and land surface maps have many applications, such as environmental management [20], flood monitoring [21], hydrological and fluvial modelling in riverine ecosystems [22,23], analysis of coastal processes [24], soil management practices, and analysis of crop yields [25–27]. Moreover, land surface maps present a background for the analysis of the correlation between the geomorphology and geology [28] and geophysical and geographic factors and processes [29].

Topographic mapping based on the digital elevation model (DEM) is considered one of the most important issues of cartography and is often used as background information for spatial analysis in the geosciences. An inherent research step in the computations of a terrain analysis based on DEM is the use of modelling methods where GIS is traditionally applied. A variety of existing GIS software can be used for the spatial data visualization and topographic analysis considered a background for both socio-geographic and physical-geographic mapping. Since the onset of the development of GIS, technical methods of cartographic visualization and approaches to terrain analysis have been constantly improved.

One of the key components in GIS software and cartographic data processing is map projection. Mapping lands evokes a distortion in their angle, area, and shape depending on the type of map projection, which may be of the conic, azimuthal, cylindrical or miscellaneous types [30]. As a result, minor distortions arise when transforming the coordinates into various types of projections. Therefore, various projections can be better adjusted to map specific study areas depending on their locations and spatial extent.

For instance, the Albers conic equal-area projection is mostly used to plot areas with large longitudinal extents, e.g., Canada and USA. The polyconic projections have true scales on the parallels represented as non-concentric circular arcs. This makes this projection class neither equal-area nor conformal with the least distortion on the parallels having their centres along a central meridian. Likewise, the Lambert conic conformal projection is suitable for regions with a W-E extent with a true scale on the two standard parallels. The equidistant conic projection keeps a balance between the conformal and equal-area types with minimised distortion over the study area and a true scale along all the meridians and standard parallels.

Large regions with a global extent are better mapped using the Lambert azimuthal equal-area or stereographic equal angle projections, while northern regions, such as Scandinavian areas, can be effectively plotted using the polar stereographic projection, where map boundaries are represented by lines of constant longitude and latitude. In this study, we used the cylindrical Mercator projection for plotting a topographic map that is conformal in type with an inserted small global map in a perspective projection.

1.2. Related Work

Various publications focused on Japan were published that, in particular, aimed to understand how the geomorphology of the Japanese Alps is linked to the geologic and surface structures [31], rock glacial processes and distribution of vegetation [32], or reflected in the adjacent bathymetry [33]. Multivariate methods of spatial analysis extract local or regional terrain features from spatial data using coordinates that locate them. Thus, a variety of analytic techniques involves methods of quantitative analysis for detecting and describing spatiotemporal information for a deeper understanding of land surface processes. These can be used as the advanced tools enabling the performance of a terrain analysis on the highly heterogeneous morphological setting of Japan, as reflected in relevant works [34–38].

By choosing a proper GIS, data handling can yield a large number of extended features. However, providing a standardised solution for selecting a proper GIS is a challenge since, in most cases, it has a similar functionality of both raster and vector data analysis. Nevertheless, whichever GIS is chosen, it requires active work in an interface menu with the manipulation of input data and modelling to present a cartographic visualization. Another point is the issue of open source availability, which has restricted access for commercial GIS. The increasing variety of GIS necessitates that alternative tools with more flexible and varied approaches to cartographic visualization and geospatial modelling should be explored. Scripting approaches can provide a substitute functional tool for cartographic visualization and geospatial modelling. Examples of scripting toolsets in cartography include the Geographic Resources Analysis Support System (GRASS) GIS [39,40] and GMT used in topographic and geomorphic mapping [41,42].

1.3. Contribution

This paper presents an application of the GMT cartographic scripting toolset [43] and the R language [44] for the spatial analysis and terrain mapping of Japan. We demonstrate the main advantages of both methods, which consist of their straightforward and logical language syntax, a scripting approach, and open-source access. We use the powerful functionality of several GMT modules for the fine-tuning of the cartographic visualization, as well as R packages to process both tabular and geospatial data directly from a console. We considered the existing cases of using R libraries for geospatial data processing and visualisation. For instance, special packages for geographic data processing include gstat [45], RStoolbox [46], terra, and raster [47].

This study makes a technical contribution to the development of cartographic methods by adapting the R language for mapping instead of the traditional GIS. We combined the GMT scripting toolset and R language for spatial data visualization and morphometric analysis in a cartographic framework, which included following general steps: (1) processing the data of the General Bathymetric Chart of the Oceans (GEBCO), Earth Topography Global Relief Model (ETOPO1) and ETOPO2 by GMT for 2D and 3D mapping by scripts; (2) the importing of Shuttle Radar Topography Mission (SRTM90) DEM by R; (3) visualising a DEM using the 'raster' package; (4) computing the topographic surface parameters of slope, aspect, and hillshade using R scripts; and (5) cartographic visualization by the 'tmap' package of R with additional cartographic elements (legend, histogram, grid, ticks, annotations, compass directions, and a bar scale).

The structure of this document is as follows. First, we discuss the key issues relevant to the morphometric modelling based on DEM in a cartographic analysis of the geospatial data. Specifically, slope, aspect, hillshade and elevation are reviewed as essential elements of the terrain analysis. Second, we present a case of the Japanese archipelago as an example for mapping and briefly outline the major geographic features and morphology of the Japanese Alps. Third, we point out that our combination framework can be easily extended to other regions with heterogeneous terrain morphology with minor modifications of the scripts. To this end, we provide technical notes on scripts, followed by a methodological description of the performed morphometric analysis using R and GMT. The full scripts are listed in the Appendix as a technical cartographic reference for similar studies. Furthermore, the methodological part demonstrates several screenshots of the performed scripting process in the RStudio environment, where maps of slope, aspect, hillshade and elevation are plotted, and the GMT for 2D and 3D modelling.

The results present the computed and visualized maps of the Japanese archipelago made by a GMT scripting toolset and in RStudio. The latter ones include the morphometric elements (slope, aspect, hillshade, and elevation) with statistical elements of data distribution as histograms and a map of prefectures of Japan visualized by the 'ggplot2' graphical package of R. A discussion of the machine-based approach to cartographic modelling and visualization follows the Results section. Finally, we provide conclusions regarding the advantages of using GMT and R in cartography compared to the state-of-the-art GIS techniques based on a standard graphical user's interface and ways of modelling geospatial data by scripting languages from the console.

2. Materials and Methods

The key methods included the GMT with diverse modules and packages of Rm, including 'tmap' for thematic mapping and 'ggplot2', 'ggmap' [48], 'maps' and 'mapdata' for the data capture and thematic regional mapping of prefectures of Japan. The package 'mapdata' contains the base location of the binary files of prefectures boundaries of Japan used by the map drawing functions. The 'raster' package was used for morphometric analysis, which also included the depending packages, such as 'sp', a package which operates with classes and methods for spatial data, and 'sf', a package operating with 'Simple Features' as object classes in an R syntax.

2.1. Study Region

The study region is Japan, shown in Figure 1.

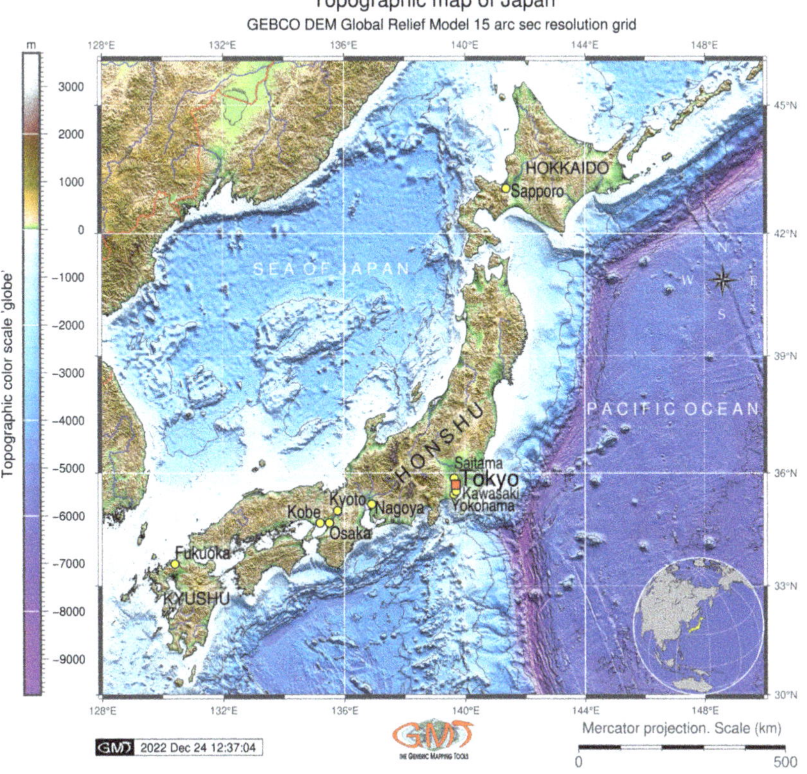

Figure 1. Topographic map of Japan. Data source: GEBCO. Software: GMT, version 6.1.1. Cartography source: authors.

The geology of Japan is characterised by an early stage of mountain range formation comprising young and active island arcs [49]. As a result of complex tectonic movements, uplift, and denudation processes, the Japanese Alps consist of a mountain range cut by river and glacial valleys, running through the main island of Honshu. During the Quaternary period, the major axis of the Japan Alps experienced more than 2000 m of uplift, which, together with the denudation processes, resulted in the exposure of the Takidani Granodiorite. The regions is divided into three major parts: the Northern, Central, and

Southern Alps. The northern region located in the Nagano, Toyama and Gifu prefectures includes the Hida Mountains with high seismicity along one of major Quaternary faults, the Atotsugawa fault, with repetitive earthquakes [50].

The geologic structure of the Hida terrane shows a complex geologic structure with gneisses formed from Permo-Carboniferous clastic sediments during a single metamorphism at ca. 250 Ma [51]. The Northern Japanese Alps are characterised by lateral and terminal moraines and outwash terraces of glacial origin. The Central Alps include the Kiso Mountains located in the Nagano prefecture [52] and are characterised by a granite structure. The highest peak of the Japanese Alps is recorded in Mt. Kita (3193 m) in the Southern Alps, or Akaishi Mountains.

The Japanese Alps had long been exploited, and their geologic structure, geomorphologic setting [53], and tectonic evolution [54] have been reconstructed. In addition to the environmental value and strong effects on the climate and vegetation setting of Japan, the Japanese Alps contribute to the economy of the country, being a source of natural mineral resources such as timber and minerals, as well as habitat for diverse species and vegetation, including medical herbs. Moreover, rice paddy fields are cultivated on the slopes of the mountains. Finally, the Alps are considered a potential source of geothermal energy [55]. All these factors make the Japanese Alps a key land surface object for nature and the society of Japan.

2.2. Datasets Preprocessing

The geospatial data processing was performed in RStudio environment, Figure 2.

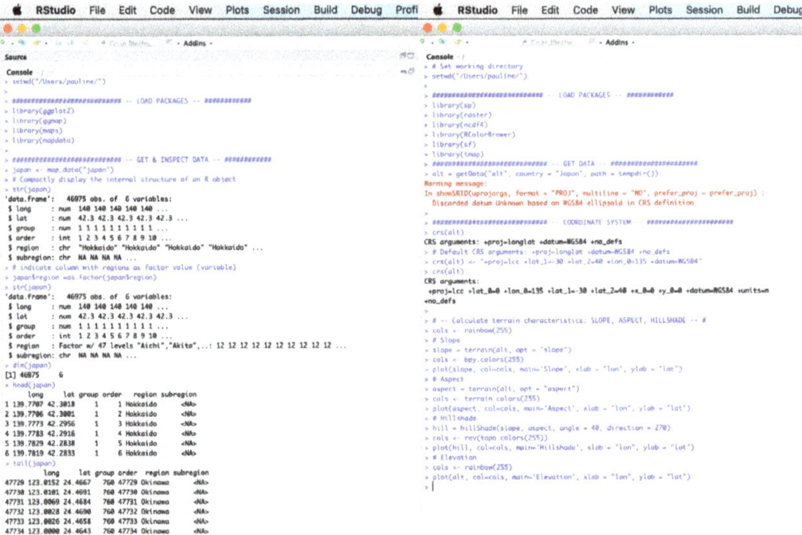

Figure 2. Geodata processing in RStudio environment. (**Left**): (1) loading packages, (2) obtaining data, (3) inspecting data.frame. (**Right**): (1) setting up coordinate system, (2) calculating terrain characteristics (slope, aspect, hillshade) by 'raster' package, (3) visualizing maps on a screen in RStudio. Source: authors.

The workflow included the following general scheme: (1) loading packages, (2) obtaining data, (3) inspecting data.frame, (4) setting up the coordinate system, (5) calculating terrain characteristics (slope, aspect, hillshade) by the 'raster' package, (6) visualizing maps on a screen in RStudio, (7) plotting cartographic aesthetics in the 'tmap' package, and (8) mapping the prefectures of Japan in the 'ggplot2' package. The data for the terrain analysis include digital elevation models (DEM) from the Shuttle Radar Topography Mission (SRTM). The SRTM is widely used in research [56] due to its acceptable resolution (30 m × 30 m) and open-source availability: the National Aeronautics and Space Administration (NASA) Shuttle Radar Topography Mission (SRTM) provided DEMs for over 80% of the land surface on the Earth. The slope, aspect, hillshade and elevation maps were plotted using the RStudio environment [57] using the 'raster' package as implemented by [58]. Specifically, the following workflow was used for data pre-processing. The data were captured using the 'getData()' function of the 'raster' package of R from the available geospatial datasets of the University of California, Davis campus, CA, U.S.

The data were then reprojected to the Lambert conformal conic projection (LCC) by the following function: $crs(alt) < -" + proj = lcc + lat\_1 = -30 + lat\_2 = 40 + lon\_0 = 140 + datum = WGS84"$. The coordinate system is an important aspect of geographic data which is implemented in the R environment by a PROJ library [59]. It should be specially mentioned that the SRTM shows elevation data not as a bare-earth model but as a surface, which includes dense canopy forests and built-up areas in the estimation of the terrain. Although such nuances might be worrying in hydrological modelling, the SRTM is generally acceptable for country-level mapping, as in our case for mapping Japan using SRTM DEM.

2.3. Methods

2.3.1. 3D Modelling by GMT Scripts

Scripting by Generic Mapping Tools (GMT) followed the existing cartographic experience [60] with the use of a 3D package for the terrain modelling of Japan, as seen in Figure 3.

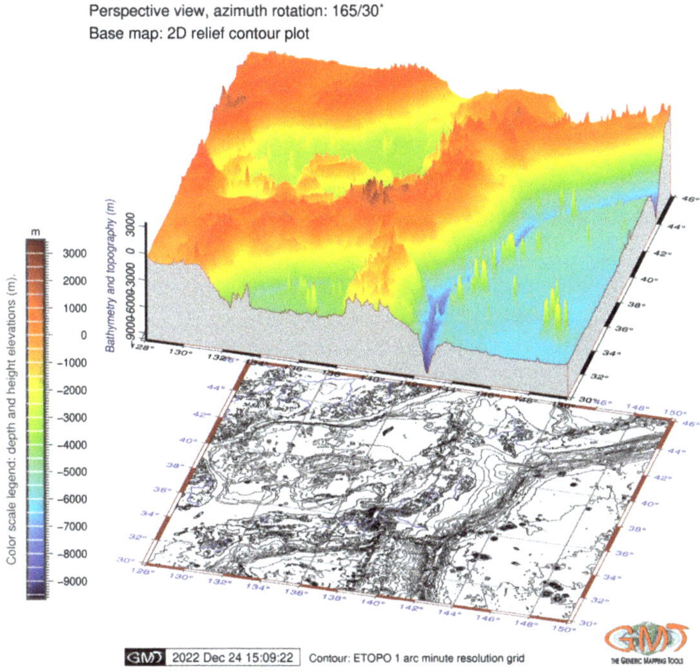

Figure 3. A 3D model of the land surface of Japan. Plotting is performed with rotation of 165/30° based on Earth topography one minute (ETOPO1) grid representing global relief. Software: GMT, version 6.1.1. Map source: authors.

It should be noted that 3D modelling is an important approach to data processing, with many applications in engineering and natural sciences [61–64]. Most 3D modelling approaches in mapping address the problem of the visualization of the land surface in the forms of multiple views of the terrain by 3D data acquisition, e.g., light detection and ranging (LiDAR) [65,66]. The high cost of special hardware for obtaining LiDAR data and substantial manual processing reduce the operational flexibility of such approaches and limit their availability as practical applications for topographic mapping .

In this part, we present 3D terrain modelling based on the GMT open soft toolset. Such approach makes 3D mapping a more practical routine and enables many applications in Earth science with advanced cartographic data visualization of the terrain. Here, the 3D shapes of the terrain are presented in perspective projections for data interpolated by splines with heights obtained from the 3D coordinates of the raster grids of the ETOPO1 and ETOPO2 datasets. The differences in the local ruggedness of the terrain as represented by ETOPO1 and ETOPO2 are visualised and compared in 3D perspective plots with a rotated azimuth view. The algorithms included the 'grdview' module of GMT, as shown in scripts in the Appendix A.

2.3.2. Mapping the Prefectures of Japan

To illustrate the location of various prefectures of Japan, we used the "ggplot2" packages of R for plotting the regions of prefectures using arguments in the dataset for visualization. The data were captured by the 'maps' package of R [67] and 'mapdata', which provide the map databases, including the data on Japan: its prefectures, areas, etc. The advantage of this approach consists of the integrated use of 'mapdata' with the 'ggplot2' package of R. The 'ggplot2' package is a common graphical package designed for general purpose scientific visualization; however, it is applicable for cartographic purposes as well. Here, the main elements, i.e., the polygons of the prefectures in Japan, were plotted using the *'geom_polygon'* function, which operates with data frames containing the coordinates of polygons and values associated with each of them:

```
geom_polygon(data=japan, aes(x=long, y=lat, fill=region, group=group)
```

The annotations of the axes were added using the 'xlab' and 'ylab' functions. The color scales were defined in the RColorBrewer and extended from the default, fixed number of colors to the number of prefectures in Japan, which was inspected by the 'length' function, as follows:

```
length(unique(japan$region))
```

Afterwards, the number of colors in the color palette was defined using the 'colourCount' function:

```
colourCount = length(unique(japan$region))
```

Following that, the color palette was expanded to the required number (47 prefectures of Japan) by the use of the 'colorRampPalette' function, as follows:

```
colorRampPalette(brewer.pal(name="Spectral", n = 8))(47).
```

Now, when the color palette was adjusted to the dataset, the map of prefectures of Japan was colored by assigning each individual color to each of the 47 prefectures, as follows:

```
scale_fill_manual(values = getPalette(colourCount)).
```

Other, additional cartographic elements included adding the titles, subtitles and captions by the 'labs' function, which enables the modification of the labels, annotations and captions. The rest of the aesthetics were added using *theme*() function, as shown in Figure 4, which provides an illustration of the script on the left and the resulting output map on the right part of the menu. These include, among others, plotting the legend, its orientation and annotations, defining the text size and font characteristics, selecting the color of background, and choosing the types of grid ticks and the frequency of breaks on a cartographic grid. The final map is shown in Figure 4.

The map of the prefectures of Japan (Figure 4) presented the visualization of the 47 prefectures made using the 'ggplot2' package of R, as well as the 'mapdata' and 'maps' packages used for data capture and processing. Each prefecture was colored as single-colored polygons modified from the default palette of R ('Spectral') using the palette extension by the 'colorRampPalette' function.

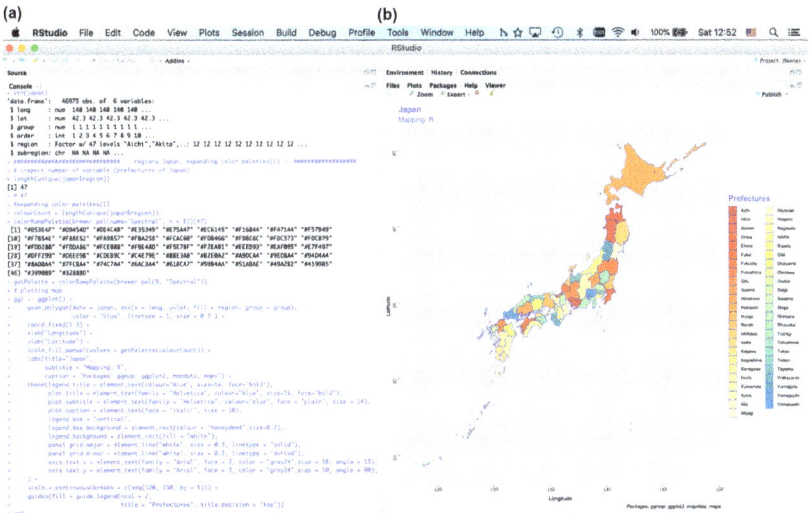

Figure 4. Mapping the prefectures of Japan. (**a**) Script used for mapping by 'ggplot2', 'ggmap', 'maps' and 'mapdata' packages. (**b**) Map of the prefectures of Japan. Source: authors.

2.3.3. Mapping Morphological Features

Modeling slope, aspect, and hillshade is a technique for visualising terrain determined by DEM as a numerical data source. Here, the slope and the aspect were plotted first as major features. Following that, we visualised the hillshade, which is a derivative from the slope and an aspect of the elevation. The parameters of hillshade were adjusted and visually changed using various illuminating positions of the light source as degrees of simulated sun angle rotation. The four derivatives of the SRTM90 DEM grid were modelled and visualized using R as follows: slope, aspect, hillshade, and elevation.

The slope gradient, aspect and hillshade were defined by the algorithm of the 'raster' package of R at any point of the input raster grid (SRTM90) using local neighbourhood analysis. These variables are automatically defined by the machine using values of altitude (elevation) and its derivatives at or around each cell point on a raster grid representing the land surface. The 'alt' has a formal class 'RasterLayer' of package 'raster' with 12 slots.

Slope

The terrain characteristics of slope were calculated and modelled using the 'raster' package of R using the following sequence of commands. First, the color palette of R was created for visualization: $cols < -rainbow(255)$. Second, the slope and aspect were computed by following: $slope = terrain(alt, opt = "slope")$. Third, the slope was visualised: $plot(slope, col = cols, main = 'Slope', xlab = "lon", ylab = "lat")$. Here, 'alt' represents the abbreviation of 'altitude'.

Aspect

The aspect was modelled using the function 'terrain' with the option 'aspect' as follows: $aspect = terrain(alt, opt = "aspect")$, followed by the definition of colors: $cols < -terrain.colors(255)$. Afterwards, using the same function $plot()$ as for the slope modeling, the map of the aspect was plotted using the following snippet of code: $plot(aspect, col = cols, main = 'Exposure', xlab = "lon", ylab = "lat")$. The full script of R is provided in the Appendix A.5.

Hillshade

The topographic hillshade represents a remarkable phenomenon in cartography that occurs when shadow caused by the elevation cue three-dimensional shape perception. Various techniques exist to produce a monochrome 3D view of a terrain relief with the relative position of the artificial illumination representing the shadows from the natural sun. The effects from this shading are well reflected in a highly rugged terrain such as Japan, where high-elevation land surface uses strong luminance contrast, while low heights use low contrast. In attempts to visualise reliefs using the shading effects of a monochrome palette, efforts have been made since the onset of the GIS development and are considered in this work. The computation of hillshade was carried out using the following algorithm: $hill = hillShade(slope, aspect, angle = 40, direction = 270)$. The color palette was applied using the code snippet: $cols < -topo.colors(255)$. Plotting the map was carried out using the plot function as described above.

Cartographic Processing

Mapping the morphological parameters in the 'tmap' package [68] included the processing of the computed raster layers by a specifically designed cartographic package where more elements could be added to the layout and more control over their aesthetics was available. The package used a function *tmap_style*, where the general style of the layout was defined. Then, *tm_shape* was used to control the main data: the name of the raster, title, and subtitle. The *tm_raster* function was applied to process the color scale of the numeric variables (values of slopes in degrees), labels, and other cartographic details in the legend; see Figure 5.

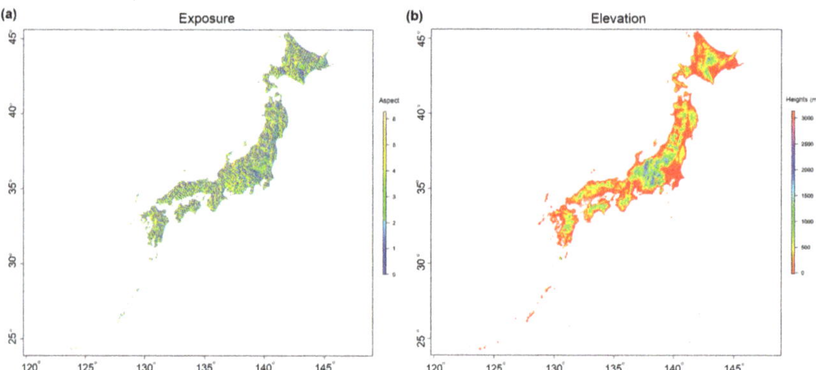

Figure 5. (**a**) Visualized plots of slope exposure. (**b**): Elevation of the terrain of Japan by 'raster' package of R. Source: authors.

The *tm_scale_bar* and *tm_compass* functions show the parameters of the annotations of the auxiliary elements on the map. Thus, the *tm_layout* function controls the variety of the cartographic aesthetics necessary for proper visualization, such as the fonts and ticks or color of panel labels. The elements of the map affect the perception of the cartographic layout, which is crucial for overall visual appreciation by the reader. Therefore, map elements were adjusted using relevant functions of the package. The final map was visualized in an RStudio environment by calling the raster for inspection (here, 'map1') and saved using the *tmap_save* function: $tmap\_save(map1, "Slope\_Japan.jpg", height = 7)$. The same procedure was repeated with all the other three files (slope, aspect, and hillshade). The script visualized in RStudio is presented in Figures where the left parts demonstrate the script and the right part shows the output maps.

3. Results

The results of the geospatial modelling of Japan are structured into the 2D and 3D plots made using GMT, the regional mapping of prefectures performed by R, and morphological mapping (slope, aspect, and hillshade sections). Various R packages were used in the workflow, as described above and presented in the Appendices. The slope gradient, being a derivative of the altitude, presents a scale-dependent variable which changes according to the reduced or increased DEM resolution. Since the input data of this research are a 30*m SRTM DEM, the results of the slope modelling refer to the given resolution.

Marked physiographic variations in the Japanese Alps control the type and distribution of morphometric parameters in the following two regards. First, the heights change between the Hokkaido, Kyushu and Honshu Islands, with the highest elevation points in the central part of the Honshu (Mt. Fuji). Therefore, the variation of topographic ruggedness and slope steepness is primarily controlled by the geographic location, with the largest difference between the extreme highest and lowest points in the Central Alps. Second, the largest earthquake recorded in Japan, the Tohoku earthquake recorded ca. 371 km NE of Tokyo in 2011 (Miyagi prefecture, see the map in Figure 4), affected the topography of the country and increased slope instability and the risk of landslides.

The comparison of the exposure (Figure 5, left) and elevation maps (Figure 5, right) shows the trends in the North-South-West-East directions with regard to the topographic altitude of the land surface. The functional options related to the visualization by RStudio are discussed in previous sections. The presented maps are based on the ETOPO1 and ETOPO2 data with 1 and 2 arc second resolution, respectively, used for 3D modelling by GMT, as shown on the surface plot of the land relief of Japan with varied rotation. The 3D mesh model with isolines drawn on top of the surface is based on ETOPO2, and the grey-shaded topographic 'waterfall' plot based on the ETOPO2 grid with a view rotation of 115/30° as shown in Figure 6.

A 3D map with a view rotation of 65/30°is presented in Figure 7. The SRTM90 DEM, which shows the morphometric models, represents the slope, aspect, and hillshade relief in the land surface of Japan. The number of pixels (over 16,000 on a raster grid) was the greatest in the 'gentle' slope level (yellow color) compared to the others: 12,000 pixels for the 'moderate' slope (orange color), over 10,000 for the 'strong' slope (light red color), and 8000 for the 'very strong' slope (magenta color); three bins are covered by the class 'extreme' slope (purple color), and the rest (blue color) are represented by the less than 3000 pixels group, that is, the steepest mountain slopes.

The slope map shows the steepness of the mountain sides in the Japanese Alps in the Northern, Central and Southern Alps. The highs and hills in the raster grids are visualised in different sub-regions of Japan. The slope directions (Figure 8) revealed the following variations in the data: 'gentle', 'moderate', 'strong', 'very strong', 'extreme', and 'steep' slopes of the mountainous regions of Japan.

The aspect map (Figure 9) demonstrates that slope orientation, according to the compass direction (W-E-S-N), differs in various parts of the mountains chains of the Japanese Alps.

Thus, in the central part of the area (around 36° N), many slopes have a primarily west orientation (as shown by a red color), which is correlated to the geographic distribution of the Central Alps. However, in other parts, the southern (yellow color) and eastern (orange color) orientations demonstrate the prevailing values. Similarly, the northern orientation (green color) of the slopes contributes the least in the examined dataset.

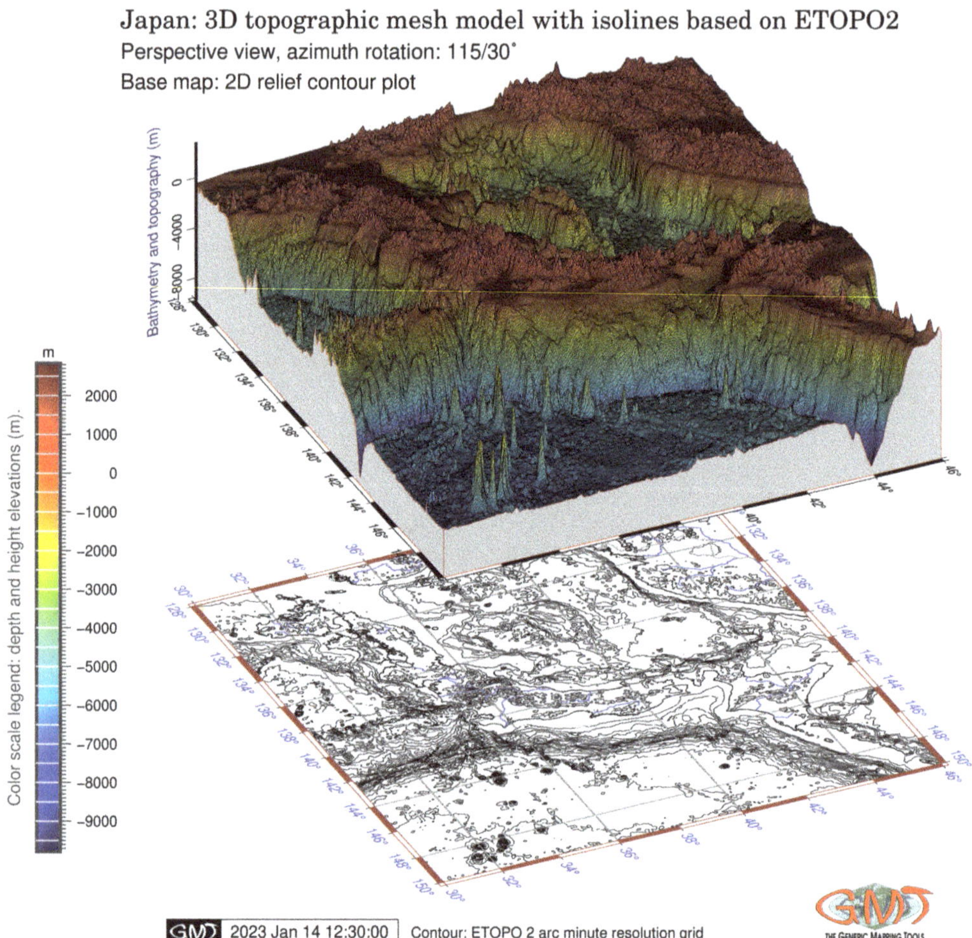

Figure 6. Japan: 3D topographic mesh model with isolines drawn on top of surface based on ETOPO2, with rotation of 115/30°. Software: GMT, version 6.1.1. Map source: authors.

Furthermore, the relationship of the slope aspect and elevation values indicating the altitude of the mountains might be the points of correlation. This should be emphasised since the impact of the geomorphic patterns on the morphometric characteristics helps detect the trends, showing the relief in the Japanese Alps. The hillshade map (Figure 10) shows a complex model based on the previously created maps of slope and aspect.

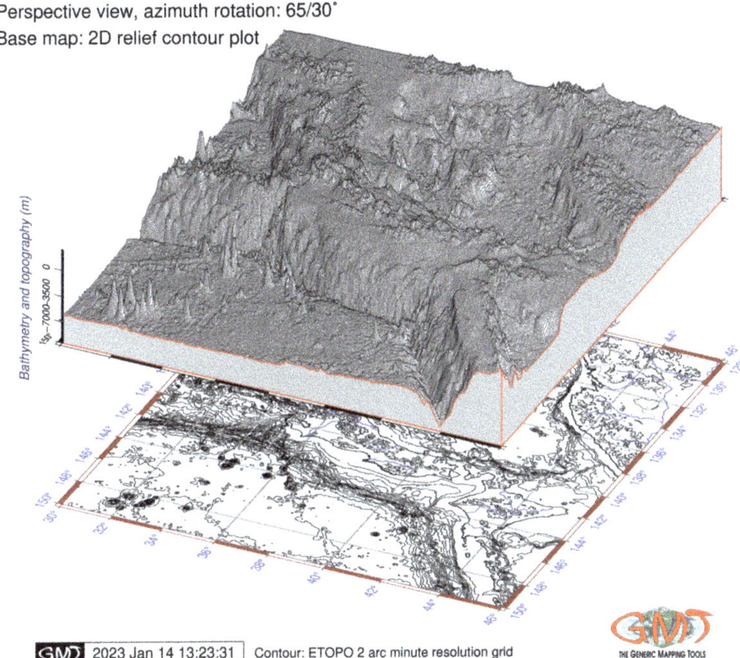

Figure 7. Japan: 3D grayshaded topographic waterfall plot based on ETOPO2 with rotation of 65/30°. Software: GMT, version 6.1.1. Map source: authors.

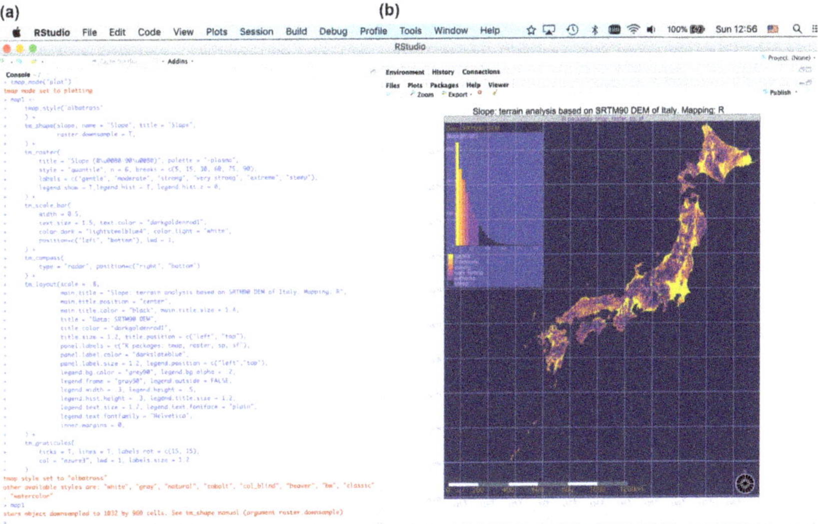

Figure 8. Slope steepness in Japan. (**a**) Script for plotting by 'raster' and 'tmap' packages. (**b**) Map in RStudio. Source: authors.

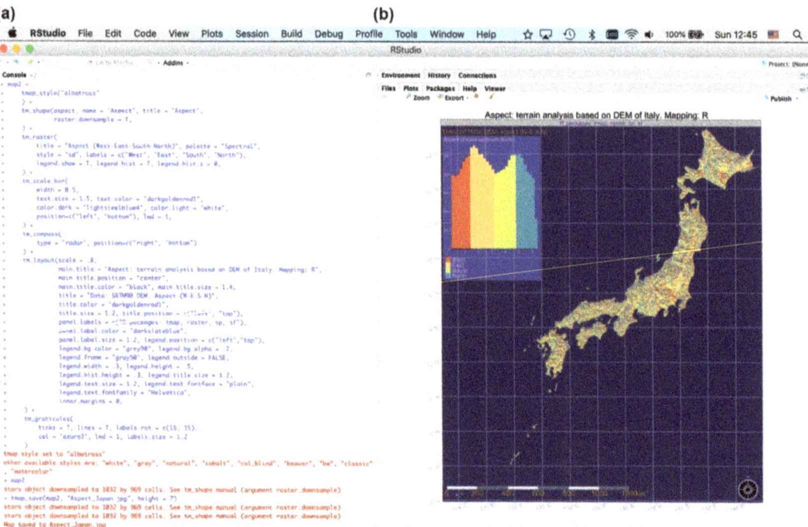

Figure 9. Aspect exposure of slopes in Japan. (**a**) Script by 'raster' and 'tmap' packages. (**b**) Map prepared in RStudio. Source: authors.

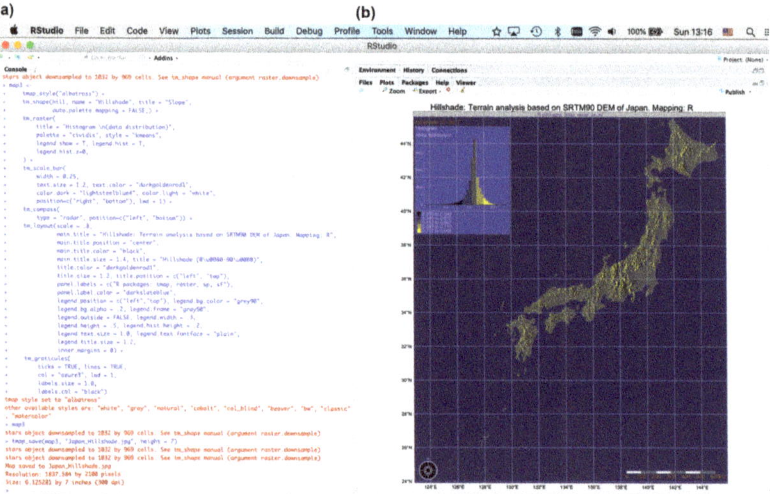

Figure 10. Terrain hillshade in Japan. (**a**) Script used for plotting using 'raster' and 'tmap' packages of R. (**b**) Map in RStudio. Source: authors.

The levels of illumination were set at angle = 40°and azimuthal direction = 270°. The elevation map was used as a basis for the relief overlaid by the hill shading. The results of the regional mapping of prefectures performed by R are shown in Figure 11.

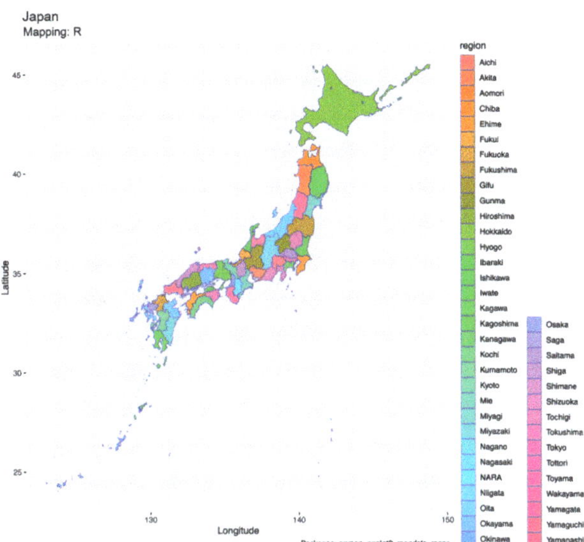

Figure 11. Japan: map of prefectures. R, v. 4.2.2 (RStudio v. 2022.12.0+353). Map source: authors.

4. Discussion

The results of the SRTM90 DEM modelling identified major morphometric characteristics, such as slope, aspect and hillshade and demonstrated the uneven elevation in the topographic maps made using R scripting. Using a variety of R packages implies both the processing of data, exploration of data frames, manipulation of its structure and spatial data analysis (e.g., using the 'raster' package), and the cartographic aesthetic visualization of the layouts supported by specially designed packages such as 'tmap', 'maps' and 'mapdata'.

The analysis of topographic attributes and DEM is of fundamental importance for reconstructing the genesis and development of landforms and, more generally, the geological setting of a specific area. The new ways of measuring, sensing, and analysing relief morphology carried out in Japan suggest that the range of the relief is reinforced by the high-elevation landforms, as well as external factors such as geomorphological processes and the climate, which contribute to eroding and modifying them. In fact, a strong correlation between geomorphology, topography and climate is commonly known, as also reported in Japan [69]. As proof of this, they pointed out that the climatic changes since the Late Glacial period have been responsible for the passage from the glacial to the permafrost environment in the current alpine zone, leading to modifications in the geomorphological processes and in the relief of the northern Japanese Alps.

Other important findings in geomorphic studies in Japan are well summarized by [70], who noted extensive sedimentation in mountain piedmonts and coastal fluvial plains and abundant sediment in steep watersheds. They furthermore pointed out the occurrence of hydro-geomorphological events in the areas of earthquakes and volcanic eruptions, as well as the post-glacial development of hillslope and flood processes along alluvial fans strongly controlled by the climate. In addition to the environmental factors, the relief directly influences the social-economic patterns through the possibility of road and building constructions and the potential triggering of landslides, which depend on the rock properties and slope steepness.

The geomorphic processes are largely driven by the gravity of the Earth and controlled by the slope steepness of the relief. As a result, the intensity of the surface processes, such as landslides, is affected by the displacement gradients connected to the geometric curvature and ruggedness of the relief. In such a way, practical applications of land surface maps in 2D and 3D representations enable the evaluation of the slope steepness quantitatively. In turn, the information retrieved from the calculated rates at which landslides are dismantling mountain slopes can be used for hazard and risk assessment for practical purposes of engineering geology, as well as for estimating potential slope instability [71]. Furthermore, other applications of land surface maps include the evaluation of environmental risks caused by climatic factors, such as precipitation intensity and the repeatability of rainfall or downpours [72,73] and visualising water surface dynamics in the context of flood and drought applications [74]. Finally, risk assessment and management in mountainous regions should consider other natural events, such as earthquakes or typhoons [75–77].

Due to the importance of topographic data visualization and the need for updated maps based on high-resolution data, there are many research reports on methods of topographic and bathymetric mapping and DEM applications for geospatial mapping. The latter includes, for instance, land surface classification based on DEM, reliefs, and geomorphological modelling using various GIS and geomorphometric computations of slope and aspect, as well as issues concerning the visualising methods of hillshade and DEM [78–80]. Since morphometric studies should always be supported by a spatial topographic representation, the questions of technical tools of mapping always remain actual for DEM-based studies. Our research contributed to this topic and showed that the integrated use of GMT and R in morphometric studies is an effective approach both for thematic mapping and for spatial analysis.

Graphics and maps, when created well, can provide eye-catching and detailed information on the land surface. This is why graphical approaches and methodologies of cartographic visualization are of high importance in Earth sciences. Despite a specific language syntax and approaches to data analysis that require mastering the tool, the GMT scripting toolset and R language are shown to be very promising tools for geospatial visualization, morphological modelling, and mapping, supporting the research in various aspects due to their different functionality.

Cartographic data visualization, regardless of what GIS software is used, is an integral part of the complex workflow of geospatial research. To mention some steps in a simplified process of geospatial data analysis using cartographic tasks, this includes data capture, preprocessing, projecting, analysis of content of the datasets, modelling variables, visualising and plotting the maps with controlled layout, and others. Applying the methods of scripts extended to the machine learning approaches to a cartographic workflow facilitates the process of spatial analysis. Specifically, it helps to increase the accuracy and precision of the final results and maps and significantly decreases the time of data processing due to the automation of geospatial data processing. Moreover, we have shown that the use of scripts and programming methods are more efficient and therefore more suitable for cartographic workflow due to repeatability. Our methods could be easily extended to also use other datasets and geographic extent of data. In fact, the computational procedure adopted here to investigate the setting of reliefs and mapping land features in 2D or 3D models could be applied to any target scales, be it global downscaled modelling or upscaling to regional (prefecture-level) and local (city-level) terrain models.

5. Conclusions

In this paper, we have designed adaptive script-based algorithms for plotting morphologic maps using GMT and R that are able to visualise maps accurately and effectively. Our two algorithms provide a tradeoff between the computational approach of R for geospatial data processing and the cartographic performance of GMT. For future relevant works, we suggest that one chooses the appropriate algorithm based on their mapping goals and available dataset. Both programs have free access as open-source tools and extensive functionality for geographic data analysis and visualization. The presented methods can be used and adopted to other regions and areas with changed relevant coordinates and modified attributes for 3D modelling (e.g., elevation range). We have shown that the use of both these tools for mapping is a very effective approach for spatial analysis that can be used instead of or besides GIS, as a complimentary script-based method.

We furthermore identified the existing challenges such as the need to use, install, and load a variety of packages in R to fit in various tasks, such as: geomorphometry, classic mapping, spatial analysis, data capture, and data conversion. In contrast, GMT enables us to perform all the steps of a cartographic workflow in the same session by calling necessary modules. On the one hand, using scripts can be an easier approach compared to GIS, but on the other hand, it can be a challenge, since it requires finding a suitable package, installing and activating it, and using it by applying its specific functions and syntax using GMT or R.

The integration of our method based on the GMT and R algorithms for processing various geospatial data is straightforward since many data formats can smoothly be imported and processed both by GMT and R. Furthermore, R has certain embedded datasets that we used for plotting the prefectures of Japan and for visualising physical features of relief. Working on grey-level images for 3D modelling by GMT enabled us to detail the land surface features and demonstrated the high accuracy of the ETOPO grids. With respect to our work, which exploited script-based mapping, we have designed and presented a series of thematic maps on Japan that can be applied for other regions and countries.

As we have shown in this study, the use of GMT and R for 2D and 3D mapping in geographic analysis is highly effective and recommended, especially for perspective visualization of the terrain. However, we should also notice that this method is not yet as popular as traditional GIS. This can be explained by the non-trivial approach of using scripts and programming techniques in mapping and cartographic data processing, as well as certain skills required for coding and mastering the syntax of both the tools. Nevertheless, the improved workflow of mapping and results suggests that both GMT and R are effective for geosciences and recommended for mapping purposes.

Author Contributions: Supervision, conceptualization, methodology, software, resources, funding acquisition, and project administration, O.D.; writing—original draft preparation, methodology, software, data curation, visualization, formal analysis, validation, writing—review and editing, and investigation, P.L. All authors have read and agreed to the published version of the manuscript.

Funding: The publication was funded by the Editorial Office of Land, Multidisciplinary Digital Publishing Institute (MDPI) with the publication fee waived for this manuscript. This project was supported by the Federal Public Planning Service Science Policy or Belgian Science Policy Office, Federal Science Policy—BELSPO (B2/202/P2/SEISMOSTORM).

Institutional Review Board Statement: Not applicable.

Informed Consent Statement: Not applicable.

Data Availability Statement: Not applicable.

Acknowledgments: The authors thank the anonymous reviewers of Land for reading, suggestions and comments that improved an earlier version of this manuscript.

Conflicts of Interest: The authors declare no conflict of interest.

Abbreviations

The following abbreviations are used in this manuscript:

DEM	Digital Elevation Model
ENVI GIS	ENvironment for Visualizing Images GIS
ETOPO	Earth Topography Global Relief Model
GEBCO	General Bathymetric Chart of the Oceans
GIS	Geographic Information System
GMT	Generic Mapping Tools
ILWIS GIS	Integrated Land and Water Information System GIS
LCC	Lambert Conformal Conic
LiDAR	Light Detection and Ranging
NASA	National Aeronautics and Space Administration
SRTM	Shuttle Radar Topography Mission
WGS	World Geodetic System

Appendix A. GMT Scripts

Appendix A.1. GMT Script for 2D Mapping of Japan

Listing 1. GMT script for topographic map.

```
#!/bin/sh
# Purpose: shaded relief raster map of Japan using GEBCO 15 arc sec
# GMT modules: gmtset, gmtdefaults, grdcut, makecpt, grdimage, psscale,
    grdcontour, psbasemap, gmtlogo, psconvert
# Extract a subset of GEBCO for the Japan trench area
gmt grdcut GEBCO_2019.nc -R128/150/30/46 -Gjp_relief.nc
gmt grdcut ETOPO1_Ice_g_gmt4.grd -R128/150/30/46 -Gjp_relief1.nc
gmt grdgdal -Ainfo jp_relief.nc
# actual_range={-9759.701171875,3700.7421875}
exec bash
# Make color palette
gmt makecpt -Cglobe.cpt -V -T-9759/3700 > myocean.cpt
# Generate a file
ps=Topo_JP.ps
# Make raster image
gmt grdimage jp_relief.nc -Cmyocean.cpt -R128/150/30/46 -JM16c -P -I+a15+ne0
    .75 -Xc -K > $ps
# Add color legend
gmt psscale -Dg125.0/30+w15.0c/0.4c+v+o0.3/0i+ml -R -J -Cmyocean.cpt \
    --FONT_LABEL=10p,0,black --FONT_ANNOT_PRIMARY=8p,0,black \
    -Bg2000f100a1000+l"Topographic color scale 'globe'" \
    -I0.2 -By+1m -O -K >> $ps
# Add isolines
gmt grdcontour jp_relief1.nc -R -J -C2000 -W0.1p -O -K >> $ps
# Add coastlines, borders, rivers
gmt pscoast -R -J -Ia/thinner,blue -Na -N1/thin,red -W0.1p -Df -O -K >> $ps
# Add grid
gmt psbasemap -R -J --MAP_FRAME_AXES=wESN --FORMAT_GEO_MAP=ddd:mm:ssF \
    --FONT_TITLE=14p,0,black --MAP_TITLE_OFFSET=0.8c \
    -Bpxg8f2a4 -Bpyg6f3a3 -Bsxg4 -Bsyg3 \
    -B+t"Topographic map of Japan" -O -K >> $ps
# Add projection scale
gmt psbasemap -R -J --FONT=10p,0,dimgray --MAP_TITLE_OFFSET=0.3c \
    -Lx14c/-0.5i+c50+w500k+l"Mercator projection. Scale (km)"+f \
    -UBL/-5p/-40p -O -K >> $ps
# Add directional rose
gmt psbasemap -R -J \
    --FONT=9p,Palatino-Roman,white --MAP_TITLE_OFFSET=0.3c \
    -Tdx14.6c/9.3c+w0.3i+f2+l+o0.15i -O -K >> $ps
# Texts
gmt pstext -R -J -N -O -K \
-F+f11p,0,white+jLB >> $ps << EOF
133 41 S E A   O F   J A P A N
144.05 37.5 P A C I F I C   O C E A N
EOF
gmt pstext -R -J -N -O -K \
-F+f11p,0,black+jLB -Gwhite@70 >> $ps << EOF
142 43.4 HOKKAIDO
130 32.5 KYUSHU
EOF
gmt pstext -R -J -N -O -K \
-F+f13p,0,black+jLB+a-320 -Gwhite@70 >> $ps << EOF
137.9 35.7 H O N S H U
EOF
# -R128/150/30/46
gmt pstext -R -J -N -O -K \
-F+f11p,21,black+jLB -Gwhite@70 >> $ps << EOF
139.53 35.0 Yokohama
EOF
gmt psxy -R -J -Sc -W0.5p -Gyellow -O -K << EOF >> $ps
139.63 35.44 0.20c
EOF
gmt pstext -R -J -N -O -K \
```

```
62  -F+f12p,21,black+jLB -Gwhite@70 >> $ps << EOF
63  135.50 34.25 Osaka
64  EOF
65  gmt psxy -R -J -Sc -W0.5p -Gyellow -O -K << EOF >> $ps
66  135.50 34.69 0.20c
67  EOF
68  # repeated with relevant coordinates for all the cities
69  gmt pstext -R -J -N -O -K \
70  -F+f15p,0,black+jLB -Gwhite@70 >> $ps << EOF
71  139.79 35.66 Tokyo
72  EOF
73  gmt psxy -R -J -Ss -W0.5p -Gred -O -K << EOF >> $ps
74  139.69 35.69 0.30c
75  EOF
76  # insert global map
77  gmt psbasemap -R -J -O -K -DjBR+w3.2c+stmp >> $ps
78  read x0 y0 w h < tmp
79  gmt pscoast --MAP_GRID_PEN_PRIMARY=thinnest,lightgray --MAP_FRAME_PEN=thin,
        white -Rg -JG140/37N/$w -Da -Gseashell3 -A2000 -Bga -Wfaint -EJP+gyellow
        -Sroyalblue3 -O -K -X$x0 -Y$y0 >> $ps
80  gmt psxy -R -J -O -K -T   -X-${x0} -Y-${y0} >> $ps
81  # Add GMT logo
82  gmt logo -Dx6.7/-1.8+o0.1i/0.1i+w2c -O -K >> $ps
83  # Add subtitle
84  gmt pstext -R0/10/0/15 -JX10/10 -X0.5c -Y8.2c -N -O \
85      -F+f10p,0,black+jLB >> $ps << EOF
86  3.0 11.0 GEBCO DEM Global Relief Model 15 arc sec resolution grid
87  EOF
88  # Convert to image file using GhostScript
89  gmt psconvert Topo_JP.ps -A0.2c -E720 -Tj -Z
```

Appendix A.2. GMT Script for 3D Mapping of Surface Plot of Japan

Listing 2. GMT script for 3D surface plot of Japan.

```
1   #!/bin/sh
2   # Purpose: 3D surface grid plot, 165/30 azimuth, from ETOPO1 for Japan
3   # GMT modules: grdcut, grd2cpt, grdcontour, pscoast, grdview, logo, psconvert
4   # Cut grid
5   gmt grdcut ETOPO1_Ice_g_gmt4.grd -R128/150/30/46 -Gjp_relief1.nc
6   gdalinfo -stats jp_relief1.nc
7   # Minimum=-9651.000, Maximum=3481.000, Mean=-2429.672, StdDev=2497.488
8   gmt makecpt -Cturbo.cpt -V -T-9651/3481 > myocean.cpt
9   # generate a file
10  ps=JP_3D.ps
11  # -B1/1NESW
12  gmt grdcontour jp_relief1.nc -JM10c -R128/150/30/46 \
13      -p165/30 -C500 --FONT_ANNOT_PRIMARY=8p,0,blue --MAP_FRAME_AXES=WESN \
14      --MAP_FRAME_PEN=brown --FORMAT_GEO_MAP=ddd:mm:ss \
15      --MAP_GRID_PEN_PRIMARY=thin,dimgray \
16      -Gd3c -Y3c -Bpxg4f2.0a2.0 -Bpyg4f2a2.0 -Bsxg1 -Bsyg1 \
17      -U/-0.5c/-1c/"Contour: ETOPO 1 arc minute resolution grid" -P -K > $ps
18  #Add coastlines, borders, rivers
19  gmt pscoast -R -J -p165/30 -P -Ia/thinner,blue \
20      -Na -N1/thin,gray -W0.1p -Df -O -K >> $ps
21  #-Bpxg2f0.5a1 -Bpyg2f0.5a1 -Bsxg2 -Bsyg1
22  # add color legend
23  gmt psscale -Dg122.0/30.0+w8.0c/0.4c+v+o0.0/0.5c+ml \
24      -R -J -Cmyocean.cpt \
25      --FONT_LABEL=8p,0,dimgray --FONT_ANNOT_PRIMARY=7p,0,black --
        MAP_ANNOT_OFFSET=0.1c \
26      -Bg500f100a1000+l"Color scale legend: depth and height elevations (m)." \
27      -I0.2 -By+lm -O -K >> $ps
28  # Add 3D
29  gmt grdview jp_relief1.nc -J -R -JZ3.0c -Cmyocean.cpt \
30      -p165/30 -Qs -N-9651+glightgray \
31      -Wm0.07p -Wf0.1p,red \
32      -B2.0/2.0/3000:"Bathymetry and topography (m)":ESwZ -S5 -Y5.0c \
```

```
33      --FORMAT_GEO_MAP=ddd:mm:ss --FONT_LABEL=8p,0,darkblue --
        FONT_ANNOT_PRIMARY=8p,0,black \
34      --MAP_FRAME_PEN=black -O -K >> $ps
35 # Add GMT logo
36 gmt logo -Dx10.5/-5.5+o0.0c/-0.5c+w2c -O -K >> $ps
37 # Add title
38 gmt pstext -R0/10/0/10 -Jx1 -X-0.8c -Y0.0c -N -O -K \
39   -F+f12p,25,black+jLB >> $ps << EOF
40 -0.5 9.0 Japan: 3D topographic surface plot based on ETOPO1
41 EOF
42 gmt pstext -R0/10/0/10 -Jx1 -X0.0c -Y0.0c -N -O\
43     -F+f10p,0,black+jLB >> $ps << EOF
44 -0.5 8.5 Perspective view, azimuth rotation: 165/30\232
45 -0.5 8.0 Base map: 2D relief contour plot
46 EOF
47 # Convert to image file using GhostScript (portrait orientation, 720 dpi)
48 gmt psconvert JP_3D.ps -A1.2c -E720 -Tj -P -Z
```

Appendix A.3. Modified GMT Script for 3D Mesh Model of Japan (Lines 29–34 of the Previous Script)

Listing 3. GMT script for 3D mesh model of Japan.

```
1 #!/bin/sh
2 # Add 3D
3 gmt grdview jp_relief2.nc -J -R -JZ3.0c -Cmyocean.cpt \
4     -p115/30 -Qsm -N-9651+glightgray \
5     -Wm0.07p -Wf0.1p,red \
6     -B2.0/2.0/3000:"Bathymetry and topography (m)":ESwZ -S5 -Y5.0c \
7     --FORMAT_GEO_MAP=ddd:mm:ss \
8     --FONT_LABEL=8p,0,darkblue \
9     --FONT_ANNOT_PRIMARY=8p,0,black \
10    --MAP_FRAME_PEN=black -O -K >> $ps
11
```

Appendix A.4. GMT Script for 3D Grayscale 'Waterfall' Model of Japan

Listing 4. GMT script for 3D grayscale 'waterfall' model of Japan.

```
1 #!/bin/sh
2 # Purpose: 3D grayscale 'waterfall' model of Japan, 65/30 azimuth, from
       ETOPO2 for Japan
3 # GMT modules: grdcut, grd2cpt, grdcontour, pscoast, grdview, logo, psconvert
4 exec bash
5 # Cut grid
6 gmt grdcut ETOPO2v2g_f4.nc -R128/150/30/46 -Gjp_relief2.nc
7 gdalinfo -stats jp_relief2.nc
8 # actual_range={-9787,2832}
9 # Add 3D
10 ps=JP_3D_wf.ps
11 # -B1/1NESW
12 gmt grdcontour jp_relief2.nc -JM10c -R128/150/30/46 \
13     -p65/30 -C500 --FONT_ANNOT_PRIMARY=8p,0,blue --MAP_FRAME_AXES=WESN \
14     --MAP_FRAME_PEN=brown --FORMAT_GEO_MAP=ddd:mm:ss \
15     --MAP_GRID_PEN_PRIMARY=thin,dimgray \
16     -Gd3c -Y3c -Bpxg4f2.0a2.0 -Bpyg4f2a2.0 -Bsxg1 -Bsyg1 \
17     -U/-0.5c/-1c/"Contour: ETOPO 2 arc minute resolution grid" -P -K > $ps
18 #Add coastlines, borders, rivers
19 gmt pscoast -R -J -p65/30 -P -Ia/thinner,blue \
20     -Na -N1/thin,gray -W0.1p -Df -O -K >> $ps
21 # Add 3D
22 gmt grdview jp_relief2.nc -J -R -JZ2.0c -Cmyocean.cpt \
23     -p65/30 -Qmx -N-9787+glightgray -Wm0.07p -Wf0.5p,red \
24     -B2.0/2.0/4000:"Bathymetry and topography (m)":ESwZ -S5 -Y3.0c \
25     --FORMAT_GEO_MAP=ddd:mm:ss --FONT_LABEL=8p,0,darkblue \
26     --FONT_ANNOT_PRIMARY=8p,0,black --MAP_FRAME_PEN=black -O -K >> $ps
27 # Add GMT logo
28 gmt logo -Dx10.5/-3.5+o0.0c/-0.5c+w2c -O -K >> $ps
```

```
29  # Add title
30  gmt pstext -R0/10/0/10 -Jx1 -X-0.8c -Y0.0c -N -O -K \
31  -F+f12p,25,black+jLB >> $ps << EOF
32  -0.5 8.5 Japan: 3D perspective waterfall plot based on gridded ETOPO2 data
33  EOF
34  gmt pstext -R0/10/0/10 -Jx1 -X0.0c -Y0.0c -N -O\
35      -F+f10p,0,black+jLB >> $ps << EOF
36  -0.5 8.0 Perspective view, azimuth rotation: 65/30\232
37  -0.5 7.5 Base map: 2D relief contour plot
38  EOF
39  # Convert to image file using GhostScript (portrait orientation, 720 dpi)
40  gmt psconvert JP_3D_wf.ps -A1.2c -E720 -Tj -P -Z
41
```

Appendix A.5. R Script for Plotting Prefectures of Japan

Listing 5. R script for plotting prefectures of Japan.

```
1   setwd("/Users/pauline/")
2   # load packages
3   library(showtext)
4   library(ggplot2)
5   library(ggmap)
6   library(maps)
7   library(mapdata)
8   # get and inspect the data
9   japan <- map_data("japan")
10  # Compactly display the internal structure of an R object
11  str(japan)
12  # indicate column with regions as factor value (variable)
13  japan$region =as.factor(japan$region)
14  str(japan)
15  dim(japan)
16  head(japan)
17  tail(japan)
18  # check up available fonts
19  library(showtext)
20  font_families()
21  font_paths()
22  font_files()
23  # regions Japan, expanding color palettes(1)
24  # inspect number of variable (prefectures of Japan)
25  length(unique(japan$region))
26  # 47
27  #expanding color palettes(1)
28  colourCount = length(unique(japan$region))
29  colorRampPalette(brewer.pal(name="Spectral", n = 8))(47)
30  getPalette = colorRampPalette(brewer.pal(9, "Spectral"))
31  # plotting map
32  gg1 <- ggplot() +
33      geom_polygon(data = japan, aes(x = long, y=lat, fill = region, group =
        group),
34          color = "blue", linetype = 1, size = 0.2 ) +
35      coord_fixed(1.3) +
36      xlab("Longitude") +
37      ylab("Latitude") +
38      scale_fill_manual(values = getPalette(colourCount)) +
39      labs(title="Japan",
40          subtitle = "Mapping: R",
41          caption = "Packages: ggmap, ggplot2, mapdata, maps") +
42      theme(legend.title = element_text(colour="blue", size=16, face="bold"),
43          plot.title = element_text(family = "Chalkboard", colour="blue", size
        =16, face="bold"),
44          plot.subtitle = element_text(family = "Chalkboard", colour="blue",
        face = "plain", size = 14),
45          plot.caption = element_text(face = "italic", size = 10),
46          legend.box = "vertical",
47          legend.box.background = element_rect(colour = "honeydew4",size=0.2),
48          legend.background = element_rect(fill = "white"),
```

```r
49          panel.grid.major = element_line("white", size = 0.3, linetype = "
       solid"),
50          panel.grid.minor = element_line("white", size = 0.2, linetype = "
       dotted"),
51          axis.text.x = element_text(family = "Arial", face = 3, color = "
       gray24",size = 10, angle = 15),
52          axis.text.y = element_text(family = "Arial", face = 3, color = "
       gray24",size = 10, angle = 90),
53          ) +
54      scale_x_continuous(breaks = c(seq(120, 150, by = 5))) +
55      guides(fill = guide_legend(ncol = 2,
56          title = "Prefectures", title.position = "top"))
57  gg1
58  # regions Japan, expanding color palettes (2nd variant)
59  length(unique(japan$region))
60  # 47
61  nb.cols <- 47
62  mycolors <- colorRampPalette(brewer.pal(8, "Set1"))(nb.cols)
63  # Create a ggplot with 47 colors
64  # Use scale_fill_manual
65  gg1 <- ggplot() +
66      geom_polygon(data = japan, aes(x = long, y=lat, fill = region, group=
       group),
67          color = "blue", linetype = 1, size = 0.2 ) +
68      coord_fixed(1.3) +
69      xlab("Longitude") +
70      ylab("Latitude") +
71      scale_fill_manual(values = mycolors) +
72      labs(title="Japan",
73          subtitle = "Mapping: R",
74          caption = "Packages: ggmap, ggplot2, mapdata, maps") +
75      theme(legend.title = element_text(colour="blue", size=16, face="bold"),
76          plot.title = element_text(family="AquaKana", face="bold", colour=
       "blue", size=16)) +
77      guides(fill = guide_legend(ncol = 2,
78          title = "Prefectures", title.position = "top"))
79  gg1
80  # Other filling, e.g., transparent
81  ggplot() +
82  geom_polygon(data = japan, aes(x=long, y = lat, group = group), fill = NA,
       color = "red") +
83  coord_fixed(1.3)
84  ggplot() +
85  geom_polygon(data = france, aes(x=long, y = lat, group = group), fill = NA,
       color = "red") +
86  coord_fixed(1.3)
87  # color
88  gg2 <- ggplot() +
89      geom_polygon(data = japan, aes(x=long, y = lat, group = group),
90          fill = "pink", color = "blue", linetype = 1, size = 0.2) +
91      coord_fixed(1.3) +
92      xlab("Longitude") +
93      ylab("Latitude") +
94      labs(title="Japan",
95          subtitle = "Mapping: R",
96          caption = "Packages: ggmap, ggplot2, mapdata, maps")
97  gg2
```

Appendix A.6. R Script for Terrain Mapping of Japan

Listing 6. R script for terrain maps.

```r
1  # Set working directory
2  setwd("/Users/pauline/")
3  # load packages
4  library(sp)
5  library(raster)
6  library(ncdf4)
7  library(RColorBrewer)
```

```
 8  library(sf)
 9  library(tmap)
10  # check up available fonts
11  library(showtext)
12  font_families()
13  font_paths()
14  font_files()
15  # get the data
16  alt = getData("alt", country = "Japan", path = tempdir())
17  # coordinate system
18  crs(alt) <- "+proj=longlat +datum=WGS84 +no_defs"
19  # Default CRS arguments: +proj=longlat +datum=WGS84 +no_defs
20  crs(alt) <- "+proj=lcc +lat_1=-30 +lat_2=40 +lon_0=140 +datum=WGS84"
21  crs(alt)
22  # -- Calculate terrain characteristics: SLOPE, ASPECT, HILLSHADE -- #
23  cols <- rainbow(255)
24  # Slope
25  slope = terrain(alt, opt = "slope")
26  cols <- bpy.colors(255)
27  plot(slope, col=cols, main='Slope', xlab = "lon", ylab = "lat")
28  # Aspect
29  aspect = terrain(alt, opt = "aspect")
30  cols <- terrain.colors(255)
31  plot(aspect, col=cols, main='Exposure', xlab = "lon", ylab = "lat")
32  # Hillshade
33  hill = hillShade(slope, aspect, angle = 40, direction = 270)
34  cols <- rev(topo.colors(255))
35  cols <- topo.colors(255)
36  plot(hill, col=cols, main='Hillshade', xlab = "lon", ylab = "lat")
37  # Elevation
38  cols <- rainbow(255)
39  plot(alt, col=cols, main='Elevation', xlab = "lon", ylab = "lat")
40  mymaps <- tmap_arrange(slope, alt)
41  mymaps
```

Appendix A.7. R Script for Mapping Slope of Japan

Listing 7. R script for mapping slope map of Japan.

```
 1  # Set working directory
 2  setwd("/Users/pauline/")
 3  # load packages
 4  library(sp)
 5  library(raster)
 6  library(ncdf4)
 7  library(RColorBrewer)
 8  library(sf)
 9  library(tmap)
10  # tmaptools::palette_explorer()
11  # initial mode: "plot"
12  # current.mode <- tmap_mode("plot")
13  # slope
14  tmap_mode("plot")
15  map1 <-
16      tmap_style("albatross"
17          ) +
18      tm_shape(slope, name = "Slope", title = "Slope",
19          raster.downsample = T,
20          ) +
21      tm_raster(
22          title = "Slope (0\u00B0-90\u00B0)", palette = "-plasma",
23          style = "quantile", n = 6, breaks = c(5, 15, 30, 60, 75, 90),
24          labels = c("gentle", "moderate", "strong", "very strong", "extreme",
      "steep"),
25          legend.show = T,legend.hist = T, legend.hist.z = 0,
26          ) +
27      tm_scale_bar(
28          width = 0.5,
29          text.size = 1.5, text.color = "darkgoldenrod1",
```

```
30      color.dark = "lightsteelblue4", color.light = "white",
31      position=c("left", "bottom"), lwd = 1,
32      ) +
33  tm_compass(
34      type = "radar", position=c("right", "bottom")
35      ) +
36  tm_layout(scale = .8,
37      main.title = "Slope: terrain analysis based on SRTM90 DEM of Italy.
    Mapping: R",
38      main.title.position = "center",
39      main.title.color = "black", main.title.size = 1.4,
40      title = "Data: SRTM90 DEM",
41      title.color = "darkgoldenrod1",
42      title.size = 1.2, title.position = c("left", "top"),
43      panel.labels = c("R packages: tmap, raster, sp, sf"),
44      panel.label.color = "darkslateblue",
45      panel.label.size = 1.2, legend.position = c("right","bottom"),
46      legend.bg.color = "grey90", legend.bg.alpha = .2,
47      legend.frame = "gray50", legend.outside = FALSE,
48      legend.width = 0.9, legend.height = .5,
49      legend.hist.height = 0.3, legend.title.size = 1.2,
50      legend.text.size = 0.6, legend.text.fontface = "plain",
51      legend.text.fontfamily = "Helvetica",
52      inner.margins = 0.1,
53      ) +
54  tm_graticules(
55      ticks = T, lines = T, labels.rot = c(15, 15),
56      col = "azure3", lwd = 1, labels.size = 1.2
57      )
58  # plot map
59  map1
60  tmap_save(map1, "Slope_Japan.jpg", height = 7)
```

Appendix A.8. R Script for Mapping Aspect of the Terrain in Japan

Listing 8. R script for mapping aspect of the terrain in Japan.

```
1   # Aspect
2   tmap_mode("plot")
3   map2 <-
4       tmap_style("albatross"
5           ) +
6       tm_shape(aspect, name = "Aspect", title = "Aspect",
7           raster.downsample = T,
8           ) +
9       tm_raster(
10          title = "Aspect (West-East-South-North)", palette = "Spectral",
11          style = "sd", labels = c("West", "East", "South", "North"),
12          legend.show = T, legend.hist = T, legend.hist.z = 0,
13          ) +
14      tm_scale_bar(
15          width = 0.5,
16          text.size = 1.5, text.color = "darkgoldenrod1",
17          color.dark = "lightsteelblue4", color.light = "white",
18          position=c("left", "bottom"), lwd = 1,
19          ) +
20      tm_compass(
21          type = "radar", position=c("right", "bottom")
22          ) +
23      tm_layout(scale = .8,
24          main.title = "Aspect: terrain analysis based on SRTM90 DEM of Italy.
    Mapping: R",
25          main.title.position = "center",
26          main.title.color = "black", main.title.size = 1.4,
27          title = "Data: SRTM90 DEM. Aspect (W-E-S-N)",
28          title.color = "darkgoldenrod1",
29          title.size = 1.2, title.position = c("left", "top"),
30          panel.labels = c("R packages: tmap, raster, sp, sf"),
31          panel.label.color = "darkslateblue",
```

```
32          panel.label.size = 1.2, legend.position = c("left","top"),
33          legend.bg.color = "grey90", legend.bg.alpha = .2,
34          legend.frame = "gray50", legend.outside = FALSE,
35          legend.width = .3, legend.height = .5,
36          legend.hist.height = .3, legend.title.size = 1.2,
37          legend.text.size = 1.2, legend.text.fontface = "plain",
38          legend.text.fontfamily = "Helvetica",
39          inner.margins = 0,
40          ) +
41      tm_graticules(
42          ticks = T, lines = T, labels.rot = c(15, 15),
43          col = "azure3", lwd = 1, labels.size = 1.2
44          )
45  # plot the map
46  map2
47  tmap_save(map2, "Aspect_Japan.jpg", height = 7)
```

Appendix A.9. R Script for Mapping Hillshade in the Terrain of Japan

Listing 9. R script for mapping hillshade in the terrain of Japan.

```
1   # hillshade
2   # tmaptools::palette_explorer()
3   tmap_mode("plot")
4   map3 <-
5       tmap_style("albatross") +
6       tm_shape(hill, name = "Hillshade", title = "Slope",
7           auto.palette.mapping = FALSE,) +
8       tm_raster(
9           title = "Histogram \n(data distribution)",
10          palette = "cividis", style = "kmeans",
11          legend.show = T, legend.hist = T,
12          legend.hist.z=0,
13          ) +
14      tm_scale_bar(
15          width = 0.25,
16          text.size = 1.2, text.color = "darkgoldenrod1",
17          color.dark = "lightsteelblue4", color.light = "white",
18          position=c("right", "bottom"), lwd = 1) +
19      tm_compass(
20          type = "radar", position=c("left", "bottom")) +
21      tm_layout(scale = .8,
22          main.title = "Hillshade: Terrain analysis based on SRTM90 DEM of
        Japan. Mapping: R",
23          main.title.position = "center",
24          main.title.color = "black",
25          main.title.size = 1.4, title = "Hillshade (0\u00B0-90\u00B0)",
26          title.color = "darkgoldenrod1",
27          title.size = 1.2, title.position = c("left", "top"),
28          panel.labels = c("R packages: tmap, raster, sp, sf"),
29          panel.label.color = "darkslateblue",
30          legend.position = c("left","top"), legend.bg.color = "grey90",
31          legend.bg.alpha = .2, legend.frame = "gray50",
32          legend.outside = FALSE, legend.width = .3,
33          legend.height = .5, legend.hist.height = .2,
34          legend.text.size = 1.0, legend.text.fontface = "plain",
35          legend.title.size = 1.2,
36          inner.margins = 0) +
37      tm_graticules(
38          ticks = TRUE, lines = TRUE,
39          col = "azure3", lwd = 1,
40          labels.size = 1.0,
41          labels.col = "black")
42  # plot the map
43  map3
44  tmap_save(map3, "Japan_Hillshade.jpg", height = 7)
```

Appendix A.10. R Script for Mapping Elevation Heights over Japan

Listing 10. R script for mapping elevation heights over Japan.

```
# elevation
tmap_mode("plot")
map4 <-
    tmap_style("white") +
    tm_shape(alt, name = "Elevation") +
    tm_raster(
        palette = terrain.colors(10),
        title = "Elevation (m asl)",
        legend.show = TRUE) +
    tm_scale_bar(
        width = 0.25,
        text.size = 0.5,
        text.color = "black",
        color.dark = "black",
        color.light = "white",
        position=c("left", "bottom"),
        lwd = 1) +
    tm_compass(position=c("left", "bottom")) +
    tm_layout(scale = .8,
        legend.position = c("left","top"),
        legend.bg.color = "grey90",
        legend.bg.alpha = .2,
        legend.frame = "gray50")
# plot the map
map4
tmap_save(map4, "Japan_Elevation.jpg", height = 7)
```

References

1. Iwahashi, J.; Kamiya, I.; Matsuoka, M. Regression analysis of Vs30 using topographic attributes from a 50-m DEM. *Geomorphology* **2010**, *117*, 202–205. [CrossRef]
2. Shoyama, K. Assessment of Land-Use Scenarios at a National Scale Using Intensity Analysis and Figure of Merit Components. *Land* **2021**, *10*, 379. [CrossRef]
3. Ayalew, L.; Yamagishi, H. The application of GIS-based logistic regression for landslide susceptibility mapping in the Kakuda-Yahiko Mountains, Central Japan. *Geomorphology* **2005**, *65*, 15–31. [CrossRef]
4. Babiker, I.S.; Mohamed, M.A.; Hiyama, T.; Kato, K. A GIS-based DRASTIC model for assessing aquifer vulnerability in Kakamigahara Heights, Gifu Prefecture, central Japan. *Sci. Total Environ.* **2005**, *345*, 127–140. [CrossRef]
5. Ohta, R.; Matsushi, Y.; Matsuzaki, H. Use of terrestrial cosmogenic 10Be to quantify anthropogenic sediment yield from mountainous watersheds: Application in reconstructing environmental change in the Tanakami Mountains, central Japan. *Geomorphology* **2022**, *405*, 108201. [CrossRef]
6. Ikemi, H. Geologically constrained changes to landforms caused by human activities in the 20th century: A case study from Fukuoka Prefecture, Japan. *Appl. Geogr.* **2017**, *87*, 115–126. [CrossRef]
7. Iwahashi, J.; Kamiya, I.; Yamagishi, H. High-resolution DEMs in the study of rainfall- and earthquake-induced landslides: Use of a variable window size method in digital terrain analysis. *Geomorphology* **2012**, *153–154*, 29–38. [CrossRef]
8. Nakayama, T.; Osako, M. Development of a process-based eco-hydrology model for evaluating the spatio-temporal dynamics of macro- and micro-plastics for the whole of Japan. *Ecol. Model.* **2023**, *476*, 110243. [CrossRef]
9. Hanaoka, K.; Nakaya, T.; Yano, K.; Inoue, S. Network-based spatial interpolation of commuting trajectories: application of a university commuting management project in Kyoto, Japan. *J. Transp. Geogr.* **2014**, *34*, 274–281. [CrossRef]
10. Thongthammachart, T.; Araki, S.; Shimadera, H.; Matsuo, T.; Kondo, A. Incorporating Light Gradient Boosting Machine to land use regression model for estimating NO2 and PM2.5 levels in Kansai region, Japan. *Environ. Model. Softw.* **2022**, *155*, 105447. [CrossRef]
11. Alifu, H.; Vuillaume, J.F.; Johnson, B.A.; Hirabayashi, Y. Machine-learning classification of debris-covered glaciers using a combination of Sentinel-1/-2 (SAR/optical), Landsat 8 (thermal) and digital elevation data. *Geomorphology* **2020**, *369*, 107365. [CrossRef]
12. Carrasco, L.; Fujita, G.; Kito, K.; Miyashita, T. Historical mapping of rice fields in Japan using phenology and temporally aggregated Landsat images in Google Earth Engine. *ISPRS J. Photogramm. Remote Sens.* **2022**, *191*, 277–289. [CrossRef]
13. Naghibi, S.A.; Hashemi, H.; Pradhan, B. APG: A novel python-based ArcGIS toolbox to generate absence-datasets for geospatial studies. *Geosci. Front.* **2021**, *12*, 101232. [CrossRef]
14. Farag, T.; Sobh, M.; Mizunaga, H. 3D constrained gravity inversion to model Moho geometry and stagnant slabs of the Northwestern Pacific plate at the Japan Islands. *Tectonophysics* **2022**, *829*, 229297. [CrossRef]

15. Lemenkova, P. Handling Dataset with Geophysical and Geological Variables on the Bolivian Andes by the GMT Scripts. *Data* **2022**, *7*, 74. [CrossRef]
16. Hara, Y.; Oki, S.; Uchiyama, Y.; Ito, K.; Tani, Y.; Naito, A.; Sampei, Y. Plant Diversity in the Dynamic Mosaic Landscape of an Agricultural Heritage System: The Minabe-Tanabe Ume System. *Land* **2021**, *10*, 559. [CrossRef]
17. Gomez, C.; Hayakawa, Y.; Obanawa, H. A study of Japanese landscapes using structure from motion derived DSMs and DEMs based on historical aerial photographs: New opportunities for vegetation monitoring and diachronic geomorphology. *Geomorphology* **2015**, *242*, 11–20.
18. Kim, M.; Rupprecht, C.D.D.; Furuya, K. Residents' Perception of Informal Green Space—A Case Study of Ichikawa City, Japan. *Land* **2018**, *7*, 102. [CrossRef]
19. Sasaki, K.; Hotes, S.; Ichinose, T.; Doko, T.; Wolters, V. Hotspots of Agricultural Ecosystem Services and Farmland Biodiversity Overlap with Areas at Risk of Land Abandonment in Japan. *Land* **2021**, *10*, 1031. [CrossRef]
20. Otani, S.; Kurosaki, Y.; Kurozawa, Y.; Shinoda, M. Dust Storms from Degraded Drylands of Asia: Dynamics and Health Impacts. *Land* **2017**, *6*, 83. [CrossRef]
21. Hooke, J.M. Changing landscapes: Five decades of applied geomorphology. *Geomorphology* **2020**, *366*, 106793. [CrossRef]
22. Siakeu, J.; Oguchi, T.; Aoki, T.; Esaki, Y.; Jarvie, H.P. Change in riverine suspended sediment concentration in central Japan in response to late 20th century human activities. *CATENA* **2004**, *55*, 231–254. [CrossRef]
23. Nakayama, T. For improvement in understanding eco-hydrological processes in mire. *Ecohydrol. Hydrobiol.* **2013**, *13*, 62–72.
24. Ito, S.; Onitsuka, T.; Kuroda, H.; Hasegawa, N.; Fukuda, H.; Gouda, H.; Akino, H.; Sonoki, S.; Endo, K.; Takayama, T.; et al. Evaluation of seafloor environmental characteristics of harvesting ground of a kelp Saccharina longissima using GIS in the Pacific coastal area of eastern Hokkaido, Japan. *Reg. Stud. Mar. Sci.* **2022**, *55*, 102527. [CrossRef]
25. Tabuchi, K.; Murakami, T.; Okudera, S.; Furihata, S.; Sakakibara, M.; Takahashi, A.; Yasuda, T. Predicting potential rice damage by insect pests using land use data: A 3-year study for area-wide pest management. *Agric. Ecosyst. Environ.* **2017**, *249*, 4–11. [CrossRef]
26. Priya, S.; Shibasaki, R. National spatial crop yield simulation using GIS-based crop production model. *Ecol. Model.* **2001**, *136*, 113–129. [CrossRef]
27. Sasai, T.; Nakai, S.; Setoyama, Y.; Ono, K.; Kato, S.; Mano, M.; Murakami, K.; Miyata, A.; Saigusa, N.; Nemani, R.R.; et al. Analysis of the spatial variation in the net ecosystem production of rice paddy fields using the diagnostic biosphere model, BEAMS. *Ecol. Model.* **2012**, *247*, 175–189. [CrossRef]
28. Matsu'ura, T.; Kimura, H.; Komatsubara, J.; Goto, N.; Yanagida, M.; Ichikawa, K.; Furusawa, A. Late Quaternary uplift rate inferred from marine terraces, Shimokita Peninsula, northeastern Japan: A preliminary investigation of the buried shoreline angle. *Geomorphology* **2014**, *209*, 1–17. [CrossRef]
29. Imaizumi, F.; Nishiguchi, T.; Matsuoka, N.; Trappmann, D.; Stoffel, M. Interpretation of recent alpine landscape system evolution using geomorphic mapping and L-band InSAR analyses. *Geomorphology* **2018**, *310*, 125–137. [CrossRef]
30. Ghaderpour, E. Some Equal-area, Conformal and Conventional Map Projections: A Tutorial Review. *J. Appl. Geod.* **2016**, *10*, 197–209. [CrossRef]
31. Matsuoka, N. A multi-method monitoring of timing, magnitude and origin of rockfall activity in the Japanese Alps. *Geomorphology* **2019**, *336*, 65–76. [CrossRef]
32. Fujino, M.; Sakakibara, K.; Tsujimura, M.; Suzuki, K. Influence of alpine vegetation on water storage and discharge functions in an alpine headwater of Northern Japan Alps. *J. Hydrol. X* **2023**, *18*, 100146. [CrossRef]
33. Kariya, Y. Geomorphic processes at a snowpatch hollow on Gassan volcano, northern Japan. *Permafr. Periglac. Process.* **2002**, *13*, 107–116. [CrossRef]
34. Oguchi, T. Geomorphology and GIS in Japan: Background and characteristics. *GeoJournal* **2000**, *52*, 195–202. [CrossRef]
35. Oguchi, T. Geomorphological debates in Japan related to surface processes, tectonics, climate, research principles, and international geomorphology. *Geomorphology* **2020**, *366*, 106805. [CrossRef]
36. Matsu'ura, T. Late Quaternary uplift rate inferred from marine terraces, Muroto Peninsula, southwest Japan: Forearc deformation in an oblique subduction zone. *Geomorphology* **2015**, *234*, 133–150. [CrossRef]
37. Niwa, Y.; Sugai, T. Millennial-scale vertical deformation of the Hachinohe coastal plain (NE Japan). *Geomorphology* **2021**, *389*, 107835. [CrossRef]
38. Hattanji, T.; Kodama, R.; Takahashi, D.; Tanaka, Y.; Doshida, S.; Furuichi, T. Migration of channel heads by storm events in two granitic mountain basins, western Japan: Implication for predicting location of landslides. *Geomorphology* **2021**, *393*, 107943. [CrossRef]
39. Lemenkova, P. NOAA Marine Geophysical Data and a GEBCO Grid for the Topographical Analysis of Japanese Archipelago by Means of GRASS GIS and GDAL Library. *Geomat. Environ. Eng.* **2020**, *14*, 25–45. [CrossRef]
40. Lemenkova, P. GRASS GIS for classification of Landsat TM images by maximum likelihood discriminant analysis: Tokyo area, Japan. *Geod. Glas.* **2020**, *51*, 5–25. [CrossRef]
41. Lemenkova, P. Mapping Climate Parameters over the Territory of Botswana Using GMT and Gridded Surface Data from TerraClimate. *ISPRS Int. J.-Geo-Inf.* **2022**, *11*, 473. [CrossRef]
42. Lemenkova, P. Console-Based Mapping of Mongolia Using GMT Cartographic Scripting Toolset for Processing TerraClimate Data. *Geosciences* **2022**, *12*, 140. [CrossRef]

43. Wessel, P.; Luis, J.F.; Uieda, L.; Scharroo, R.; Wobbe, F.; Smith, W.H.F.; Tian, D. The Generic Mapping Tools Version 6. *Geochem. Geophys. Geosyst.* **2019**, *20*, 5556–5564. [CrossRef]

44. R Core Team. *R: A Language and Environment for Statistical Computing*; R Foundation for Statistical Computing: Vienna, Austria, 2022.

45. Hengl, T.; Bajat, B.; Blagojević, D.; Reuter, H.I. Geostatistical modeling of topography using auxiliary maps. *Comput. Geosci.* **2008**, *34*, 1886–1899. [CrossRef]

46. Lemenkova, P.; Debeir, O. R Libraries for Remote Sensing Data Classification by K-Means Clustering and NDVI Computation in Congo River Basin, DRC. *Appl. Sci.* **2022**, *12*, 12554. [CrossRef]

47. Lemenkova, P.; Debeir, O. Satellite Image Processing by Python and R Using Landsat 9 OLI/TIRS and SRTM DEM Data on Côte d'Ivoire, West Africa. *J. Imaging* **2022**, *8*, 317. [CrossRef]

48. Kahle, D.; Wickham, H. ggmap: Spatial Visualization with ggplot2. *R J.* **2013**, *5*, 144–161. [CrossRef]

49. Sueoka, S.; Kobayashi, Y.; Fukuda, S.; Kohn, B.P.; Yokoyama, T.; Sano, N.; Hasebe, N.; Tamura, A.; Morishita, T.; Tagami, T. Low-temperature thermochronology of active arc-arc collision zone, South Fossa Magna region, central Japan. *Tectonophysics* **2022**, *828*, 229231. [CrossRef]

50. Enescu, B.; Ito, K. The 1998 Hida Mountain, Central Honshu, Japan, earthquake swarm: Double-difference event relocation, frequency–magnitude distribution and Coulomb stress changes. *Tectonophysics* **2005**, *409*, 147–157. [CrossRef]

51. Tsujimori, T.; Liou, J.; Ernst, W.; Itaya, T. Triassic paragonite- and garnet-bearing epidote-amphibolite from the Hida Mountains, Japan. *Gondwana Res.* **2006**, *9*, 167–175.

52. Sueoka, S.; Tsutsumi, H.; Tagami, T. New approach to resolve the amount of Quaternary uplift and associated denudation of the mountain ranges in the Japanese Islands. *Geosci. Front.* **2016**, *7*, 197–210. [CrossRef]

53. Kariya, Y.; Sato, G.; Komori, J. Landslide-induced terminal moraine-like landforms on the east side of Mount Shiroumadake, Northern Japanese Alps. *Geomorphology* **2011**, *127*, 156–165. [CrossRef]

54. Sato, H.; Imaizumi, T.; Yoshida, T.; Ito, H.; Hasegawa, A. Tectonic evolution and deep to shallow geometry of Nagamachi-Rifu Active Fault System, NE Japan. *Earth, Planets Space* **2002**, *54*, 1039–1043. [CrossRef]

55. Tsuchiya, N.; Yamada, R. Geological and Geophysical Perspective of Supercritical Geothermal Energy in Subduction Zone, Northeast Japan. *Procedia Earth Planet. Sci.* **2017**, *17*, 193–196. [CrossRef]

56. Rabus, B.; Eineder, M.; Roth, A.; Bamler, R. The shuttle radar topography mission—A new class of digital elevation models acquired by spaceborne radar. *ISPRS J. Photogramm. Remote Sens.* **2003**, *57*, 241–262. [CrossRef]

57. RStudio Team. RStudio: Integrated Development Environment for R; RStudio Inc., Boston, MA, USA, 2017. Available online: https://www.RStudio.com/ (accessed on 23 December 2022).

58. Hijmans, R.J. raster: Geographic Data Analysis and Modeling. R Package Version 2.6-7. 2017. Available online: https://CRAN.R-project.org/package=raster (accessed on 23 December 2022).

59. Evers, K.; Knudsen, T. Transformation pipelines for PROJ.4. In Proceedings of the FIG Working Week 2017, Surveying the World of Tomorrow—From Digitalisation to Augmented Reality, Helsinki, Finland, 29 May–2 June 2017; pp. 1–13.

60. Lemenkova, P.; Debeir, O. Seismotectonics of Shallow-Focus Earthquakes in Venezuela with Links to Gravity Anomalies and Geologic Heterogeneity Mapped by a GMT Scripting Language. *Sustainability* **2022**, *14*, 15966. [CrossRef]

61. Narksri, P.; Takeuchi, E.; Ninomiya, Y.; Morales, Y.; Akai, N.; Kawaguchi, N. A Slope-robust Cascaded Ground Segmentation in 3D Point Cloud for Autonomous Vehicles. In Proceedings of the 2018 21st International Conference on Intelligent Transportation Systems (ITSC), Maui, HI, USA, 4–7 November 2018; pp. 497–504. [CrossRef]

62. Matsuyama, H.; Aoki, S.; Yonezawa, T.; Hiroi, K.; Kaji, K.; Kawaguchi, N. Deep Learning for Ballroom Dance Recognition: A Temporal and Trajectory-Aware Classification Model With Three-Dimensional Pose Estimation and Wearable Sensing. *IEEE Sens. J.* **2021**, *21*, 25437–25448. [CrossRef]

63. Watanabe, K.; Hiroi, K.; Kamiyama, T.; Sano, H.; Tsukamoto, M.; Katagiri, M.; Ikeda, D.; Kaji, K.; Kawaguchi, N. A Smartphone 3D Positioning Method using a Spinning Magnet Marker. *J. Inf. Process.* **2019**, *27*, 10–24. [CrossRef]

64. Wang, W.; Sakurada, K.; Kawaguchi, N. Incremental and Enhanced Scanline-Based Segmentation Method for Surface Reconstruction of Sparse LiDAR Data. *Remote Sens.* **2016**, *8*, 967. [CrossRef]

65. Kasai, M.; Ikeda, M.; Asahina, T.; Fujisawa, K. LiDAR-derived DEM evaluation of deep-seated landslides in a steep and rocky region of Japan. *Geomorphology* **2009**, *113*, 57–69.

66. Kondo, H.; Toda, S.; Okumura, K.; Takada, K.; Chiba, T. A fault scarp in an urban area identified by LiDAR survey: A Case study on the Itoigawa–Shizuoka Tectonic Line, central Japan. *Geomorphology* **2008**, *101*, 731–739. [CrossRef]

67. Becker, R.A.; Wilks, A.R.; Brownrigg, R.; Minka, T.P. Maps: Draw Geographical Maps. R Package Version 2.3-2. 2013. Available online: http://CRAN.R-project.org/package=maps (accessed on 23 December 2022).

68. Tennekes, M. tmap: Thematic Maps in R. *J. Stat. Softw.* **2018**, *84*, 1–39. [CrossRef]

69. Aoyama, M. Rock glaciers in the northern Japanese Alps: palaeoenvironmental implications since the Late Glacial. *J. Quat. Sci.* **2005**, *20*, 471–484. [CrossRef]

70. Oguchi, T.; Saito, K.; Kadomura, H.; Grossman, M. Fluvial geomorphology and paleohydrology in Japan. *Geomorphology* **2001**, *39*, 3–19.

71. Fujisawa, K.; Marcato, G.; Nomura, Y.; Pasuto, A. Management of a typhoon-induced landslide in Otomura (Japan). *Geomorphology* **2010**, *124*, 150–156.

72. Tsunetaka, H. Comparison of the return period for landslide-triggering rainfall events in Japan based on standardization of the rainfall period. *Earth Surf. Process. Landforms* **2021**, *46*, 2984–2998. [CrossRef]

73. Oku, Y.; Nakakita, E. Future change of the potential landslide disasters as evaluated from precipitation data simulated by MRI-AGCM3.1. *Hydrol. Process.* **2013**, *27*, 3332–3340. [CrossRef]

74. Sogno, P.; Klein, I.; Kuenzer, C. Remote Sensing of Surface Water Dynamics in the Context of Global Change—A Review. *Remote Sens.* **2022**, *14*, 2475. [CrossRef]

75. Yamada, M.; Kumagai, H.; Matsushi, Y.; Matsuzawa, T. Dynamic landslide processes revealed by broadband seismic records. *Geophys. Res. Lett.* **2013**, *40*, 2998–3002. [CrossRef]

76. Wang, G.; Suemine, A.; Schulz, W.H. Shear-rate-dependent strength control on the dynamics of rainfall-triggered landslides, Tokushima Prefecture, Japan. *Earth Surf. Process. Landforms* **2010**, *35*, 407–416. .: 10.1002/esp.1937. [CrossRef]

77. Hirata, Y.; Chigira, M. Landslides associated with spheroidally weathered mantle of granite porphyry induced by 2011 Typhoon Talas in the Kii Peninsula, Japan. *Eng. Geol.* **2019**, *260*, 105217. [CrossRef]

78. Prima, O.D.A.; Echigo, A.; Yokoyama, R.; Yoshida, T. Supervised landform classification of Northeast Honshu from DEM-derived thematic maps. *Geomorphology* **2006**, *78*, 373–386. [CrossRef]

79. Lin, Z.; Oguchi, T. Drainage density, slope angle, and relative basin position in Japanese bare lands from high-resolution DEMs. *Geomorphology* **2004**, *63*, 159–173. [CrossRef]

80. Hayakawa, Y.S.; Oguchi, T. DEM-based identification of fluvial knickzones and its application to Japanese mountain rivers. *Geomorphology* **2006**, *78*, 90–106.

 *land*

Article

Machine Learning-Based Assessment of Watershed Morphometry in Makran

Reza Derakhshani [1,2,*], Mojtaba Zaresefat [3], Vahid Nikpeyman [1], Amin GhasemiNejad [4], Shahram Shafieibafti [2], Ahmad Rashidi [5,6], Majid Nemati [2,5] and Amir Raoof [1]

[1] Department of Earth Sciences, Utrecht University, 3584CB Utrecht, The Netherlands
[2] Department of Geology, Shahid Bahonar University of Kerman, Kerman 76169-13439, Iran
[3] Copernicus Institute of Sustainable Development, Utrecht University, 3584CB Utrecht, The Netherlands
[4] Department of Economics, Faculty of Management and Economics, Shahid Bahonar University of Kerman, Kerman 76169-13439, Iran
[5] Department of Earthquake Research, Shahid Bahonar University of Kerman, Kerman 76169-13439, Iran
[6] Department of Seismotectonics, International Institute of Earthquake Engineering and Seismology, Tehran 19537-14453, Iran
* Correspondence: r.derakhshani@uu.nl

Abstract: This study proposes an artificial intelligence approach to assess watershed morphometry in the Makran subduction zones of South Iran and Pakistan. The approach integrates machine learning algorithms, including artificial neural networks (ANN), support vector regression (SVR), and multivariate linear regression (MLR), on a single platform. The study area was analyzed by extracting watersheds from a Digital Elevation Model (DEM) and calculating eight morphometric indices. The morphometric parameters were normalized using fuzzy membership functions to improve accuracy. The performance of the machine learning algorithms is evaluated by mean squared error (MSE), mean absolute error (MAE), and correlation coefficient (R^2) between the output of the method and the actual dataset. The ANN model demonstrated high accuracy with an R^2 value of 0.974, MSE of 4.14×10^{-6}, and MAE of 0.0015. The results of the machine learning algorithms were compared to the tectonic characteristics of the area, indicating the potential for utilizing the ANN algorithm in similar investigations. This approach offers a novel way to assess watershed morphometry using ML techniques, which may have advantages over other approaches.

Keywords: watershed morphometry; fuzzy analytic hierarchy process; artificial neural networks; support vector regression; multivariate linear regression; tectonics; Makran

Citation: Derakhshani, R.; Zaresefat, M.; Nikpeyman, V.; GhasemiNejad, A.; Shafieibafti, S.; Rashidi, A.; Nemati, M.; Raoof, A. Machine Learning-Based Assessment of Watershed Morphometry in Makran. *Land* **2023**, *12*, 776. https://doi.org/10.3390/land12040776

Academic Editor: Chuanrong Zhang

Received: 14 February 2023
Revised: 25 March 2023
Accepted: 27 March 2023
Published: 29 March 2023

1. Introduction

Watershed morphometry is a crucial factor in determining the impact of tectonic processes on the landscape. By analyzing the shape and geometry of watersheds at a regional scale, we can identify the relative significance of tectonic deformation versus erosion in landscape evolution [1,2]. Understanding the impact of these geological forces on the morphology of watersheds and the development of drainage systems is essential, as it can have implications for sediment supply to river reaches and increase the risk of landslides [3]. Furthermore, by quantifying the morphotectonic situation of watersheds, we can gain insight into their evolution and assess the role of regional tectonic control in shaping their development [2,4–6]. Recent studies (e.g., [7–10]) have expanded our understanding of the relationship between morphotectonic factors and the development of drainage systems and landscapes, building upon the foundational work of Horton [11], Strahler [12], and Hack [13]. Through a quantitative assessment of watershed development in relation to regional tectonics, we can better understand the morphotectonic situations and their implications for the broader landscape [14].

The utilization of Geographic Information System (GIS), satellite remote sensing data processing, particularly Global Digital Elevation Models (DEMs), along with the application of Analytic Hierarchy Process (AHP) and Fuzzy Analytic Hierarchy Process (FAHP) techniques, can facilitate the determination of comprehensive geomorphometric analyses (e.g., [15–17]).

Due to the inhomogeneity of geological landscapes and vast data, we need complicated models and methods to study geological features. Numerous complex mathematical approaches have been implemented to overcome this issue. To this end, multiple types of research have been conducted using AHP, FAHP, Technique for Order Preference by Similarity to Ideal Solution (TOPSIS), and machine learning algorithms. Some of them are combined (hybrid methods) to take advantage of the best parts of each and obtain the best results [18–20].

The use of a Geographic Information System (GIS), a digital database management system specifically designed for handling large-scale and spatially diverse data from various sources, has the potential to reveal the drainage patterns of watersheds [21,22] and is well-suited for advanced zoning applications.

A notable benefit of the Analytic Hierarchy Process (AHP) is its foundation in pairwise comparisons, which effectively facilitate the computation of criteria weights. In addition, the AHP calculates inconsistency indices; ratios of a decision maker's inconsistency. Nevertheless, decision makers must occasionally perform a huge number of pairwise comparisons, and this condition, particularly with fuzzy AHP, makes using the AHP procedure impracticable [23]. When the researcher is confident in the certainty of the collected data, the classical Analytic Hierarchy Process (AHP) outperforms the Fuzzy Analytic Hierarchy Process (FAHP). However, in cases where the data is uncertain, the FAHP technique is advised [21]. The Fuzzy Analytic Hierarchy Process (FAHP) approach is used in this study to evaluate the weight criteria related to active tectonics based on morphometric parameters.

Artificial intelligence algorithms have attracted increasing attention in recent years as solid computational tools to simulate complicated phenomena in various academic domains [24–27]. Among the various methods of machine learning, artificial neural networks are considered the backbone of machine learning algorithms. This method's key advantages are its learning capability based on the training process, which eliminates the requirement for statistical assumptions for the source data, and its ability to cope with non-linear situations. Researchers in numerous scientific and engineering fields are interested in ANN models since they can correlate huge and complicated multi-parameter datasets without a prior understanding of the relationships between the parameters [28]. Applying machine learning in combination with GIS to analyze morphometric parameters allows us to understand the evolution of watersheds and provides a clear picture of landscape evolution [27].

A few investigations have used geomorphic indices of drainage basins by calculating their arithmetic mean or considering a weight for each index in the AHP to map out a relative tectonic activity (e.g., [15,27,29]), but no study employing artificial intelligence has been published on this subject as of yet. Considering that the morphometry of watersheds can reveal the relative tectonic activity of an area over a long period, this paper attempts to investigate a new method to combine the traditional FAHP technique with artificial intelligence algorithms as an innovation in morphotectonic evaluation. This research employs several geomorphic indices to assess the regional tectonic activity across such a broad area. The main goal of this study is to use machine learning algorithms and FAHP to identify which of the 423 drainage basins studied in the Makran subduction zone are more affected by tectonic features.

2. Materials and Methods

2.1. Study Area

The Makran subduction zone with east-west trending is located in southern Pakistan and Iran. This zone indicates a convergence zone where the Oman oceanic plate subducts beneath the Eurasian continental plate. The subduction probably started during the Late Cretaceous [30]. The western boundary of Makran is often called the Oman Line and Minab-Zendan Fault system, which runs northward and separates a highly seismic region in Zagros [31] from a region of low seismic activity in Makran to the east [32,33]. The eastern boundary of Makran is defined by a transfer zone consisting of three individual faults, namely the Ornach-Nal, Ghazaband, and Chaman faults [34,35].

In contrast to the majority of accretionary complexes in the world, the Makran accretionary wedge lacks an obvious trench [36]. The lack of a trench in this region may be because the angle of the subduction slab at the accretionary front is low, which may be caused by the presence of 7 km thick sediments with low compaction [37]. The seaward 70 km of the forearc comprises semi-consolidated and unconsolidated sediments with high pore fluid pressures and low seismic velocities. These sediments are capable of causing and failing large tsunamigenic slides [38].

The convergence rate decreases east to west along the Makran boundary [39]. In Makran, the average convergence rate is 4 cm/year, decreasing from 4.2 cm/year in the east to 3.65 cm/year in the west. Furthermore, GPS measurements suggest that the highest subduction rate of the Oman oceanic plate beneath the Eurasian plate occurs in the east at about 2.7 cm/year. In comparison, the lowest rate occurs in the west at about 1.95 cm/year [40]. Compared to other subduction zones, such as the Cascadia subduction zone at 35 mm/year [41], the Mexico subduction zone at 41 mm/year [42], the Sumatra subduction zone at 65 mm/year [43], the south Chile subduction zone at 70 mm/year [44], the Japan subduction zone at 80 mm/year [45], and the Tonga subduction zone at 160 mm/year [44], Makran is considered a relatively slow subduction zone.

Zarifi [46] states that the compressional stress direction along the Makran zone is rotating. The stress field in the eastern Makran is influenced by the collision between the Indian and Eurasian plates, while the western Makran stress field is affected by the Arabia-Eurasian collision. Eastern and western parts of the Makran subduction zone exhibit distinctive seismic behavior. The eastern region of Makran displays more seismic activity than the western region [38]. The plate boundary in western Makran lacks well-documented great instrumental and historical earthquakes. In contrast, eastern Makran was ruptured by thrust faulting during the 1945 earthquake and currently experiences earthquakes of varying magnitudes [38].

The main deformation phase from the Late Miocene to the Late Pliocene occurred by imbricating fans at the front of the Makran accretionary prism [47]. The seabed's current geometry suggests that most of these imbricated fans remain active and that the accretionary prisms continue to be impacted by ongoing deformation from some branches of the imbricated fans. Following the Pliocene, the Makran coast and the mid-slope region experienced normal faulting, uplifting, and ductile flows [47].

The Makran subduction zone, like many other subduction zones around the world, has active mud volcanoes, some of which have formed along anticlinal axes [48]. As the largest mud volcanoes in the world [48], the mud volcanoes in Makran result from mud diapirism, providing evidence for the tectonic expulsion of mud and fluids seaward of the accretionary front. This confirms that tectonic forcing plays a significant role in forming mud volcanoes in Makran [49].

2.2. Extracting the Geomorphic Parameters

In this study, the morphotectonic zoning method was implemented in two steps. The watersheds of the research region were first determined using the Japan Aerospace Exploration Agency's ALOS World 3D30m images (AW3D30), Ver.3.2 [50], with a resolution of 30 m (Figure 1). Numerous studies have demonstrated that AW3D30 DEM data can be compared to other open-source DEMs, such as SRTM and ASTER, with the advantage of being more accurate in delineating river basins and presenting drainage networks [51,52]. Next, ArcGIS Pro was used to evaluate the morphometric parameters and all physiographic indices to determine the watersheds with higher uplift. The eight indices employed in the morphotectonic process are shown in Figure 2.

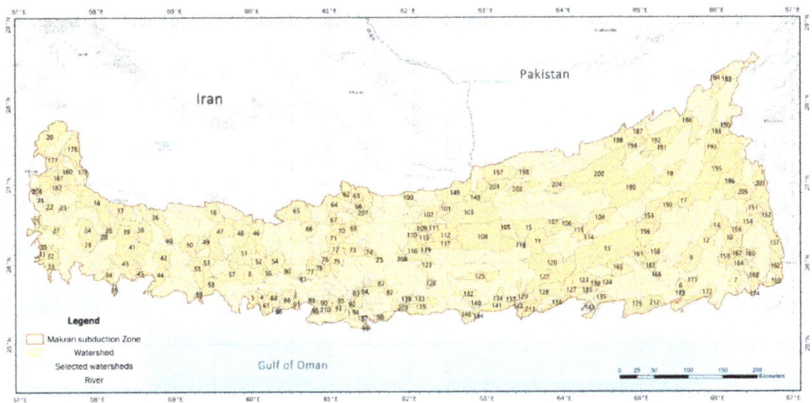

Figure 1. Location of Makran subduction zone and border of 423 extracted watersheds. The number labels show the randomly selected watershed used in AI.

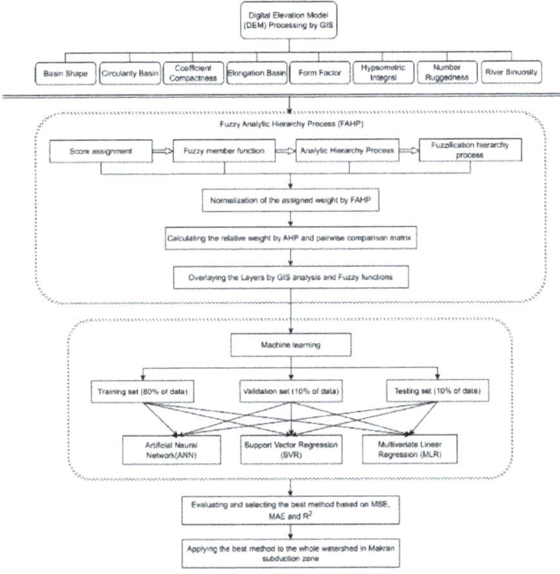

Figure 2. Decision tree developed for morphotectonic zoning of Makran region.

2.3. Calculating Criterion Weights by FAHP

The Analytical Hierarchy Process (AHP) provides a systematic approach for evaluating complex decisions by combining qualitative and quantitative factors within a single framework, resulting in a prioritized list of alternatives [53]. Despite its widespread use, there have been concerns raised by researchers regarding certain limitations of the AHP, such as ambiguity in standardizing non-commensurate criteria (i.e., criteria that cannot be compared due to differences in size, type, or scale) and the influence of personal assessments [54]. These limitations can significantly impact the results of the AHP [55–57]. To address these issues, the Fuzzy Analytical Hierarchy Process (FAHP) has been developed to overcome the limitations and flaws of the AHP [58–60]. In this context, FAHP was employed to assess the weighting criteria related to morphotectonic zoning.

Mohebbi Tafreshi et al. [61] presented a fuzzy modelling method consisting of the following steps. The initial step in fuzzy models involves standardizing parameters using a fuzzy membership function. The experts' opinions are utilized to assign fuzzy values to raw input values using a transformation function, where values close to 1 are deemed more suitable for the desired outcome, and values close to 0 are considered less suitable. The fuzzy logic extension of the ArcGIS Pro 2.8 software (version 10.8) offers various fuzzy membership functions. The selection of the fuzzification function is based on the nature, significance, and relation of each criterion to the goal. For this preliminary analysis, the Linear and Gaussian functions were chosen from the seven available fuzzy membership functions to standardize the factors.

The fuzzy linear transformation function applies a linear transformation between the minimum and maximum values specified by the user. Any value below the minimum will be assigned a 0 (definitely not a member), and any value above the maximum will be assigned a 1 (definitely a member). On the other hand, the fuzzy Gaussian function transforms the original values into a normal distribution [62].

The process of identifying the critical factors in the morphotectonic analysis was initiated by conducting a literature review, which resulted in identifying the most important indices, as shown in Figure 2. The relative significance of each parameter was then estimated through the use of the Analytic Hierarchy Process (AHP) and a constructed pairwise comparison matrix (8 × 8) based on the input factors, as determined by Saaty's scale (Table 1). It is worth noting that the likelihood of inconsistencies in pairwise comparisons increases with the number of comparisons made, as stated in reference [63]. To account for this, AHP incorporates a consistency index (CI) to evaluate the calculated weight matrix. The weight is deemed acceptable if the CI is less than 10% (Table 2). Finally, the calculated weights were normalized to a scale between 0 and 1 to facilitate the integration of the weighted map layers.

Table 1. Saaty's 1–9 scale of relative importance [63].

Intensity of Importance	Interpretation
1	Equal importance
3	Moderate importance
5	Essential
7	Extreme importance
9	Extreme importance
2, 4, 6, 8	Intermediate values between adjacent scale values

Table 2. Pairwise comparison matrix for standardizing factor scores.

Linear and Areal Aspects	Hi	Bs	Cb	Er	Rn	Rs	Cc	Ff	Score
Hypsometric integral (Hi)		0.5	0.5	2	0.5	0.5	2	2	0.136
Basin shape (Bs)			2	2	2	2	2	2	0.037
Circularity basin (Cb)				0.33	0.5	2	2	0.5	0.123
Elongation ratio (Er)					2	2	2	2	0.084
Ruggedness number (Rn)						2	3	2	0.078
River sinuosity (Rs)							0.5	0.5	0.297
CoefficientCompactness (Cc)								0.5	0.050
Form factor (Ff)									0.197
CI					0.03				

2.4. Description and Application of the Criterion

Morphometric analysis of watersheds as a prerequisite of hydrological studies would be a valuable method to make informed management choices based on a more comprehensive view of the drainage network's behavior and the morphology of the watershed.

In this study, eight linear and area-based indices of watersheds are evaluated. These indices provide a broad overview of the drainage basin network with regard to morphology and relief. These indices assess the stream network's complexity, texture, and distortion due to neotectonic disturbances [64]. The extracted values are valuable in understanding the development of the drainage network concerning lithology and landscape. These indices are explained in detail as follows.

2.4.1. Hypsometric Integral (Hi)

The hypsometric curve represents the elevation distribution concerning the drainage area at various levels, including regional and continental scales [65,66]. The hypsometric integral (Hi) calculates the uneroded volume of a basin by determining the area under the hypsometric curve [67]. This can be computed through Equation (1).

$$\text{Hi} = \frac{H_{avg} - H_{min}}{H_{max} - H_{min}} \tag{1}$$

A value of Hi greater than 0.6 signifies a tectonically active region with significant uplift and steep topography [68–70]. On the other hand, a mature drainage basin exhibits a moderate-to-low Hi value [71]. In this study, a fuzzy linear membership function was utilized, where a Hi value of 0.65 was assigned the highest weight, and the weight progressively decreased to zero as the Hi value approached 0.04.

2.4.2. Basin Shape (Bs)

The shape of drainage basins in active tectonic zones is generally elongated, but over time it tends to become more circular [67]. The Basin Shape Index (Bs) is calculated using Equation (2) [72,73] to describe this change.

$$\text{Bs} = \frac{\text{Bl}}{\text{Bw}} \tag{2}$$

This index is based on the length (Bl) and width (Bw) of the basin, measured from the headwater to the mountain ridge and at the broadest point, respectively. These values are calculated using the minimum bounding geometry script. Basins that have lower Bs values are considered more circular in shape and are usually associated with low tectonic activity. On the other hand, steep basins that have high tectonic activity are elongated [29,74]. The most significant weight is assigned to Bs values around 7.9 using the increasing fuzzy linear membership, and as the Bs value decreases to 0.16, the weight decreases until it reaches zero.

2.4.3. Circularity Basin (Cb)

The concept of the circulatory basin (Cb) was introduced by Miller [75] and Strahler [76] and it is defined as the ratio of the area of the basin (A) to the area of a circle with the same perimeter as the basin Equation (3).

$$\text{Cb} = \frac{4\pi A}{P2} \tag{3}$$

where A represents the area of the basin and P is the basin's perimeter. The circularity index ranges from 0 for a straight line to 1 for a perfect circle. The more circular the shape of the basin, the higher the value of Cb. Factors such as the length, frequency of streams (Fs), geological structures, land cover, climate, relief, and slope of the basin all play a role in determining the circulatory ratio, making it a key factor in determining the stage of a basin. Low, medium and high Cb values correspond to a basin's youthful, mature, and ageing periods [77,78]. In this context, the decreasing fuzzy linear membership assigns the lowest weight to a Cb value close to 0.3, and as the Cb value decreases to 0.03, the weight also decreases until it reaches zero.

2.4.4. Elongation Ratio (Er)

The elongation ratio (Er) measures the shape of a drainage basin, representing the ratio of the diameter of a circle with the same area as the basin to the basin's length [79]. A value of 1 indicates a perfectly circular shape, while a lower value suggests a more elongated and tectonically impacted basin (as described in Equation (4)).

$$Er = \frac{2\sqrt{\frac{A}{\pi}}}{L} \tag{4}$$

In this study, the highest and lowest Er values were found to be 0.17 and 0.06, respectively. Using the increasing fuzzy linear membership, the highest weight was assigned to an Er value close to 0.17, steadily decreasing as the Er value decreased to 0.06 and reached zero.

2.4.5. Ruggedness Number (Rn)

The Ruggedness Index, which combines two factors: relief (H) and drainage density (Dd), indicates the steepness and length of slopes. The calculation of the Ruggedness Index is carried out using Equation (5a–c).

$$Rn = Dd \cdot H \tag{5a}$$

$$Dd = \frac{\sum Li}{A} \tag{5b}$$

$$H = H_{max} - H_{min} \tag{5c}$$

Where Li is the length of the river, and A is the area of the watershed. H_{max} and H_{min} are the highest and lowest elevations of the watershed. The Rn close to 473 is allocated the highest weight utilizing the rising fuzzy linear membership. As the Rn value approaches 1, the weight drops until it reaches zero.

2.4.6. River Sinuosity (Rs)

The morphology of rivers that interact with an active fault zone tend to become uneven due to changes in slope and variations in curvature, leading to increased meandering [80]. Hence, the sinuosity of a river (Rs) can be a valuable indicator of tectonic activity in a drainage basin. A highly sinuous river suggests stability, while a straight river profile suggests ongoing tectonic activity. The formula in Equation (6) can be used to calculate the sinuosity of a river, where c represents the channel length, and v represents the straight length of the valley. Utilizing a decreasing fuzzy linear membership, the highest weight is assigned to the Rs value closest to 1.03, while the weight decreases as the Rs value approaches 2.58.

$$Rs = \frac{C}{V} \tag{6}$$

2.4.7. Compactness Coefficient (Cc)

The compactness coefficient, also known as the Gravellus Index, evaluates the basin's shape irregularity. The more irregular the shape, the higher the value of Cc [81]. This value can be a good indicator of the area's tectonic activity level Equation (7).

$$Cc = 0.2821 \frac{P}{A^{0.5}} \tag{7}$$

where P is the perimeter and A is the area of the watershed. With the use of increasing fuzzy linear membership, a Cc value close to 5.03 is given the lowest weight. As the Cc value decreases to 1.83, the weight decreases until it reaches zero.

2.4.8. Form Factor (Ff)

As Horton [11] defined, the form factor represents the relationship between a basin's area and length, squared. The equation for calculating the form factor is shown in Equation (8).

$$Ff = \frac{A}{L_b^2} \tag{8}$$

Here, A is the basin area in square kilometers, and L_b^2 is the square of the basin length. A perfectly circular basin will have a form factor value lower than 0.78. Basins with a lower form factor are considered to be more elongated and potentially influenced by tectonic activity. Using the decreasing fuzzy linear membership, a form factor value close to 0.61 will be given the lowest weight. As the form factor decreases to 0.05, the weight will increase until it reaches one.

2.5. Machine Learning Algorithms

Traditionally, geological research has faced challenges with data sources that are complex and imprecise due to the size and complexity of geological objects [82]. However, with advancements in science and technology, new methods have emerged that enhance the precision of geological data and increase the amount of available information.

Artificial intelligence (AI) is a field within computer science that focuses on developing intelligent systems created by humans. It can be separated into two types: strong AI and weak AI. Weak AI views the creation of reasoning and problem-solving machines as impossible, while strong AI aims to build machines with the ability to think and make decisions. The study of AI encompasses various areas, such as expert systems, machine learning, natural language processing, computer vision, and recommendation systems. Machine learning explicitly explores how computer systems can improve automatically through experience and the fundamental laws that govern all learning systems, including humans, organizations, and computers.

Machine learning algorithms allow us to gain new insights and capabilities. The advent of deep learning has revolutionized the field of AI by addressing the challenges faced in traditional machine learning, such as limited model options, time-consuming training processes, and the complexity of determining model parameters. As a result, deep learning has become a pivotal area of AI advancement.

Representation learning enables machines to automatically identify patterns in raw data and learn the necessary representations for detection or classification. Deep learning methods are a type of representation learning that involve multiple levels of abstraction created by combining simple, nonlinear modules. As these modules are layered, increasingly complex functions can be learned and implemented.

The advancement of big data and AI has opened new opportunities in the field of geology in recent years. Thanks to increased computing power, particularly the progress in GPU technology, the limitations of big data and AI in terms of computing have been significantly reduced. This expands the potential applications and growth prospects of geology. AI is commonly utilized for geological surveys and resource exploration, such as mineral recognition and geochemical anomaly detection. This article delves into the topic of big data and AI in geology.

2.5.1. Artificial Neural Networks (ANNs)

Artificial Neural Networks (ANNs) are computational systems that draw inspiration from functioning biological nervous systems such as the human brain. ANNs comprise many interconnected computational nodes, called neurons, that work together to learn from inputs and optimize the output.

The input, typically a multidimensional vector, is fed into the input layer and distributed to the hidden layers. These layers make decisions based on the previous layer and assess the impact of random changes on the final output, a process referred to as learning.

When multiple hidden layers are stacked, it is called deep learning. ANNs have several key features, including the ability to learn and adapt, generalize information, and process and analyze information uniformly and with error tolerance. These features make ANNs powerful tools for pattern recognition, classification, and nonlinear function estimation.

Activation functions are used in ANNs to transform the input into an output, which is then supplied as the input to the next layer. The weights of the connections between neurons in the layers are adjusted during the training phase, allowing the network to learn the patterns between inputs and outputs. ANNs can be divided into two learning models: supervised and unsupervised. In supervised learning, the network is trained with proper outputs for each input pattern. The weights are adjusted to minimize the error between the network output and the actual value. In unsupervised learning, the network discovers the relationship between the data patterns without needing actual responses.

ANNs can also be divided into two categories: recurrent and feed-forward networks. Recurrent networks have a feedback loop, while feed-forward networks do not. The neurons in each layer provide information to both the previous and the subsequent layers.

The ANNs offer these hypotheses:

- Data processing occurs in the units known as neurons. The neurons (or artificial neurons) present a model of brain neurons.
- The exchange of data is facilitated through communication between neurons.
- There is a weight for communicative ways between neurons.
- Every neuron utilizes a nonlinear function to process its inputs (weighted data), producing a specific output [82].

It is possible to identify a neural network via the communicative model between different layers of the network, the number of layers, the number of neurons, the neuron's operational function, and the learning algorithm. However, no general principle is available regarding the standard size of the network components. It is an innovative approach in most cases where the multilayered networks have a different amount of neurons in each layer, and different activation functions and various learning rates do the training of these networks. Then, it is followed by a selection of the best network. The network training in the learning phase takes place via the adjustment of weights so that outputs can be predicted or classified based on a set of inputs [83].

In this paper, the input parameters of the neural network included Basin Shape, Circularity Basin, Coefficient Compactness, Elongation Basin, Form Factor, Hypsometric Integral, River Sinuosity, Drainage Density, and Number Ruggedness, and the FAHP outputs were considered as the network output parameters. The data on these parameters were divided into training, testing, and data validation. A total of 80% of these data were used for training, 10% of data for validation and the other 10% for testing.

2.5.2. Support Vector Regression (SVR)

Considering a data set with N elements $\{(X_i,\ y_i)\}_{i=1}^{N}$, where $X_i = [x_{1,i},\ \dots,\ x_{n,i}] \in R^n$ and X_i represents the i_{th} element in a space with n dimensions, y_i ($y_i \in R$) indicates the actual value for X_i, the definition of a nonlinear function is as follows: $\varphi: R^n \rightarrow R^{nh}$. For mapping the entry data, X_i represents an R^{nh} space of high dimension known as feature space, which specifies the nonlinear transformation φ. Hence, a linear function f in a high-dimensional space, and consequently, the entry data, X_i can be related to output y_i. Equation (9) presents the linear function, i.e., SVR.

$$f(X) = W^T \cdot \varphi(X) + b \tag{9}$$

where $b \in R$ and $W \in R^n$, and $f(X)$ is the foretold value. As indicated in Equation (10), the empirical risk is minimized by the SVR.

$$R_{reg}(f) = C \sum_{i=1}^{N} \Theta_\varepsilon (y_i - f(X_i)) + \frac{1}{2} ||W^T|| \tag{10}$$

where the cost function is represented by $\Theta\varepsilon(yi - f(Xi))$, regarding the ε-SVR, as shown in Equation (11), a loss function ε-insensitive is utilized:

$$\Theta_\varepsilon(y - f(X)) = \begin{cases} |y - f(X)| - \varepsilon & \text{If} |y - f(X)| \geq \varepsilon \\ 0 & \text{In another case} \end{cases} \tag{11}$$

The nonlinear function φ is determined using Θ_ε in the R^{nh} space for finding a function with the ability to fit present training data with a deviation equal to or below ε. Using the mentioned function, the training error is minimized between the data training, and Equation (12) provides the function ε-insensitive [84].

$$\min_{W,b,\xi^*,\xi} R_{reg}(W, \xi^*, \xi) = \frac{1}{2} W^T W + C \sum_{i=1}^{N} (\xi_i^* + \xi_i) \tag{12}$$

The training errors of $f(X)$ and Y are punished by Equation (12) via the function ε-insensitive. Using the parameter C, the compromise between the points meeting condition $|f(X) - y| \geq \varepsilon$ in Equation (11) and the model complexity (vector W) is determined. With $C \to \infty$, a small model margin is observed that is adjusted to the data. When $C \to 0$, there is a large model margin, which is why it softens. Lastly, ξ_i represents errors more minor than $-\varepsilon$ and ξ_i^* indicates training errors larger than ε.

For solving the regression problem, the internal product of Equation (9) can be replaced by kernel $K()$ functions. Thus, this operation can be performed in a higher dimension by low-dimensional space data input regardless of knowledge of the transformation φ [85].

2.5.3. Multivariate Linear Regression (MLR)

The regression method is applied to two theories. Firstly, regression analysis is typically utilized for prediction and forecasting, and its application significantly overlaps the machine learning field. Secondly, it is possible to use regression analysis sometimes for determining causal relationships between the dependent and independent variables. Regression alone presents just relationships between a fixed dataset of different variables and a dependent variable.

Based on the regression models, the dependent variables are predicted by the independent variables. The value of the dependent 'y' variable is estimated by regression analysis because of the range of independent variable 'x' values. This articlediscusses polynomial and linear regression, which fit better into the predictive model. Regression could be multiple regression or a simple linear regression [86].

Simple linear regression is a statistical method used to model the relationship between a dependent variable and a single independent variable. An equation represents it, $y = \beta 0 + \beta 1x + \varepsilon$, where y is the dependent variable, x is the independent variable, $\beta 0$ and $\beta 1$ are coefficients, and ε represents the error term. The goal of simple linear regression is to determine the strength and direction of the relationship between the dependent and independent variables and estimate the effect of the independent variable on the dependent variable [86].

Multivariate linear regression is a statistical method for predicting the result of an answer variable, which uses some explanatory variables, as shown in Equation (13). Multiple Linear Regression (MLR) aims to establish a linear association between the dependent variable y and one or more independent variables x, which will then be analyzed.

$$\hat{\beta} = \left(X^T X\right)^{-1} X^T y \tag{13}$$

2.6. Integrating the FAHP and ML Algorithms

We first used the FAHP model to generate a target database suitable for our ML. This database served as the basis for training and testing the algorithms. Next, we randomly selected 212 out of 423 watersheds across our study area for training and testing the algorithms. By randomly selecting the watersheds, we ensured a representative sample of the study area. Here, raw data from eight different morphometric indices were used to train and test the machine learning algorithms, as shown in Figure 2. These indices were selected based on their relevance and importance in the morphotectonic characterization of watersheds. We then used these raw data as input features for the machine learning algorithms to generate predictions for the target database. Using a representative sample of the study area and including relevant input features, we aimed to create a distinct training network and improve the accuracy and precision of our predictions and our cost-effectiveness. Three distinct ML algorithms, namely, ANNs, SVR, and MLR methods, were employed for each set. It is important to note that the training and testing set includes 212 watershed indices. After optimizing and obtaining the best algorithms using statistical equations, the method was generalized for the whole domain.

To assess the effectiveness of the ANNs, SVR, and MLR techniques, Mean Squared Error (MSE), Mean Absolute Error (MAE), and Correlation Coefficient (R^2) were utilized as performance metrics. These metrics are described in Equation (14a–c) [82].

$$\text{MSE} = \frac{1}{N} \sum_{i=1}^{N} (y_i - \hat{y}_i)^2 \tag{14a}$$

$$\text{MAE} = \frac{1}{N} \sum_{i=1}^{N} |y_i - \hat{y}_i| \tag{14b}$$

$$R^2 = 1 - \frac{\sum_i (y_i - \hat{y}_i)^2}{\sum_i (y_i - \overline{y})^2} \tag{14c}$$

We used ANN for regression here and utilized the supervised method. Also, we use L2 (ridge) regularization to avoid overfitting. To have a faster converging, we need zero-centered data points, and each dimension should be scaled according to its standard deviation. Thus we normalize our data.

Figure 3 shows the performance of the ANN method for the training and validation sets. From epoch 200 and on, the trend was deemed acceptable. After completing the learning process, the ANN model demonstrated high accuracy with an R^2 value of 0.974, MSE of 4.14×10^{-6}, and MAE of 0.00151.

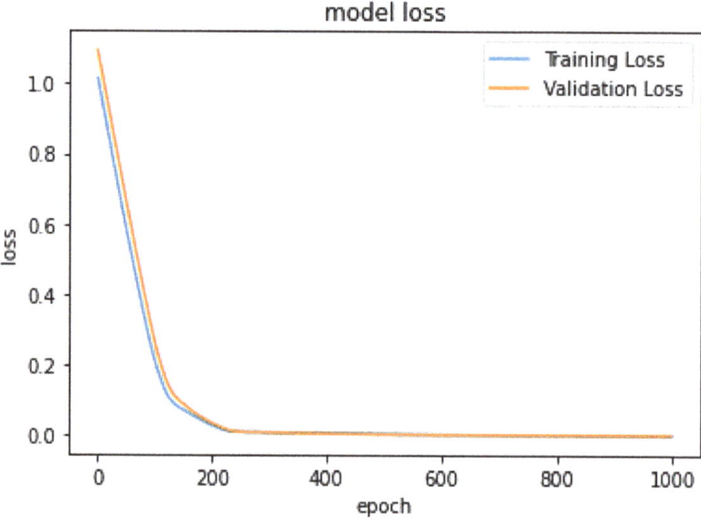

Figure 3. Training and validation loss for ANN model.

After developing and training the SVR model with the value of 0.947, MSE and MAE obtained 8.12×10^{-6} and 1.94×10^{-3}, respectively.

After developing and training the MLR model with the value of 0.967, MSE and MAE obtained 5.06×10^{-6} and 1.61×10^{-3}, respectively.

According to the simulation results obtained, there is a good agreement between the results obtained from ANN, SVR and MLR models but ANN model outperformed the other models. All three algorithms demonstrated similar error patterns in our study, with samples 16 and 96 displaying particularly large prediction errors. We have further investigated the reason for these high prediction errors and found that the location of these two samples may have contributed to the discrepancies. Specifically, sample 16 is located in the northernmost part of the study area, where geological and climatic conditions differ significantly from the rest of the watersheds. Similarly, sample 96 is situated in the southernmost part of the study area, which also has distinct geological and climatic conditions compared to the majority of the watersheds.

These differences in geological and climatic conditions may have played a role in the higher prediction errors observed for samples 16 and 96. Our findings suggest that when analyzing and predicting data in areas with distinct geological and climatic conditions, particular attention should be paid to samples that are situated in the outlier regions of the study area. By doing so, we may better understand and account for the differences in geological and climatic conditions and obtain more accurate predictions.

Overall, our study highlights the importance of considering the spatial distribution of data and the potential impact of varying geological and climatic conditions when making predictions. Further research may be necessary to determine how these findings can be applied in other contexts or how they can inform the development of more accurate predictive models.

3. Results and Discussion

ML techniques use algorithms to learn from data and make predictions or decisions without being explicitly programmed. Meanwhile, the fuzzy technique assigns degrees of truth to statements or rules, making it useful in assessing morphometry in watersheds where data such as rainfall patterns or soil properties can be difficult to quantify accurately. These techniques are useful in engineering problems where the relationship between input and output variables is complex and challenging to model analytically, such as landslides and mass movements [87,88].

However, both techniques have limitations in practical applications. Fuzzy logic relies heavily on expert knowledge to define rules and membership functions, resulting in models that are challenging to interpret and validate. ML techniques are limited by the quality and quantity of available data and the choice of algorithm and parameters used. Overfitting or underfitting data can result in poor predictive performance on new data. Successful application requires careful consideration of their limitations and appropriate use of available data and expertise.

Our study divided the dataset into three groups: a training dataset that accounted for 80% of the collected data, a validation dataset that accounted for 10%, and a testing dataset that comprised the remaining 10%. The performance of the machine learning methods on both the training and testing data is depicted in Figures 4–6. These figures compared the target (FAHP output) and output (predicted by algorithms) values and calculated the model error. Numerical results closely matched those obtained from the ANN, SVR, and MLR methods (Table 3). However, the ANN model showed superior performance compared to the other methods.

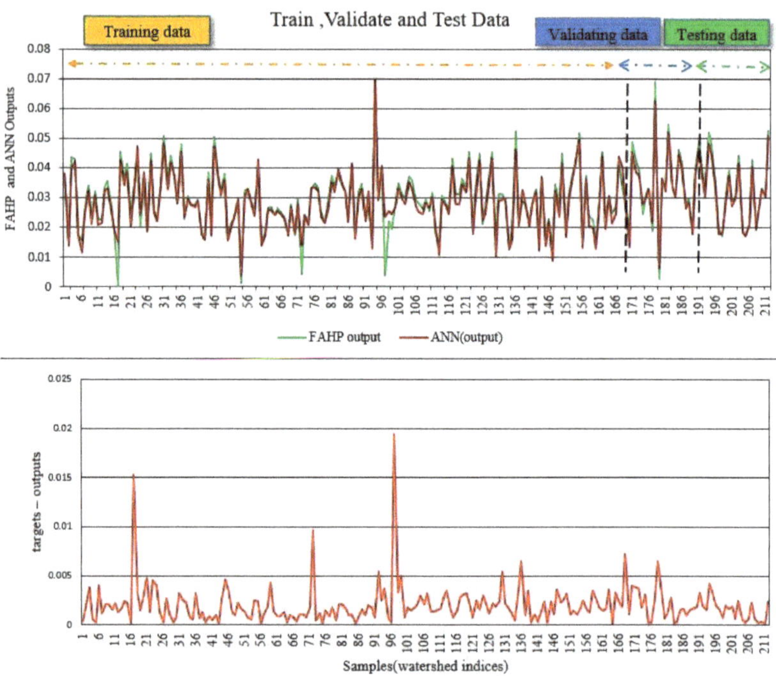

Figure 4. ANN performance for training, validating and testing data.

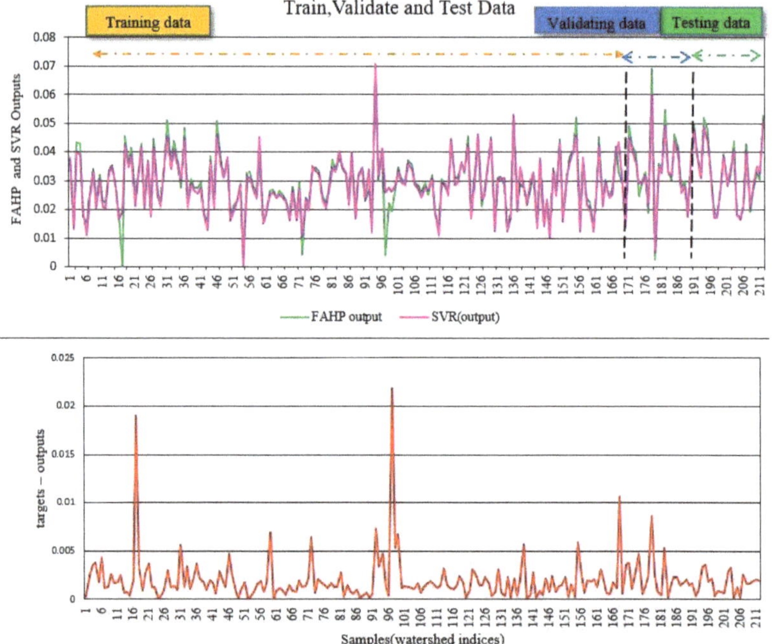

Figure 5. SVR performance for training, validating and testing data.

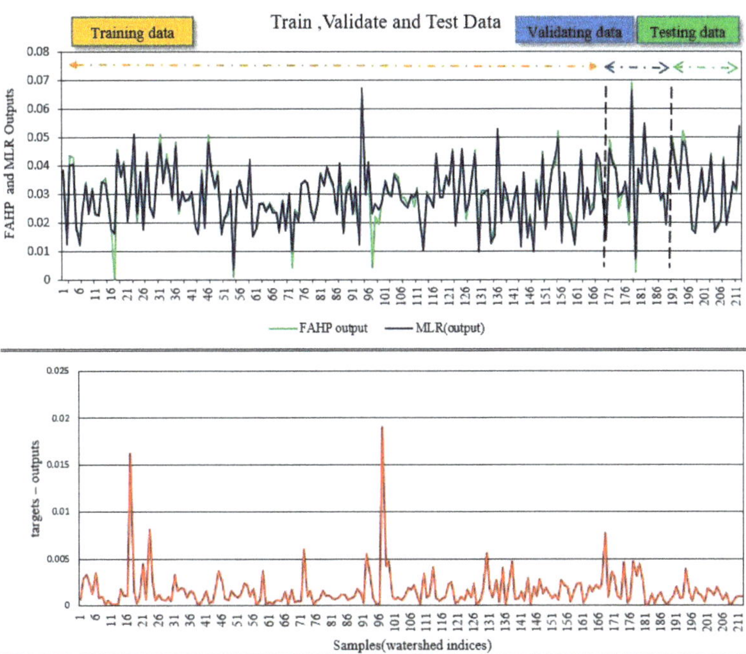

Figure 6. MLR performance for training, validating, and testing data.

Table 3. Performance of the machine learning algorithms.

Methods	MSE	MAE	R^2
ANN	4.14×10^{-6}	0.00151	0.974
SVR	8.12×10^{-6}	0.00194	0.947
MLR	5.06×10^{-6}	0.00161	0.967

We investigated the potential of coupled artificial intelligence algorithms and FAHP to predict the watershed's behavior in response to the region's tectonics. After selecting the best from among the ANN, SVR and MLR methods, the model was applied to whole watersheds in the Makran Subduction zone (Figure 7).

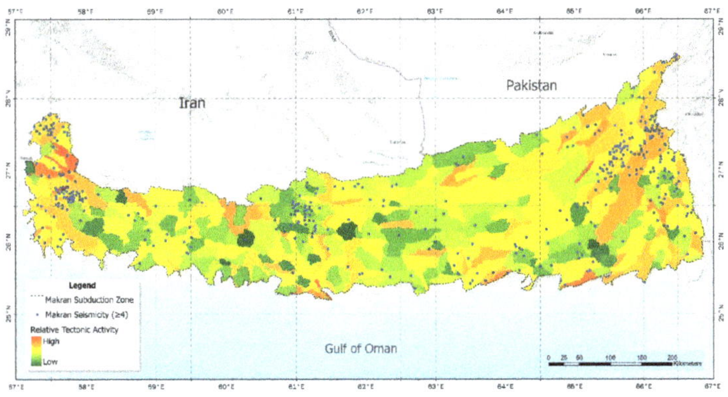

Figure 7. The result of applying the best method for whole watersheds in Makran subduction zone.

Mapping the results acquired by artificial intelligence algorithms for analog interpretation and comparing them with tectonic features and seismic events indicates the acceptable accuracy of the algorithms used. As shown in Figure 7, the findings measured for each watershed are illustrated in the range of low- to high levels of tectonic activity, which correlates well with the tectonic situation. On the west side of the figure, the location of the Minab fault zone is in good agreement with the area marked on the map as a zone with relatively high tectonic activity. Additionally, this region experiences the effects of two unique geological occurrences: the convergence of the Iranian and Arabian tectonic plates in the west and the subduction of the Oman sea plate beneath the Iranian continent in the east, both ongoing processes. On the eastern side of the region, it is observed that tectonic activity is estimated to be relatively high based on morphometric parameters.

This area aligns with the Chaman fault zone in Pakistan and marks the eastern boundary where the Oman oceanic plate is subducting beneath the Iranian continental plate. Moreover, instrumental epicenters also confirm the higher activity of this zone. The noteworthy point in this figure is the presence of instrumental epicenters in the central region, which are consistent with the results obtained using AI approaches, albeit with a slight shift to the left.

It is affected by the inclination of seismic faults, earthquakes, the geometry of the subducted oceanic plate, and instrumental error. The development of morphometric indicators can be affected by various factors, including geological heterogeneity, land use, and climate conditions. While high erosion rates in certain geological layers may contribute to the lack of clear development of tectonic indicators in some watersheds, it is important to note that this is only one potential explanation among many. Other factors, such as high levels of sedimentation, variations in precipitation patterns, and land use changes, could also contribute to the observed variability in technical indicator development. Additionally, interactions between multiple factors may contribute to the observed patterns.

Furthermore, it should be noted that seismic instrument data belong to only the last hundred years, whereas the morphometric features date back several million years which makes it more reliable. To better understand the underlying causes of the variability, it may be necessary to conduct further analyses that consider these different factors and potential interactions. It is also important to consider the limitations of the data and methods used in the study, as these may affect the accuracy and reliability of the technical indicators.

4. Conclusions

In conclusion, the morphometric analysis of watersheds in a region has the potential to reveal insights into long-term tectonic activity, and the present study aims to develop a novel approach to enhance this analysis. The proposed method leverages advanced machine learning techniques to reduce computational cost and time, especially in the case of large-scale regions. In this study, we examined the morphometry of watersheds in the Makran subduction zone using three artificial intelligence methods: ANN, SVR, and MLR. The results of our analysis indicate that the ANN method is the most accurate of the three, with a value of 0.974. The MSE and MAE values of the ANN method were also found to be lower compared to the other methods. Based on these findings, it can be concluded that the ANN approach can be effectively utilized in the morphometric analysis of watersheds and provides better results than previous techniques.

Author Contributions: Conceptualization, R.D. and M.Z.; methodology, M.Z. and V.N.; software, A.G.; validation, R.D., M.Z. and V.N.; formal analysis, A.R. (Ahmad Rashidi) and M.N.; investigation, S.S.; resources, R.D., M.Z. and A.R. (Ahmad Rashidi); data curation, R.D. and V.N.; writing—original draft preparation, R.D., M.Z., A.G. and V.N.; writing—review and editing, R.D., M.Z. and V.N.; supervision, R.D and A.R. (Amir Raoof); project administration, R.D. All authors have read and agreed to the published version of the manuscript.

Funding: This research received no external funding.

Data Availability Statement: Not applicable.

Conflicts of Interest: The authors declare no conflict of interest.

References

1. Christophe, O.; Olaniyi Ti, D. Applications of Geographical Information Systems (GIS) for Spatial Decision Support in Eco-Tourism Development. *Environ. Res. J.* **2010**, *4*, 187–194. [CrossRef]
2. Segura, F.S.; Pardo-Pascual, J.E.; Rosselló, V.M.; Fornós, J.J.; Gelabert, B. Morphometric Indices as Indicators of Tectonic, Fluvial and Karst Processes in Calcareous Drainage Basins, South Menorca Island, Spain. *Earth Surf. Process. Landf.* **2007**, *32*, 1928–1946. [CrossRef]
3. Juez, C.; Garijo, N.; Hassan, M.A.; Nadal-Romero, E. Intraseasonal-to-Interannual Analysis of Discharge and Suspended Sediment Concentration Time-Series of the Upper Changjiang (Yangtze River). *Water Resour. Res.* **2021**, *57*, e2020WR029457. [CrossRef]
4. Mesa, L.M. Morphometric Analysis of a Subtropical Andean Basin (Tucumán, Argentina). *Environ. Geol.* **2006**, *50*, 1235–1242. [CrossRef]
5. Kermani, A.F.; Derakhshani, R.; Bafti, S.S. Data on Morphotectonic Indices of Dashtekhak District, Iran. *Data Brief.* **2017**, *14*, 782–788. [CrossRef]
6. Rahbar, R.; Bafti, S.S.; Derakhshani, R. Investigation of the Tectonic Activity of Bazargan Mountain in Iran. *Sustain. Dev. Mt. Territ.* **2017**, *9*, 380–386. [CrossRef]
7. Aher, P.D.; Adinarayana, J.; Gorantiwar, S.D. Quantification of Morphometric Characterization and Prioritization for Management Planning in Semi-Arid Tropics of India: A Remote Sensing and GIS Approach. *J. Hydrol.* **2014**, *511*, 850–860. [CrossRef]
8. Salvany, J.M. Tilting Neotectonics of the Guadiamar Drainage Basin, SW Spain. *Earth Surf. Process. Landf.* **2004**, *29*, 145–160. [CrossRef]
9. Javed, A.; Khanday, M.Y.; Rais, S. Watershed Prioritization Using Morphometric and Land Use/Land Cover Parameters: A Remote Sensing and GIS Based Approach. *J. Geol. Soc. India* **2011**, *78*, 63–75. [CrossRef]
10. Rashidi, A.; Abbassi, M.-R.; Nilfouroushan, F.; Shafiei, S.; Derakhshani, R.; Nemati, M. Morphotectonic and Earthquake Data Analysis of Interactional Faults in Sabzevaran Area, SE Iran. *J. Struct. Geol.* **2020**, *139*, 104147. [CrossRef]
11. Horton, R.E. Erosional Development of Streams and Their Drainage Basins; Hydrophysical Approach to Quantitative Morphology. *Geol. Soc. Am. Bull* **1945**, *56*, 275–370. [CrossRef]
12. Strahler, A.N. Hypsometric (Area-Altitude) Analysis of Erosional Topography. *Geol. Soc. Am. Bull* **1952**, *63*, 1117–1142. [CrossRef]
13. Hack, J.T. Studies of Longitudinal Stream Profiles in Virginia and Maryland. In *USGS Professional Paper*; US Government Printing Office: Washington, DA, USA, 1957; Volume 294.
14. Ribolini, A.; Spagnolo, M. Drainage Network Geometry versus Tectonics in the Argentera Massif (French-Italian Alps). *Geomorphology* **2008**, *93*, 253–266. [CrossRef]
15. Bemis, S.P.; Micklethwaite, S.; Turner, D.; James, M.R.; Akciz, S.T.; Thiele, S.; Bangash, H.A. Ground-Based and UAV-Based Photogrammetry: A Multi-Scale, High-Resolution Mapping Tool for Structural Geology and Paleoseismology. *J. Struct. Geol.* **2014**, *69*, 163–178. [CrossRef]
16. Ozdemir, H.; Bird, D. Evaluation of Morphometric Parameters of Drainage Networks Derived from Topographic Maps and DEM in Point of Floods. *Environ. Geol.* **2009**, *56*, 1405–1415. [CrossRef]
17. Chorowicz, J.; Dhont, D.; Gündogdu, N. Neotectonics in the Eastern North Anatolian Fault Region (Turkey) Advocates Crustal Extension: Mapping from SAR ERS Imagery and Digital Elevation Model. *J. Struct. Geol.* **1999**, *21*, 511–532. [CrossRef]
18. Azarafza, M.; Azarafza, M.; Akgün, H.; Atkinson, P.M.; Derakhshani, R. Deep Learning-Based Landslide Susceptibility Mapping. *Sci. Rep.* **2021**, *11*, 24112. [CrossRef]
19. Ghosh, M.; Gope, D. Hydro-Morphometric Characterization and Prioritization of Sub-Watersheds for Land and Water Resource Management Using Fuzzy Analytical Hierarchical Process (FAHP): A Case Study of Upper Rihand Watershed of Chhattisgarh State, India. *Appl. Water Sci.* **2021**, *11*, 17. [CrossRef]
20. Kumar, R.; Dwivedi, S.B.; Gaur, S. A Comparative Study of Machine Learning and Fuzzy-AHP Technique to Groundwater Potential Mapping in the Data-Scarce Region. *Comput. Geosci.* **2021**, *155*, 104855. [CrossRef]
21. Zaresefat, M.; Ahrari, M.; Reza Shoaei, G.; Etemadifar, M.; Aghamolaie, I.; Derakhshani, R. Identification of Suitable Site-Specific Recharge Areas Using Fuzzy Analytic Hierarchy Process (FAHP) Technique: A Case Study of Iranshahr Basin (Iran). *Air Soil Water Res.* **2022**, *15*, 11786221211063849. [CrossRef]
22. Zaresefat, M.; Derakhshani, R.; Nikpeyman, V.; GhasemiNejad, A.; Raoof, A. Using Artificial Intelligence to Identify Suitable Artificial Groundwater Recharge Areas for the Iranshahr Basin. *Water* **2023**, *15*, 1182. [CrossRef]
23. Önüt, S.; Efendigil, T.; Soner Kara, S. A Combined Fuzzy MCDM Approach for Selecting Shopping Center Site: An Example from Istanbul, Turkey. *Expert. Syst. Appl.* **2010**, *37*, 1973–1980. [CrossRef]
24. Bui, D.T.; Shahabi, H.; Shirzadi, A.; Chapi, K.; Pradhan, B.; Chen, W.; Khosravi, K.; Panahi, M.; Bin Ahmad, B.; Saro, L. Land Subsidence Susceptibility Mapping in South Korea Using Machine Learning Algorithms. *Sensors* **2018**, *18*, 2464. [CrossRef]
25. Corsini, A.; Cervi, F.; Ronchetti, F. Weight of Evidence and Artificial Neural Networks for Potential Groundwater Spring Mapping: An Application to the Mt. Modino Area (Northern Apennines, Italy). *Geomorphology* **2009**, *111*, 79–87. [CrossRef]

26. Naghibi, S.A.; Pourghasemi, H.R. A Comparative Assessment between Three Machine Learning Models and Their Performance Comparison by Bivariate and Multivariate Statistical Methods in Groundwater Potential Mapping. *Water Resour. Manag.* **2015**, *29*, 5217–5236. [CrossRef]
27. Arabameri, A.; Saha, S.; Roy, J.; Chen, W.; Blaschke, T.; Bui, D.T. Landslide Susceptibility Evaluation and Management Using Different Machine Learning Methods in the Gallicash River Watershed, Iran. *Remote Sens.* **2020**, *12*, 475. [CrossRef]
28. Sarangi, A.; Madramootoo, C.A.; Enright, P.; Prasher, S.O.; Patel, R.M. Performance Evaluation of ANN and Geomorphology-Based Models for Runoff and Sediment Yield Prediction for a Canadian Watershed. *Curr. Sci.* **2005**, *89*, 2022–2033.
29. El Hamdouni, R.; Irigaray, C.; Fernández, T.; Chacón, J.; Keller, E.A. Assessment of Relative Active Tectonics, Southwest Border of the Sierra Nevada (Southern Spain). *Geomorphology* **2008**, *96*, 150–173. [CrossRef]
30. Dykstra, J.D.; Birnie, R.W. Reconnaissance Geologic Mapping in Chagai Hills, Baluchistan, Pakistan, by Computer Processing of Landsat Data. *Am. Assoc. Pet. Geol. Bull.* **1979**, *63*, 1490–1503. [CrossRef]
31. Kamali, Z.; Nazari, H.; Rashidi, A.; Heyhat, M.R.; Khatib, M.M.; Derakhshani, R. Seismotectonics, Geomorphology and Paleoseismology of the Doroud Fault, a Source of Seismic Hazard in Zagros. *Appl. Sci.* **2023**, *13*, 3747. [CrossRef]
32. Derakhshani, R.; Farhoudi, G. Existence of the Oman Line in the Empty Quarter of Saudi Arabia and Its Continuation in the Red Sea. *J. Appl. Sci.* **2005**, *5*, 745–752. [CrossRef]
33. Ghanbarian, M.A.; Derakhshani, R. The Folds and Faults Kinematic Association in Zagros. *Sci. Rep.* **2022**, *12*, 8350. [CrossRef]
34. Regard, V.; Hatzfeld, D.; Molinaro, M.; Aubourg, C.; Bayer, R.; Bellier, O.; Yamini-Fard, F.; Peyret, M.; Abbassi, M. The Transition between Makran Subduction and the Zagros Collision: Recent Advances in Its Structure and Active Deformation. *Geol. Soc. Lond. Spec. Publ.* **2010**, *330*, 43–64. [CrossRef]
35. Lawrence, R.D.; Khan, S.H.; Nakata, T. Chaman Fault, Pakistan-Afghanistan. *Ann. Tecton.* **1992**, *6*, 196–223.
36. Mokhtari, M.; Abdollahie Fard, I.; Hessami, K. Structural Elements of the Makran Region, Oman Sea and Their Potential Relevance to Tsunamigenisis. *Nat. Hazards* **2008**, *47*, 185–199. [CrossRef]
37. Kopp, C.; Fruehn, J.; Flueh, E.R.; Reichert, C.; Kukowski, N.; Bialas, J.; Klaeschen, D. Structure of the Makran Subduction Zone from Wide-Angle and Reflection Seismic Data. *Tectonophysics* **2000**, *329*, 171–191. [CrossRef]
38. Byrne, D.E.; Sykes, L.R.; Davis, D.M. Great Thrust Earthquakes and Aseismic Slip along the Plate Boundary of the Makran Subduction Zone. *J. Geophys. Res. Solid Earth* **1992**, *97*, 449–478. [CrossRef]
39. DeMets, C.; Gordon, R.G.; Argus, D.F.; Stein, S. Current Plate Motions. *Geophys. J. Int.* **1990**, *101*, 425–478. [CrossRef]
40. Vernant, P.; Nilforoushan, F.; Hatzfeld, D.; Abbassi, M.R.; Vigny, C.; Masson, F.; Nankali, H.; Martinod, J.; Ashtiani, A.; Bayer, R. Present-Day Crustal Deformation and Plate Kinematics in the Middle East Constrained by GPS Measurements in Iran and Northern Oman. *Geophys. J. Int.* **2004**, *157*, 381–398. [CrossRef]
41. Wong, I.G. Low Potential for Large Intraslab Earthquakes in the Central Cascadia Subduction Zone. *Bull. Seismol. Soc. Am.* **2005**, *95*, 1880–1902. [CrossRef]
42. Cruz, G.; Wyss, M. Large Earthquakes, Mean Sea Level, and Tsunamis along the Pacific Coast of Mexico and Central America. *Bull. Seismol. Soc. Am.* **1983**, *73*, 553–570. [CrossRef]
43. Gahalaut, V.K.; Catherine, J.K. Rupture Characteristics of 28 March 2005 Sumatra Earthquake from GPS Measurements and Its Implication for Tsunami Generation. *Earth Planet. Sci. Lett.* **2006**, *249*, 39–46. [CrossRef]
44. Bevis, M.; Taylor, F.W.; Schutz, B.E.; Recy, J.; Isacks, B.L.; Helu, S.; Singh, R.; Kendrick, E.; Stowell, J.; Taylor, B.; et al. Geodetic Observations of Very Rapid Convergence and Back-Arc Extension at the Tonga Arc. *Nature* **1995**, *374*, 249–251. [CrossRef]
45. Kawasaki, I.; Asai, Y.; Tamura, Y. Space-Time Distribution of Interplate Moment Release Including Slow Earthquakes and the Seismo-Geodetic Coupling in the Sanriku-Oki Region along the Japan Trench. *Tectonophysics* **2001**, *330*, 267–283. [CrossRef]
46. Zarifi, Z. Unusual Subduction Zones: Case Studies in Colombia and Iran. Ph.D. Thesis, The University of Bergen, Bergen, Norway, 2006.
47. Grando, G.; McClay, K. Morphotectonics Domains and Structural Styles in the Makran Accretionary Prism, Offshore Iran. *Sediment. Geol.* **2007**, *196*, 157–179. [CrossRef]
48. Snead, R.E. Recent Morphological Changes along the Coast of West Pakistan. *Ann. Assoc. Am. Geogr.* **1967**, *57*, 550–565. [CrossRef]
49. Wiedicke, M.; Neben, S.; Spiess, V. Mud Volcanoes at the Front of the Makran Accretionary Complex, Pakistan. *Mar. Geol.* **2001**, *172*, 57–73. [CrossRef]
50. JAXA ALOS Global Digital Surface Model "ALOS World 3D—30 m" (AW3D30). 2023. Available online: https://www.eorc.jaxa.jp/ALOS/en/dataset/aw3d30/aw3d30_e.htm (accessed on 23 March 2023).
51. Florinsky, I.V.; Skrypitsyna, T.N.; Luschikova, O.S. Comparative Accuracy of the AW3D30 DSM, ASTER GDEM, and SRTM1 DEM: A Case Study on the Zaoksky Testing Ground, Central European Russia. *Remote Sens. Lett.* **2018**, *9*, 706–714. [CrossRef]
52. Liu, K.; Song, C.; Ke, L.; Jiang, L.; Pan, Y.; Ma, R. Global Open-Access DEM Performances in Earth's Most Rugged Region High Mountain Asia: A Multi-Level Assessment. *Geomorphology* **2019**, *338*, 16–26. [CrossRef]
53. D'Apuzzo, L.; Marcarelli, G.; Squillante, M. Analysis of Qualitative and Quantitative Rankings in Multicriteria Decision Making. *New Econ. Windows* **2009**, *7*, 157–170. [CrossRef]
54. Darko, A.; Chan, A.P.C.; Ameyaw, E.E.; Owusu, E.K.; Pärn, E.; Edwards, D.J. Review of Application of Analytic Hierarchy Process (AHP) in Construction. *Int. J. Constr. Manag.* **2019**, *19*, 436–452. [CrossRef]
55. Lin, Q.; Wang, D. Facility Layout Planning with SHELL and Fuzzy AHP Method Based on Human Reliability for Operating Theatre. *J. Healthc. Eng.* **2019**, *2019*, 8563528. [CrossRef] [PubMed]

56. Musumba, G.W.; Wario, R.D. Towards Fuzzy Analytical Hierarchy Process Model for Performance Evaluation of Healthcare Sector Services. In *Information and Communication Technology for Development for Africa*; Mekuria, F., Nigussie, E., Tegegne, T., Eds.; Springer International Publishing: Cham, Switzerland, 2019; pp. 93–118.

57. Pourebrahim, S.; Hadipour, M.; Mokhtar, M.B.; Taghavi, S. Application of VIKOR and Fuzzy AHP for Conservation Priority Assessment in Coastal Areas: Case of Khuzestan District, Iran. *Ocean Coast. Manag.* **2014**, *98*, 20–26. [CrossRef]

58. Akbar, M.A.; Alsanad, A.; Mahmood, S.; Alothaim, A. Prioritization-Based Taxonomy of Global Software Development Challenges: A FAHP Based Analysis. *IEEE Access* **2021**, *9*, 37961–37974. [CrossRef]

59. Atıcı, U.; Adem, A.; Şenol, M.B.; Dağdeviren, M. A Comprehensive Decision Framework with Interval Valued Type-2 Fuzzy AHP for Evaluating All Critical Success Factors of e-Learning Platforms. *Educ. Inf. Technol.* **2022**, *27*, 5989–6014. [CrossRef]

60. Xie, J.; Liu, B.; He, L.; Zhong, W.; Zhao, H.; Yang, X.; Mai, T. Quantitative Evaluation of the Adaptability of the Shield Machine Based on the Analytic Hierarchy Process (AHP) and Fuzzy Analytic Hierarchy Process (FAHP). *Adv. Civ. Eng.* **2022**, *2022*, 3268150. [CrossRef]

61. Mohebbi Tafreshi, G.; Nakhaei, M.; Lak, R. Land Subsidence Risk Assessment Using GIS Fuzzy Logic Spatial Modeling in Varamin Aquifer, Iran. *GeoJournal* **2019**, *86*, 1203–1223. [CrossRef]

62. Bahrani, S.; Ebadi, T.; Ehsani, H.; Yousefi, H.; Maknoon, R. Modeling Landfill Site Selection by Multi-Criteria Decision Making and Fuzzy Functions in GIS, Case Study: Shabestar, Iran. *Environ. Earth Sci.* **2016**, *75*, 1–14. [CrossRef]

63. Saaty, T.L.; Vargas, L.G. Hierarchical Analysis of Behavior in Competition: Prediction in Chess. *Behav. Sci.* **1980**, *25*, 180–191. [CrossRef]

64. Argyriou, A. A Methodology for the Rapid Identification of Neotectonic Features Using Geographical Information Systems and Remote Sensing. A Case Study from Western Crete: Greece. Ph.D. Thesis, University of Portsmouth, Portsmouth, UK, 2012.

65. Pérez-Peña, J.V.; Azor, A.; Azañón, J.M.; Keller, E.A. Active Tectonics in the Sierra Nevada (Betic Cordillera, SE Spain): Insights from Geomorphic Indexes and Drainage Pattern Analysis. *Geomorphology* **2010**, *119*, 74–87. [CrossRef]

66. Walcott, R.C.; Summerfield, M.A. Scale Dependence of Hypsometric Integrals: An Analysis of Southeast African Basins. *Geomorphology* **2008**, *96*, 174–186. [CrossRef]

67. Dehbozorgi, M.; Pourkermani, M.; Arian, M.; Matkan, A.A.; Motamedi, H.; Hosseiniasl, A. Quantitative Analysis of Relative Tectonic Activity in the Sarvestan Area, Central Zagros, Iran. *Geomorphology* **2010**, *121*, 329–341. [CrossRef]

68. Chen, Y.C.; Cheng, K.Y.; Huang, W.S.; Sung, Q.C.; Tsai, H. The Relationship between Basin Hypsometric Integral Scale Dependence and Rock Uplift Rate in a Range Front Area: A Case Study from the Coastal Range, Taiwan. *J. Geol.* **2019**, *127*, 223–239. [CrossRef]

69. Liao, Y.; Zheng, M.; Li, D.; Wu, X.; Liang, C.; Nie, X.; Huang, B.; Xie, Z.; Yuan, Z.; Tang, C. Relationship of Benggang Number, Area, and Hypsometric Integral Values at Different Landform Developmental Stages. *Land Degrad. Dev.* **2020**, *31*, 2319–2328. [CrossRef]

70. Pande, C.; Moharir, K.; Pande, R. Assessment of Morphometric and Hypsometric Study for Watershed Development Using Spatial Technology—A Case Study of Wardha River Basin in Maharashtra, India. *Int. J. River Basin Manag.* **2018**, *19*, 43–53. [CrossRef]

71. Keller, E.A.; Pinter, N. *Active Tectonics, Earthquakes, Uplift, and Landscape*, 2nd ed.; Prentice Hall: Hoboken, NJ, USA, 2002.

72. Cheng, W.; Wang, N.; Zhao, M.; Zhao, S. Relative Tectonics and Debris Flow Hazards in the Beijing Mountain Area from DEM-Derived Geomorphic Indices and Drainage Analysis. *Geomorphology* **2016**, *257*, 134–142. [CrossRef]

73. Bahrami, S.; Capolongo, D.; Mofrad, M.R. Morphometry of Drainage Basins and Stream Networks as an Indicator of Active Fold Growth (Gorm Anticline, Fars Province, Iran). *Geomorphology* **2020**, *355*, 107086. [CrossRef]

74. Faghih, A.; Samani, B.; Kusky, T.; Khabazi, S.; Roshanak, R. Geomorphologic Assessment of Relative Tectonic Activity in the Maharlou Lake Basin, Zagros Mountains of Iran. *Geol. J.* **2012**, *47*, 30–40. [CrossRef]

75. Potter, P.E. A Quantitative Geomorphic Study of Drainage Basin Characteristics in the Clinch Mountain Area, Virginia and Tennessee. *J. Geol.* **1957**, *65*, 112–113. [CrossRef]

76. Strahler, A.N. Quantitative Geomorphology of Drainage Basins and Channel Networks, Part II. In *Handbook of Applied Hydrology*; McGraw-Hill: New York, NY, USA, 1964; pp. 4–39.

77. Sreedevi, P.D.; Subrahmanyam, K.; Ahmed, S. The Significance of Morphometric Analysis for Obtaining Groundwater Potential Zones in a Structurally Controlled Terrain. *Environ. Geol.* **2005**, *47*, 412–420. [CrossRef]

78. Sreedevi, P.D.; Owais, S.; Khan, H.H.; Ahmed, S. Morphometric Analysis of a Watershed of South India Using SRTM Data and GIS. *J. Geol. Soc. India* **2009**, *73*, 543–552. [CrossRef]

79. Strahler, A.N. Quantitative Analysis of Watershed Geomorphology. *Eos Trans. Am. Geophys. Union* **1957**, *38*, 913–920. [CrossRef]

80. Sharma, A.; Singh, P.; Rai, P.K. Correction to: Morphotectonic Analysis of Sheer Khadd River Basin Using Geo-Spatial Tools. *Spat. Inf. Res.* **2018**, *26*, 405–414. [CrossRef]

81. Bendjoudi, H.; Hubert, P. The Gravelius Compactness Coefficient: Critical Analysis of a Shape Index for Drainage Basins. *Hydrol. Sci. J.* **2002**, *47*, 921–930. [CrossRef]

82. Jalaee, M.S.; Shakibaei, A.; Ghaseminejad, A.; Jalaee, S.A.; Derakhshani, R. A Novel Computational Intelligence Approach for Coal Consumption Forecasting in Iran. *Sustainability* **2021**, *13*, 7612. [CrossRef]

83. Jalaee, S.A.; Shakibaei, A.; Akbarifard, H.; Horry, H.R.; GhasemiNejad, A.; Nazari Robati, F.; Amani zarin, N.; Derakhshani, R. A Novel Hybrid Method Based on Cuckoo Optimization Algorithm and Artificial Neural Network to Forecast World's Carbon Dioxide Emission. *MethodsX* **2021**, *8*, 101310. [CrossRef]

84. Shokri, S.; Sadeghi, M.T.; Marvast, M.A.; Narasimhan, S. Improvement of the Prediction Performance of a Soft Sensor Model Based on Support Vector Regression for Production of Ultra-Low Sulfur Diesel. *Pet. Sci.* **2015**, *12*, 177–188. [CrossRef]

85. Kazem, A.; Sharifi, E.; Hussain, F.K.; Saberi, M.; Hussain, O.K. Support Vector Regression with Chaos-Based Firefly Algorithm for Stock Market Price Forecasting. *Appl. Soft Comput. J.* **2013**, *13*, 947–958. [CrossRef]

86. Maulud, D.; Abdulazeez, A.M. A Review on Linear Regression Comprehensive in Machine Learning. *J. Appl. Sci. Technol. Trends* **2020**, *1*, 140–147. [CrossRef]

87. Lacasta, A.; Juez, C.; Murillo, J.; García-Navarro, P. An Efficient Solution for Hazardous Geophysical Flows Simulation Using GPUs. *Comput. Geosci.* **2015**, *78*, 63–72. [CrossRef]

88. Huang, C.; Sun, Y.; An, Y.; Shi, C.; Feng, C.; Liu, Q.; Yang, X.; Wang, X. Three-Dimensional Simulations of Large-Scale Long Run-out Landslides with a GPU-Accelerated Elasto-Plastic SPH Model. *Eng. Anal. Bound. Elem.* **2022**, *145*, 132–148. [CrossRef]

Article

Enhancing Wind Erosion Assessment of Metal Structures on Dry and Degraded Lands through Machine Learning

Marta Terrados-Cristos *, Francisco Ortega-Fernández, Marina Díaz-Piloñeta, Vicente Rodríguez Montequín and José Valeriano Álvarez Cabal

Project Engineering Department, University of Oviedo, 33004 Oviedo, Spain
* Correspondence: marta.terrados@api.uniovi.es

Abstract: With the increasing construction activities in dry or degraded lands affected by wind-driven particle action, the deterioration of metal structures in such environments becomes a pressing concern. In the design and maintenance of outdoor metal structures, the emphasis has mainly been on preventing corrosion, while giving less consideration to abrasion. However, the importance of abrasion, which is closely linked to the terrain, should not be underestimated. It holds significance in two key aspects: supporting the attainment of sustainable development goals and assisting in soil planning. This study aims to address this issue by developing a predictive model that assesses potential material loss in these terrains, utilizing a combination of the literature case studies and experimental data. The methodology involves a comprehensive literature analysis, data collection from direct impact tests, and the implementation of a machine learning algorithm using multivariate adaptive regression splines (MARS) as the predictive model. The experimental data are then validated and cross-verified, resulting in an accuracy rate of 98% with a relative error below 15%. This achievement serves two primary objectives: providing valuable insights for anticipating material loss in new structure designs based on prospective soil conditions and enabling effective maintenance of existing structures, ultimately promoting resilience and sustainability.

Keywords: wind erosion; degraded land; metal structures; abrasion; machine learning

Citation: Terrados-Cristos, M.;
Ortega-Fernández, F.; Díaz-Piloñeta,
M.; Montequín, V.R.; Cabal, J.V.Á.
Enhancing Wind Erosion Assessment
of Metal Structures on Dry and
Degraded Lands through Machine
Learning. *Land* **2023**, *12*, 1503.
https://doi.org/10.3390/
land12081503

Academic Editor: Chuanrong
Zhang

Received: 16 June 2023
Revised: 22 July 2023
Accepted: 26 July 2023
Published: 28 July 2023

1. Introduction

Wind erosion is a natural process that involves removal, transport and deposition of coarse and fine particles, primarily sand, by the wind [1]. Differences in atmospheric pressure generate air movements capable of eroding surface materials (also known as abrasion) when velocities reach sufficient levels [2]. The scientific community has increasingly recognized the significance of wind erosion due to its impact on soil health, agricultural production, climate and structures resilience [3]. Efforts have been devoted to simulating and predicting wind-driven effects, including soil erosion, to control land degradation and implement appropriate agricultural management practices [4]. Various methods, ranging from empirical equations for average soil erosion [5,6] to advanced models predicting crop yields and conservation of natural resources [7–9], have been developed.

However, wind erosion is gaining increasing relevance in other fields that have not been extensively studied. The durability of metal structures is greatly influenced by damage caused by wind erosion, particularly in degraded areas where wind-driven particle movement is more intense [10]. While the degradation of metal structures in outdoor conditions, both chemically and physically, is directly influenced by their geographical location [11], the attention has predominantly been on studying corrosion [12–14], with less emphasis on terrain-related abrasion, which holds relevance for achieving sustainable development goals and effective land planning.

Identifying and determining suitable soils for construction would facilitate their classification, allowing for redirection to alternative uses or assigning specific wear values, aligning with the objectives of sustainable development, and minimizing material wastage. This process results in significant economic, social, and environmental losses, affecting various metal constructions.

Windblown sand transport is characterized by three types of movement based on grain diameter (d): suspension (d < 0.07 mm); saltation (0.07 < d < 0.5 mm); and creep (d > 0.5 mm) [15] (Figure 1). Among these, saltation plays a crucial role in the total mass of sand transported, driven by wind shear forces on the land surface that lead to the rebound of sand particles and horizontal sand mass flow in the downwind direction [16,17].

Figure 1. Windblown sand transport modes.

Although wind erosion can occur in all climates, it is more prevalent in semi-arid and arid environments characterized by extensive land degradation or dry conditions [18]. As a result, metal structures were historically not exposed to this problem. However, the proliferation of constructions in these areas, including new cities [19] and the development of renewable energy projects [20,21], has brought wind erosion into focus. Approximately one-fifteenth of the Earth's surface is susceptible to significant sand blowing [16] and the expansion of wind erosion-prone areas is expected due to climate change [22].

Factors influencing the movement of sand and hazardous particles by wind include specific particle size distribution, extensive plain lands without vegetation or wind barriers, high wind speeds combined with low relative humidity and elevated concentrations of total suspended particulate matter [23]. In contrast, as height increases, the negative impact of the process becomes less severe due to its inherent characteristics, as higher altitudes result in fewer particles reaching the area [24,25]. The parameters that influence erosion can be categorized into three main groups.

1. Impact conditions, which include the velocity and angle of impact;
2. Characteristics of the eroding particle, such as its size, shape, and other parameters;
3. Properties of the material being eroded, including its ductility, hardness, density, and other relevant factors.

Understanding how land conditions affect infrastructure in the long term is crucial for the design and maintenance of both new and historic buildings. The maintenance of structures in aggressive environments, such as the north-west coast of Egypt exposed to sandstorms, presents significant challenges [26]. Wind erosion implications for high-speed lines in Saudi Arabia are also garnering attention [27]. Researchers at the Inner Mongolia University of Technology have studied the impact of wind erosion on steel structure coatings in central and western regions of Mongolia affected by sandstorms [28,29]. However, the design, analysis, and evaluation of wind erosion processes are still in the early stages of study.

Common responses to wind erosion include increasing protection and coating of materials, which is prevalent in the wind and aeronautics industry, with research exploring multilayer coatings and alloys [30,31]. Prior knowledge during the design or engineering phase is essential for sustainability as it facilitates calculations that help to mitigate the economic and environmental implications of excessive material waste [32]. Other studies have focused on soil treatment solutions, such as protective barriers [33] or surface treatments [34–36], but implementing these solutions on larger surfaces is challenging. The current approach to studying wind erosion often relies on localized and case-specific investigations, compounded by a lack of standardized terminology in the literature. These factors pose challenges in unifying the research efforts and effectively addressing the issue. Therefore, it is imperative to establish methods for determining and predicting the extent of wind erosion-induced abrasion on structures to enable the implementation of appropriate preventive measures.

The objective of this study is to develop a predictive machine learning model capable of determining the erosion rate experienced by metal structures based on their geographical location. By integrating data from various sources, including existing studies and experimental data, the model aims to provide insights into potential degradation associated with the surrounding land. These insights enable to design environmentally conscious structures, optimize material usage, and extend the lifespan of metal structures through careful maintenance planning and preventive measures.

This paper presents a detailed description of the methodology employed, starting with the creation of a robust database serving as the foundation for training the predictive models. The database comprises information sourced from existing studies in the literature. Given the limited literature data available, a specific and comprehensive dataset was generated, incorporating a wider range of materials and measurable variables obtained through direct impact tests conducted in a laboratory setting. Subsequently, the modelling techniques and evaluation methods utilized throughout this study are elucidated. Finally, the results are thoroughly analysed, and the conclusions drawn from this research are presented.

2. Materials and Methods

The methodology employed in this study is outlined in Figure 2 and encompasses three key phases. The initial phase involved the creation of a database, which serves as the key point for the application of predictive algorithms that facilitate the estimation of erosion rates for specific metals under different conditions and types of terrain. Subsequently, in the second phase, the model was developed based on the analysis of the compiled data. Finally, in the third phase, the model's efficacy was evaluated through validation procedures, and the obtained results were assessed.

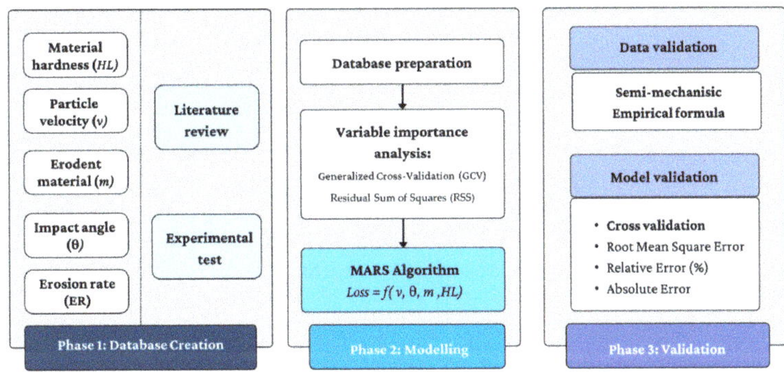

Figure 2. Overall process followed.

2.1. Phase 1: Database Creation

For this first phase, two main sources of data were used: external data derived from international literature, and internal data acquired from experimental laboratory tests. The first source involved assembling the cases and analysis of the relevant information in the literature related to the study topic. Additionally, several laboratory tests were carried out in order to expand the information with our own experimental data.

2.1.1. Literature Review

Erosion is a phenomenon influenced by multiple factors, including the properties of both the material being eroded and the material causing the erosion, as well as the conditions under which the phenomenon occurs. Table 1 summarises the most significant variables considered in the literature.

Table 1. Most significant variables of the direct impact test.

Process Parameters	Eroded Material Parameters
Impact angle [37]	Hardness [38]
Particle diameter of impacting particles [39]	Fracture toughness [38]
Impact velocity [40]	Elastic modulus [37]

However, to attempt a macroscopic approach and ensure that the model is truly useful and applicable to any case study, the variables that form the model should be readily available or easily obtainable. Therefore, the variables collected were selected based on their availability and significance according to the literature.

1. Material hardness (*HL*): Studies agree that material hardness is a highly influential variable in calculating wind erosion [41];
2. Particle velocity (*v*): It is key point to determine the force with which particles impact the structure, as abrasion increases with higher particle velocities [41];
3. Amount of erodent material (*m*): The quantity of material impacting the structure directly influences the level of abrasion [41];
4. Impact angle (*θ*): Studies have shown that for ductile materials as metallic structures, the highest abrasion damage occurs at impact angles between 15 and 30 degrees and decreases towards 90 degrees [42];
5. Erosion rate (*ER*): The majority of scientific literature describes wind erosion using the erosion ratio which is usually measured as follows (1) [43,44]:

$$ER = \frac{Mass\ of\ material\ lost\ due\ to\ erosion}{Mass\ of\ material\ eroded} \tag{1}$$

Measuring the impact in this way, instead of using mass loss, has the advantage of allowing better comparison of erosion between different materials [45]. At this point, all experimental studies in the literature that aim to characterize the effect of different parameters on erosion and erosion resistance of various materials were collected. These studies typically involve conducting tests with sand or other particles and measuring the impact [42,46].

The database consists of 778 data points. The dataset, comprising data from different laboratory tests, undergoes thorough pre-processing to handle missing values, outliers, and inconsistencies. Standardisation of measurement units is applied to facilitate meaningful comparisons, while min-max scaling rescales the variables for analysis. Categorical data are appropriately encoded, and the normalized data from various sources are integrated into a unified dataset stored as relational data in a CSV (comma-separated values) format.

2.1.2. Experimental Test

Experimental data were obtained by conducting various laboratory tests. The analysed and collected variables were the same as those identified as relevant in the literature review. The procedure for obtaining each of these experimental data is specified below.

Material Hardness (*HL*)

Hardness tests were performed on plates made of different materials using the Leeb hardness test. The Leeb hardness (*HL*) [47] relates the rebound velocity to the impact velocity of a spherical device, with a diameter of 3 mm or 5 mm (2).

$$HL = \frac{rebound\ velocity}{impact\ velocity} \times 1000 \tag{2}$$

The tests were performed according to the following standards: ASTM A956/A956m–17a, Standard Test Method for Leeb Hardness Testing of Steel Products and ISO 16859-1/2/3:2015, and Metallic materials–Leeb hardness test [48,49].

Particle Velocity (*v*)

Velocity can be adjusted based on factors such as the pressure of the compressor, atmospheric pressure, and the diameter of the nozzle. By measuring the air velocity, we can estimate the particle velocity and determine its range of values. According to studies in the literature, the relationship with the velocity of the carrier fluid itself is estimated to be less than one-third [42].

Amount of Erodent Material (*m*)

The material impacting the structure can be estimated based on the concentration of erodent material in the air (expressed in micrograms per cubic meter) (m_a), multiplied by the wind velocity (in meters per second) (v_w); the duration of impact per year (in hours) (*d*); and the surface area (in square meters) (*s*) (3). At a laboratory level, the amount of sand is determined via weighing.

$$m = m_a * v_w * d * s \tag{3}$$

Impact angle (*θ*)

The impact angle (*θ*) can be determined by comparing the orientation of the structure with the dominant wind direction. At a laboratory level, the impact angle can be set by sample's colocation.

Erosion Rate (*ER*)

Erosion rate was determined by conducting direct impact tests according to the ASTM G76-2013 standard [50]. A total of 216 tests were conducted, involving 12 different types of materials, including bare steel, stainless steel, galvanized steel, aluminium, and tinplate. Each material underwent 3 repetitions of the test. The tests were performed using 3 batches of 300 g of sand, resulting in a total of 900 g of eroding particles. Two different sizes of sand were used (150 and 300 μm).

All these tests were carried out in a sandblasting cabin (CHC60, PA, Spain) equipped with a sandblasting gun operated with ceramic nozzles. The required airflow rate of 340 L/min was achieved using a compressor (METALWORKS 458804090, PA, Spain). Figure 3 shows an outline of the testing procedure. To separate the sand into different particle sizes, a sieve shaker (CISA BA200N, PA, Spain) was employed.

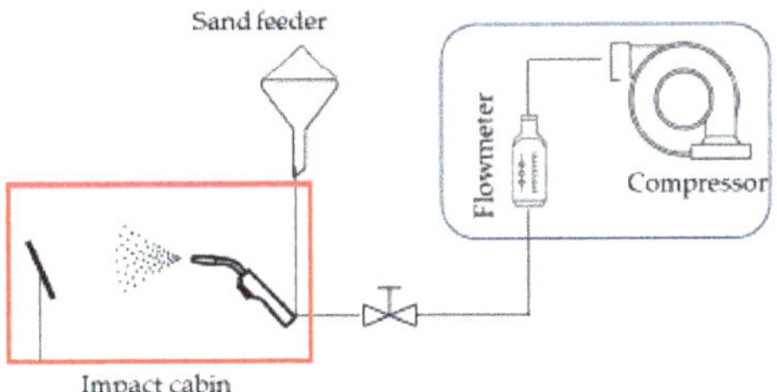

Figure 3. Schematic representation of laboratory tests performed.

The plates were weighed before and after each sand batch using Laboratory Precision Balance (Raswag AS 310 R2 PLUS, PA, Spain) to determine the mass loss. By comparing the final weight with the initial weight, the mass loss caused by the impact was determined, providing valuable information about the energy absorption capacity of the samples and the erosion ratio (*ER*).

2.2. Phase 2: Modelling

Once all the data are collected, complementing the information from the literature with experimental test, the modelling stage began. The collected data from both sources underwent a thorough cleaning and pre-processing process to ensure data quality and consistency. An exploratory analysis was conducted to understand the data structure and identify patterns. Relevant variables were selected for predictive models. Two methods are used to determine the importance of each variable in the model: generalized cross-validation (GCV) and residual sum of squares (RSS).

- Generalized cross-validation (GCV): It involves fitting the model with all variables, calculating GCV scores by temporarily excluding each variable, and ranking them based on their scores. Variables with higher GCV scores are considered more important;
- Residual sum of squares (RSS): It calculates the sum of the squared differences between the observed values and the predicted values obtained by the model. The RSS represents the overall amount of unexplained variation in the data. A lower RSS indicates a better fit of the model to the data.

The database was then prepared for model construction by partitioning the data and handling missing values. These steps ensured the integrity of the data and facilitated the construction of accurate predictive models.

The modelling stage is carried out using the MARS algorithm (multivariate adaptive regression splines). This algorithm is an effective tool for constructing accurate and robust predictive models from complex datasets. MARS algorithm enables the identification of nonlinear and nonparametric relationships among variables, which is particularly useful in the study of direct impact where relationships can be highly nonlinear. This machine learning technique combines linear regression with non-linear functions called splines. It begins by constructing an initial linear model and then adds splines to capture non-linear relationships in the data. It uses an iterative approach to improve the fit and selects the most relevant variables [51]. Ultimately, a flexible model is obtained that combines both linear and non-linear terms to predict a continuous response variable [52].

The MARS algorithm is capable of predicting the amount of material that can be lost due to abrasion, as shown in Equation (4) in the following form:

$$Loss\ (g) = f(v,\ \theta,\ m,\ HL) \tag{4}$$

where

- v : Particle velocity (m/s);
- θ: Impact angle (°);
- m: Mass of sand (g);
- HL: Material hardness.

2.3. Phase 3: Validation

Validating the obtained results is crucial to ensure the reliability and generalizability of the developed models. In this methodology, two validation phases are conducted: data validation and model validation.

2.3.1. Data Validation

To validate the obtained results, it is proposed to employ an empirical semi-mechanistic erosion equation [37]. This formula is based on theoretical principles and physical laws related to direct impact. By comparing the data with the values calculated, the consistency and validity of the obtained results can be evaluated.

The erosion damage is caused by two mechanisms: cutting (ER_C) (5) and deformation (ER_D) (6). Therefore, the total erosion damage is given by the sum of both terms.

$$ER_C = \begin{cases} C_1 F_s \dfrac{U^{2.41}\sin(\theta)[2\,K\cos(\theta)-\sin(\theta)]}{2K^2} & \theta < tan^{-1}(K) \\[2ex] C_1 F_s \dfrac{U^{2.41}\cos^2(\theta)}{2} & \theta > tan^{-1}(K) \end{cases} \tag{5}$$

$$ER_D = C_2 F_s \frac{(U\sin(\theta) - U_{tsh})^2}{2} \tag{6}$$

where

- U_{tsh} is the threshold velocity below which deformation is negligible;
- F_s is the angularity factor of the particle, ranging from 0.25 for completely rounded particles to 1 for very angular particles. In this case, Fs was considered as 0.5;
- K is the ratio between the contact area in the x-direction and the contact area in the y-direction of the particle with the material. In most materials eroded by sand, it is 0, so is the ratio used in this study;
- C is the cutting constant, which depends on the hardness of the material. It has been shown to be proportional to the inverse square root of materials hardness [42];
- U is the initial velocity of the particle. According to experimental studies, the average relationship between particle velocity and gas velocity is 3.1739 [42];
- θ is the impact angle, considered perpendicular in this case.

2.3.2. Model Validation

Cross-validation is a widely used technique for evaluating the performance of predictive models. In this context, the dataset is divided into training (75%) and testing (25%) subsets. The model is trained using the training subset, and its performance is evaluated using the testing subset. This process is repeated several times (6 blocks), alternating the training and testing subsets, and an average performance measure is calculated to assess the model's generalization capability, based on the following.

- The root mean square error (RMSE) measures the average magnitude of the residuals (differences between predicted and actual values). A lower RMSE indicates a better fit between the model and the observed data;
- Relative error measures the percentage difference between the predicted and actual values, providing insight into the relative accuracy of the model's predictions;
- Absolute error represents the absolute difference between predicted and actual values, giving an indication of the magnitude of the prediction errors;
- Mean directly compares the values, indicating the overall bias of the model.

3. Results

The results are presented in detail throughout the different phases of the proposed methodology.

3.1. Phase 1: Database Creation

After an exhaustive study of the scientific literature and analysis of direct impact tests from research such as [46,53,54], the data and variables that align with the context of the object of this study are collected, analysed, identified, and added. A total of 778 initial data points were collected before eliminating and cleaning the database. The collected parameters and the range of values studied are summarised in the following Table 2.

Table 2. Values in the study variables: range, mean and standard deviation (Sd).

Material Hardness [-]			Particle Velocity [m/s]			Amount of Erodent Material [g]			Impact Angle [°]		
Range	Mean	Sd	Range	Mean	Sd	Range	Mean	Sd	Range	Mean	Sd
395–710	193.64	87.49	9.2–32.56	19.35	7.29	300–1800	670.65	343.25	15–90	49.22	25.71

The distribution of these variables is shown in the form of box plots in Figure 4.

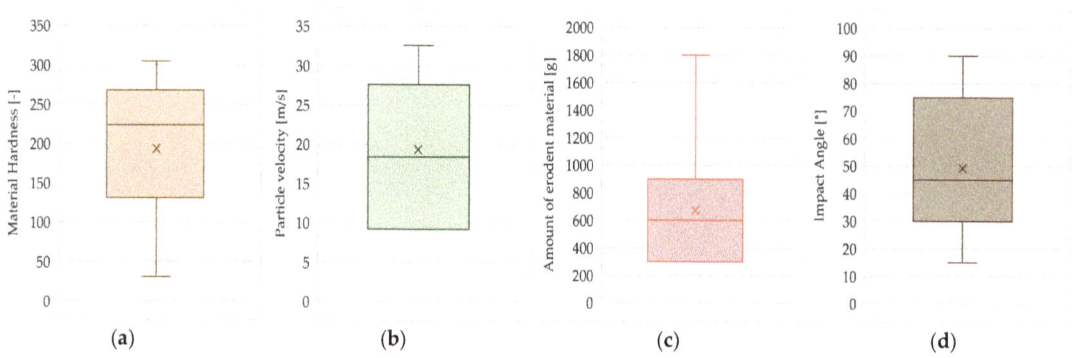

Figure 4. Variable distribution: (**a**) material hardness, (**b**) particle velocity, (**c**) amount of erodent material, (**d**) impact angle.

On the other hand, the experimental tests were conducted under normal pressure and temperature conditions. The eroding material particles, in this case sand, had diameters of 150 μm and 300 μm and were propelled at a velocity ranging between 13 and 14 m/s.

Upon the completion of the impact tests, clear surface deformation was observed in the samples. Furthermore, evident surface changes were measured, indicating the influence on the structure and external appearance of the samples, suggesting the need for further detailed analysis. Some examples of the experimental test results are shown in Figure 5.

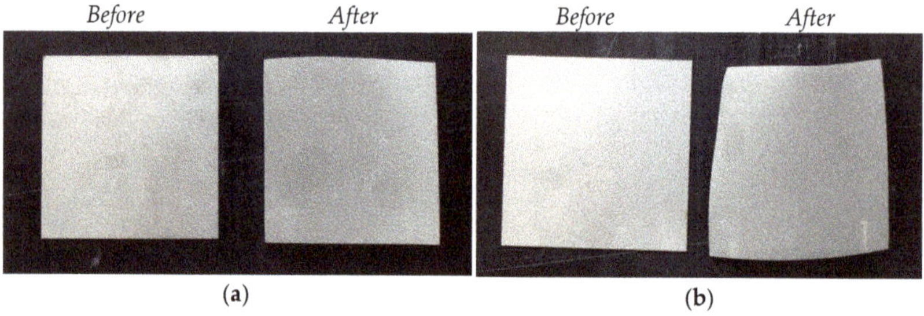

Figure 5. Metal samples before and after direct impingement tests. (**a**) Stainless steel, (**b**) galvanized steel.

It was observed that some plates, such as aluminium, showed mass gains of up to 0.05%. This phenomenon can be attributed not only to the absence of significant wear but also to the embedding of sand particles in the material. This phenomenon was also observed in tinplate samples. The remaining plates exhibited mass losses ranging from approximately 0.20% to 0.30%, except for galvanized steel, which showed losses of 0.99%.

The radial chart in Figure 6 displays the average values of each of the 12 materials under different test conditions. Mass loss after impact for the three defined amounts of sand, as well as the total mass loss, is shown in four different colours. In this following chart, the axes extend outward from the centre and the magnitude of the mass loss is represented on each axis using dots or lines.

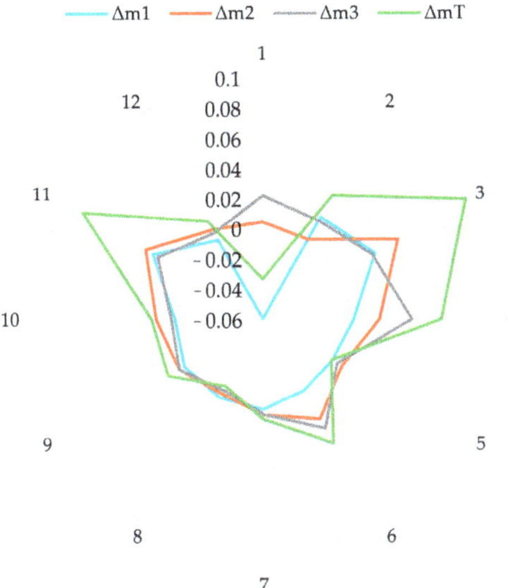

Figure 6. Representation of the average mass change in each study condition.

By comparing the mass losses among the different amounts of sand (300 g (Δm1), 600 g (Δm2) and 900 g (Δm3)), patterns or trends can be identified. The chart shows that as the amount of sand increases, the mass loss also increases, except for materials where sand particles become embedded due to their low hardness. Additionally, the chart presents the total mass loss as a consolidated measure across all amounts of sand.

On the other hand, Figure 7 provides information about the distribution and variability of hardness values. It can be observed that the majority of values are within a close range, with a single outlier, corresponding to aluminium.

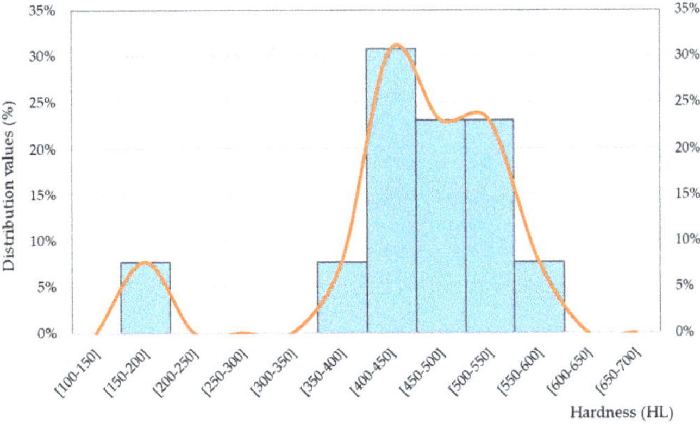

Figure 7. Representation of the hardness distribution of the tested materials.

The Pearson correlation coefficient obtained between mass change and hardness is 0.28, indicating a moderate positive correlation between hardness data and mass loss. The *p*-value of 0.361 suggests that this correlation is not statistically significant at a significance level of 0.05. It is important to note that other factors or variables not considered in this analysis could have a more relevant influence on the results. Therefore, further comprehensive studies are recommended to better understand the nature and strength of the relationship between the variables in question.

3.2. Phase 2: Modelling

Once the database is prepared, the predictive algorithm is applied to create a model for predicting the material loss (mass loss) that a metal structure will experience under those conditions.

Two methods are used to determine the importance of each variable in the model: generalized cross-validation (GCV) and residual sum of squares (RSS). The most significant variables, in order, are shown in Table 3.

Table 3. Importance of each variable determined via GCV and RSS.

	GCV	RSS
Velocity	100	100
Impact Angle	76.2	76.2
Amount of Sand	62.9	62.9
Material Hardness	27	27.8

Velocity of impact is the most relevant factor according to both methods. Furthermore, the values obtained for each of the variables according to the two methods are similar and coherent with each other. Hence, these variables can be deemed as valid and integrated into the predictive model.

3.3. Phase 3: Validation

Figure 8 displays the results after validating the data obtained empirically through experimental trials and the data calculated using well-established equations in the scientific community. The dashed line represents the ideal situation for these values. Each set of experiments samples is represented by a unique colour. It can be observed that there are no significant deviations between the theoretical and practical values, and the differences are acceptable ($R^2 = 0.9207$). Therefore, these results can be considered valid and incorporated into the predictive model.

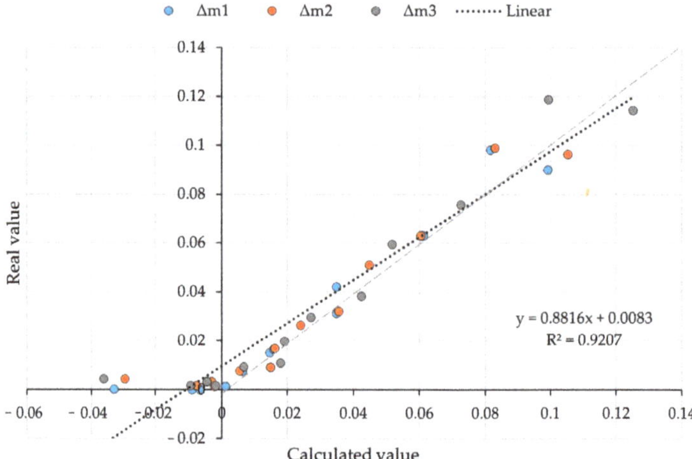

Figure 8. Comparison between experimentally and theoretically obtained results.

In Figure 9, the predicted values are represented on the vertical axis, while the actual values are shown on the horizontal axis.

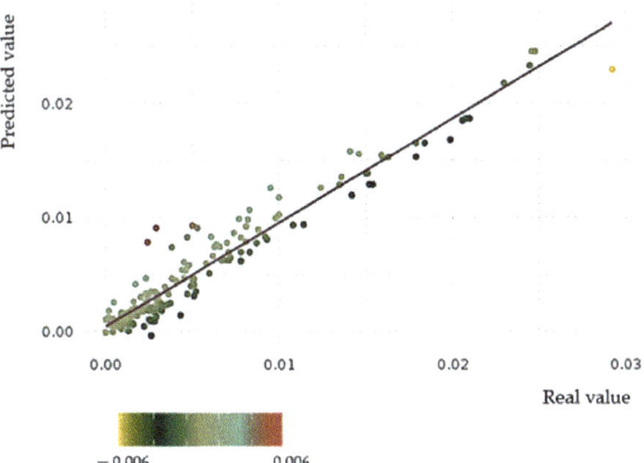

Figure 9. Abrasion model results.

Ideally, the points in this plot should be distributed along the diagonal line, indicating an exact correspondence between the model's predictions and the actual values. In this case, a high correlation is observed between the predicted and actual values, as most of the points are close to the diagonal line ($R^2 = 0.9083$). This demonstrates that the MARS model is capable of generating accurate estimations of mass loss based on the study parameters.

The proximity of the points to the diagonal line also suggests that the model generalizes well, meaning it can provide accurate predictions even for data not used during the model's training. This ability to generalize is essential to ensure the applicability and reliability of the model in practical situations.

The residuals represent the differences between the predicted and actual values of mass loss based on the study parameters. In a precise and reliable model, the residuals should be randomly distributed around zero and show no systematic trend.

In Figure 10, a homogeneous distribution of residuals around zero is observed, indicating that the MARS model can capture the variability in the data, adequately adjusting to the patterns of mass loss.

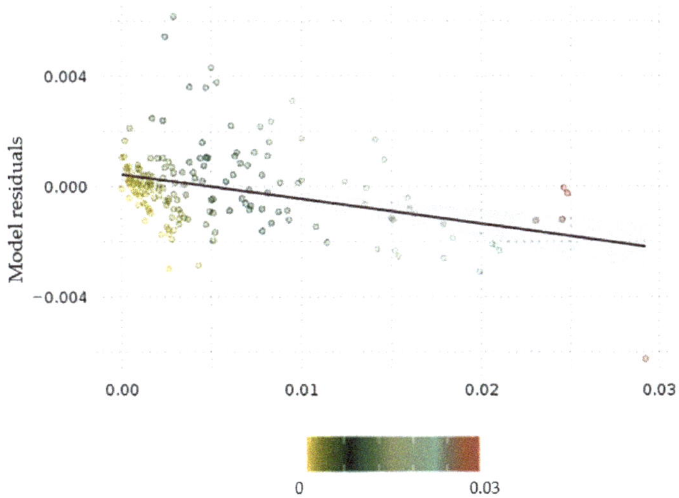

Figure 10. Residual plot of the MARS model.

The root mean square error (RMSE) used in this case to measure the differences between the predicted values of the model and the actual values has a value of 0.005587. Table 4 shows a comparison between the relative error, absolute error, the percentage predicted through the model, and an example of what it would be using the mean value.

Table 4. Relative error, absolute error, and mean error of the model.

Relative Error (%)	Absolute Error	Mean (%)	Model (%)
1	0.000292	1.12	25.7
5	0.00146	15.08	77.09
10	0.00292	38.55	94.41
14	0.004088	58.1	97.77
20	0.00584	85.47	98.88
25	0.0073	86.59	100
Inf	Inf	100	100

These results provide an assessment of the model's performance in predicting the abrasion values. The RMSE value indicates the average difference between the predicted and actual values, with lower values indicating better accuracy. The table presents the relative and absolute errors for different percentages, comparing the model's predictions to the mean value. It can be observed that the model's predictions have significantly lower errors compared to using the mean value, demonstrating its effectiveness in estimating the abrasion values. For a relative error of less than 15%, the model shows an efficiency of 98% accuracy.

4. Discussion

4.1. Interpretation of Results

The results of this study highlight the importance of considering the conditions and characteristics of the surrounding terrain when designing and maintaining outdoor metal structures on dry and degraded lands. This study emphasizes that wind erosion can lead to significant degradation of metal structures in such environments, a factor often overlooked during the design process. The developed predictive model incorporating data from various sources provides valuable insights into the potential material degradation and erosion experienced by these structures. The findings underscore the significance of including terrain-related parameters as essential factors in the design and maintenance practices for outdoor metal structures.

The compilation of a comprehensive database from the existing literature and the inclusion of experimental data from direct impingement tests on metal plates subjected to high-pressure air and sand impacts the study's findings. The experimental tests revealed mass losses ranging from 0.20% to 0.99% for different metal plates. It was interesting to observe that certain plates, such as aluminium and various types of tinplate, showed mass gains, likely due to minimal wear and the embedding of sand particles. These observations underscore the complexity of abrasion processes and highlight the need for a more nuanced understanding of material responses under different impact conditions.

4.2. Implications and Applications

The study's implications are significant for the construction industry and outdoor metal structure maintenance. By incorporating information about the land and drylands circumstances and environmental factors into the design process, engineers and designers can better anticipate and mitigate potential material loss and degradation. Understanding the impact of wind-driven particle action on metal structures will facilitate more informed decision-making in product development and material selection, ultimately leading to more durable and resilient structures.

The developed predictive model using the multivariate adaptive regression splines (MARS) algorithm holds great promise for practical applications. The model's accuracy in predicting material mass loss based on parameters such as hardness, impact angle, impact velocity, and sand quantity makes it a valuable tool in assessing material performance and durability under different impact conditions. Designers and engineers can use this model to optimize the design of metal structures and select appropriate materials, considering the specific environmental conditions they will be exposed to. Moreover, the model's efficiency of 98% accuracy for a relative error of less than 15% indicates its reliability and suitability for real-world applications.

4.3. Limitations and Future Research

The present study offers valuable insights into the relationship between terrain conditions and material degradation, focusing on outdoor metal structures in a controlled environment. Although this study acknowledges certain limitations, it could be further enhanced to explore the significance of its findings in dryland regions, where the impact of environmental factors is more pronounced.

One aspect that could be clarified is how dryland conditions were specifically modelled in the lab. Understanding the methodology used to replicate these conditions would add depth to the study and provide insight into the relevance of the findings to real-world desert environments.

To enhance the study's applicability, future research should consider in situ challenges that may be encountered in actual deserts. Factors such as extreme temperature fluctuations, the presence of abrasive particles in winds, and limited water resources for structure maintenance can significantly affect material degradation in dryland areas.

Moreover, investigating the long-term performance of the predictive model under cyclic weather patterns and varying wind velocities in dryland conditions would provide valuable information about its practical reliability.

Overall, expanding the study to encompass a broader range of dryland scenarios and addressing the in situ challenges faced in actual deserts would contribute to a more comprehensive understanding of material degradation in these regions.

5. Conclusions

With the proliferation of constructions on dry and degraded lands, it is crucial to consider the conditions and characteristics of the surrounding terrain when designing and maintaining outdoor metal structures due to the potential problems caused by wind erosion. However, these parameters are often overlooked during the design process. To address this issue, this study emphasizes the importance of incorporating information about land circumstances in the design and maintenance of metal structures exposed to outdoor conditions.

By developing a predictive model that considers data from diverse sources, it provides valuable insights into the potential degradation and erosion experienced by such structures. The findings underscore the need to include terrain-related parameters as essential factors in the design and maintenance practices for outdoor metal structures.

A comprehensive database was compiled from the existing literature and supplemented with experimental data collected for this study. The tests evaluated the mass loss experienced by metal plates subjected to high-pressure air and sand impacts using direct impingement tests. Sample plates exhibited mass losses ranging from 0.20% to 0.99%. Notably, some plates, such as aluminium and different types of tinplate, showed mass gains, likely due to minimal wear and sand particle embedding.

Based on the literature review and experimental data, a predictive model was developed using the multivariate adaptive regression splines (MARS) algorithm. This model accurately predicted material mass loss based on parameters such as hardness, impact angle, impact velocity, and sand quantity. The practical application of the MARS model was demonstrated in assessing the material performance and durability under different impact conditions, aiding in informed decision-making for product development and material selection. For a relative error of less than 15%, the model shows an efficiency of 98% accuracy.

Future research should focus on studying the influence of wind speed and its parameterization in this context, further enhancing our understanding of material degradation, and enabling more precise modelling and predictions.

Author Contributions: Conceptualization, M.T.-C. and F.O.-F.; methodology, M.D.-P.; validation, V.R.M. and J.V.Á.C.; writing—original draft preparation, M.D.-P. and M.T.-C.; writing—review and editing, M.T-C. and F.O-F. All authors have read and agreed to the published version of the manuscript.

Funding: This research was funded by the Council of Science, Innovation, and University through FICYT for the realization of R + D + i network projects with grant number AYUD/2021/57418. Additionally, The APC was funded by the Council of Science, Innovation, and University of the Principality of Asturias with grant number AYUD/2021/50953.

Data Availability Statement: The data used to support the findings of this study are available from the corresponding author upon request.

Conflicts of Interest: The authors declare no conflict of interest.

References

1. Wei, X.; Wu, X.; Wang, D.; Wu, T.; Li, R.; Hu, G.; Zou, D.; Bai, K.; Ma, X.; Liu, Y.; et al. Spatiotemporal variations and driving factors for potential wind erosion on the Mongolian Plateau. *Sci. Total. Environ.* **2023**, *862*, 160829. [CrossRef] [PubMed]
2. Zobeck, T.M.; Van Pelt, R.S.; Hatfield, J.L.; Sauer, T.J. Wind Erosion. In *Soil Management: Building a Stable Base for Agriculture*; John Wiley & Sons, Ltd.: Hoboken, NJ, USA, 2015; pp. 209–227. [CrossRef]
3. Webb, N.P.; Kachergis, E.; Miller, S.W.; McCord, S.E.; Bestelmeyer, B.T.; Brown, J.R.; Chappell, A.; Edwards, B.L.; Herrick, J.E.; Karl, J.W.; et al. Indicators and benchmarks for wind erosion monitoring, assessment and management. *Ecol. Indic.* **2020**, *110*, 105881. [CrossRef]
4. Jarrah, M.; Mayel, S.; Tatarko, J.; Funk, R.; Kuka, K. A review of wind erosion models: Data requirements, processes, and validity. *Catena* **2020**, *187*, 104388. [CrossRef]
5. Woodruff, N.P.; Siddoway, F.H. A Wind Erosion Equation. *Soil Sci. Soc. Am. J.* **1965**, *29*, 602–608. [CrossRef]
6. Williams, J.R.; Jones, C.A.; Dyke, P.T. A Modeling Approach to Determining the Relationship between Erosion and Soil Productivity. *Am. Soc. Agric. Biol. Eng.* **1984**, *27*, 0129–0144. [CrossRef]
7. Liu, B.; Qu, J.; Ning, D.; Han, Q.; Yin, D.; Du, P. WECON: A model to estimate wind erosion from disturbed surfaces. *Catena* **2019**, *172*, 266–273. [CrossRef]
8. Böhner, J.; Schäfer, W.; Conrad, O.; Gross, J.; Ringeler, A. The WEELS model: Methods, results and limitations. *Catena* **2003**, *52*, 289–308. [CrossRef]
9. Hong, C.; Chenchen, L.; Xueyong, Z.; Huiru, L.; Liqiang, K.; Bo, L.; Jifeng, L. Wind erosion rate for vegetated soil cover: A prediction model based on surface shear strength. *Catena* **2020**, *187*, 104398. [CrossRef]
10. Xu, Y.; Liu, L.; Zhou, Q.; Wang, X.; Tan, M.Y.; Huang, Y. An Overview of Major Experimental Methods and Apparatus for Measuring and Investigating Erosion-Corrosion of Ferrous-Based Steels. *Metals* **2020**, *10*, 180. [CrossRef]
11. Savill, T.; Jewell, E.; Barker, P. Development of Techniques and Non-Destructive Methods for In-Situ Performance Monitoring of Organically Coated Pre-Finished Cladding Used in the Construction Sector. In *Electrochemical Society Meeting Abstracts*; The Electrochemical Society, Inc.: Pennington, NJ, USA, 2022; p. 1016. [CrossRef]
12. Laukkanen, A.; Lindgren, M.; Andersson, T.; Pinomaa, T.; Lindroos, M. Development and validation of coupled erosion-corrosion model for wear resistant steels in environments with varying pH. *Tribol. Int.* **2020**, *151*, 106534. [CrossRef]
13. Terrados-Cristos, M.; Ortega-Fernández, F.; Alonso-Iglesias, G.; Díaz-Piloneta, M.; Fernández-Iglesias, A. Corrosion Prediction of Weathered Galvanised Structures Using Machine Learning Techniques. *Materials* **2021**, *14*, 3906. [CrossRef] [PubMed]
14. Zhang, Y.; Ayyub, B.M.; Fung, J.F. Projections of corrosion and deterioration of infrastructure in United States coasts under a changing climate. *Resilient Cities Struct.* **2022**, *1*, 98–109. [CrossRef]
15. Kok, J.F.; Parteli, E.J.R.; Michaels, T.I.; Karam, D.B. The physics of wind-blown sand and dust. *Rep. Prog. Phys.* **2012**, *75*, 106901. [CrossRef] [PubMed]
16. Raffaele, L.; Bruno, L. Windblown sand action on civil structures: Definition and probabilistic modelling. *Eng. Struct.* **2019**, *178*, 88–101. [CrossRef]
17. Shao, Y. (Ed.) Integrated Wind-Erosion Modelling. In *Physics and Modelling of Wind Erosion, in Atmospheric and Oceanographic Sciences Library*; Springer: Dordrecht, The Netherlands, 2008; pp. 303–360. [CrossRef]
18. Wang, W.; Samat, A.; Ge, Y.; Ma, L.; Tuheti, A.; Zou, S.; Abuduwaili, J. Quantitative Soil Wind Erosion Potential Mapping for Central Asia Using the Google Earth Engine Platform. *Remote. Sens.* **2020**, *12*, 3430. [CrossRef]
19. Al-Sayed, A.; Al-Shammari, F.; Alshutayri, A.; Aljojo, N.; Aldhahri, E.; Abouola, O. The Smart City-Line in Saudi Arabia: Issue and Challenges. *Postmod. Openings* **2022**, *13*, 15–37. [CrossRef]
20. Hunold, C.; Leitner, S. 'Hasta la vista, baby!' The Solar Grand Plan, environmentalism, and social constructions of the Mojave Desert. *Environ. Polit.* **2011**, *20*, 687–704. [CrossRef]

21. Vo, T.T.E.; Je, S.-M.; Jung, S.-H.; Choi, J.; Huh, J.-H.; Ko, H.-J. Review of Photovoltaic Power and Aquaculture in Desert. *Energies* **2022**, *15*, 3288. [CrossRef]
22. Parteli, E.J.R. Predicted expansion of sand deserts. *Nat. Clim. Chang.* **2022**, *12*, 967–968. [CrossRef]
23. Wiesinger, F.; Sutter, F.; Fernández-García, A.; Wette, J.; Wolfertstetter, F.; Hanrieder, N.; Schmücker, M.; Pitz-Paal, R. Sandstorm erosion on solar reflectors: Highly realistic modeling of artificial aging tests based on advanced site assessment. *Appl. Energy* **2020**, *268*, 114925. [CrossRef]
24. Dentoni, V.; Grosso, B.; Pinna, F.; Lai, A.; Bouarour, O. Emission of Fine Dust from Open Storage of Industrial Materials Exposed to Wind Erosion. *Atmosphere* **2022**, *13*, 320. [CrossRef]
25. Shi, X. Numerical prediction on erosion damage caused by wind-blown sand movement. *Eur. J. Environ. Civ. Eng.* **2014**, *18*, 550–566. [CrossRef]
26. El-Sherbiny, Y.M. Erosive wear of different facade finishing materials. *HBRC J.* **2018**, *14*, 431–437. [CrossRef]
27. Carrascal, I.; Casado, J.; Diego, S.; Polanco, J. Dynamic behaviour of high-speed rail fastenings in the presence of desert sand. *Constr. Build. Mater.* **2016**, *117*, 220–228. [CrossRef]
28. Hao, Y.-H.; Li, Y. Erosion-behaviors of the coating on steel structure eroded at low erosion-angle in sandstorm. *Mocaxue Xuebao/Tribology* **2013**, *33*, 343–348. [CrossRef]
29. Hao, Y.-H.; Ren, Y.; Duan, G.-L.; Zhu, M.-X.; Feng, Y.-J. Erosion mechanism and evaluation of steel structure coating eroded under sandstorm environment. *Jianzhu Cailiao Xuebao/J. Build. Mater.* **2014**, *34*, 357–363. [CrossRef]
30. Cao, X.; He, W.; Liao, B.; Zhou, H.; Zhang, H.; Tan, C.; Yang, Z. Sand particle erosion resistance of the multilayer gradient TiN/Ti coatings on Ti6Al4V alloy. *Surf. Coat. Technol.* **2018**, *365*, 214–221. [CrossRef]
31. Dalili, N.; Edrisy, A.; Carriveau, R. A review of surface engineering issues critical to wind turbine performance. *Renew. Sustain. Energy Rev.* **2009**, *13*, 428–438. [CrossRef]
32. Coelho, L.B.; Zhang, D.; Van Ingelgem, Y.; Steckelmacher, D.; Nowé, A.; Terryn, H. Reviewing machine learning of corrosion prediction in a data-oriented perspective. *Npj Mater. Degrad.* **2022**, *6*, 8. [CrossRef]
33. Bruno, L.; Horvat, M.; Raffaele, L. Windblown sand along railway infrastructures: A review of challenges and mitigation measures. *J. Wind. Eng. Ind. Aerodyn.* **2018**, *177*, 340–365. [CrossRef]
34. Almajed, A.; Lemboye, K.; Arab, M.G.; Alnuaim, A. Mitigating wind erosion of sand using biopolymer-assisted EICP technique. *Soils Found.* **2020**, *60*, 356–371. [CrossRef]
35. Meng, H.; Gao, Y.; He, J.; Qi, Y.; Hang, L. Microbially induced carbonate precipitation for wind erosion control of desert soil: Field-scale tests. *Geoderma* **2020**, *383*, 114723. [CrossRef]
36. Shi, Y.; Shi, Z. Ultrasonic surface treatment for improving wind-blown sand erosion resistance of cementitious materials. *Wear* **2020**, *460–461*, 203185. [CrossRef]
37. Khanouki, H.A. Development of Erosion Equations for Solid Particle and Liquid Droplet Impact. Ph.D. Thesis, University of Tulsa, Tulsa, OK, USA, 2015.
38. Bouledroua, O.; Meliani, M.H.; Azari, Z.; Sorour, A.; Merah, N.; Pluvinage, G. Effect of Sandblasting on Tensile Properties, Hardness and Fracture Resistance of a Line Pipe Steel Used in Algeria for Oil Transport. *J. Fail. Anal. Prev.* **2017**, *17*, 890–904. [CrossRef]
39. Pastore, G.; Baird, T.; Vermeesch, P.; Bristow, C.; Resentini, A.; Garzanti, E. Provenance and recycling of Sahara Desert sand. *Earth-Sci. Rev.* **2021**, *216*, 103606. [CrossRef]
40. Zheng, X.; Bo, T. Representation model of wind velocity fluctuations and saltation transport in aeolian sand flow. *J. Wind. Eng. Ind. Aerodyn.* **2022**, *220*, 104846. [CrossRef]
41. Oka, Y.; Okamura, K.; Yoshida, T. Practical estimation of erosion damage caused by solid particle impact: Part 1: Effects of impact parameters on a predictive equation. *Wear* **2005**, *259*, 95–101. [CrossRef]
42. Arabnejad, H.; Mansouri, A.; Shirazi, S.; McLaury, B. *Evaluation of Solid Particle Erosion Equations and Models for Oil and Gas Industry Applications*; SPE: San Antonio, TX, USA, 2015. [CrossRef]
43. Wiesinger, F.; Sutter, F.; Wolfertstetter, F.; Hanrieder, N.; Fernández-García, A.; Pitz-Paal, R.; Schmücker, M. Assessment of the erosion risk of sandstorms on solar energy technology at two sites in Morocco. *Sol. Energy* **2018**, *162*, 217–228. [CrossRef]
44. Harsha, A.; Bhaskar, D.K. Solid particle erosion behaviour of ferrous and non-ferrous materials and correlation of erosion data with erosion models. *Mater. Des.* **2008**, *29*, 1745–1754. [CrossRef]
45. Bingley, M.; O'flynn, D. Examination and comparison of various erosive wear models. *Wear* **2005**, *258*, 511–525. [CrossRef]
46. Huttunen-Saarivirta, E.; Kinnunen, H.; Tuiremo, J.; Uusitalo, M.; Antonov, M. Erosive wear of boiler steels by sand and ash. *Wear* **2014**, *317*, 213–224. [CrossRef]
47. Çelik, S.B.; Çobanoğlu, I.; Koralay, T.; Gireson, K. Investigation of the Leeb hardness test in rapid characterisation of rock cores with particular emphasis on the effect of length to diameter ratio. *Int. J. Min. Reclam. Environ.* **2023**, *37*, 524–543. [CrossRef]
48. *ASTM A956/A956m-17a*; Standard Test Method for Leeb Hardness Testing of Steel Products. ASTM: West Conshohocken, PA, USA, 2022.
49. *ISO 16859-1/2/3:2015*; Metallic Materials—Leeb Hardness Test. ISO: Geneva, Switzerland, 2015.
50. *ASTM G76-2013*; Standard Test Method for Conducting Erosion Tests by Solid Particle Impingement Using Gas Jets. ASTM: West Conshohocken, PA, USA, 2013.
51. Friedman, J.H. Multivariate Adaptive Regression Splines. *Ann. Stat.* **1991**, *19*, 1–67. [CrossRef]

52. Naser, A.H.; Badr, A.H.; Henedy, S.N.; Ostrowski, K.A.; Imran, H. Application of Multivariate Adaptive Regression Splines (MARS) approach in prediction of compressive strength of eco-friendly concrete. *Case Stud. Constr. Mater.* **2022**, *17*, e01262. [CrossRef]
53. Praveen, A.S.; Sarangan, J.; Suresh, S.; Subramanian, J.S. Erosion wear behaviour of plasma sprayed NiCrSiB/Al$_2$O$_3$ composite coating. *Int. J. Refract. Met. Hard Mater.* **2015**, *52*, 209–218. [CrossRef]
54. Mathapati, M.; Ramesh, M.; Doddamani, M. High temperature erosion behavior of plasma sprayed NiCrAlY/WC-Co/cenosphere coating. *Surf. Coat. Technol.* **2017**, *325*, 98–106. [CrossRef]

 *land*

Article

The Development of an Experimental Framework to Explore the Generative Design Preference of a Machine Learning-Assisted Residential Site Plan Layout

Pei Sun [1], Fengying Yan [1,*], Qiwei He [1] and Hongjiang Liu [2]

[1] School of Architecture, Tianjin University, Tianjin 300072, China; sun_814224622@tju.edu.cn (P.S.); fslight@tju.edu.cn (Q.H.)
[2] China Architecture Design & Research Group, Beijing 100044, China; hongjiang@tju.edu.cn
*   Correspondence: fengying@tju.edu.cn; Tel.: +86-139-2030-9555

Abstract: Generative design based on machine learning has become an important area of application for artificial intelligence. Regarding the generative design process for residential site plan layouts (hereafter referred to as "RSPLs"), the lack of experimental demonstration begs the question: what are the design preferences of machine learning? In this case, all design elements of the target object need to be extracted as much as possible to conduct experimental studies to produce scientific experimental results. Based on this, the Pix2pix model was used as the test case for Chinese residential areas in this study. An experimental framework of "extract-translate-machine-learning-evaluate" is proposed, combining different machine and manual computations, as well as quantitative and qualitative evaluation techniques, to jointly determine which design elements and their characteristic representations are machine learning design preferences in the field of RSPL. The results show that machine learning can assist in optimizing the design of two particular RSPL elements to conform to residential site layout plans: plaza paving and landscaped green space. In addition, two other major elements, public facilities and spatial structures, were also found to exhibit more significant design preferences, with the largest percentage increase in the number of changes required after machine learning. Finally, the experimental framework established in this study compensates for the lack of consideration that all design elements of a residential area simultaneously utilize the same methodological framework. This can also assist planners in developing solutions that better meet the expectations of residents and can clarify the potential and advantageous directions for the application of machine learning-assisted RSPL.

Keywords: machine learning; generative design preference; planning design elements; Pix2pix model; residential site layout planning; experimental framework

Citation: Sun, P.; Yan, F.; He, Q.; Liu, H. The Development of an Experimental Framework to Explore the Generative Design Preference of a Machine Learning-Assisted Residential Site Plan Layout. *Land* **2023**, *12*, 1776. https://doi.org/10.3390/land12091776

Academic Editor: Chuanrong Zhang

Received: 18 July 2023
Revised: 5 September 2023
Accepted: 8 September 2023
Published: 13 September 2023

1. Introduction

With the growth of computer science, machine learning-based generative design has become popular. This gives us new ways to learn about the generative design process for RSPLs. Generative design is performed using a computer that generates new design solutions in a given design space structure via random noise sampling. Machine learning, as a data-driven approach, is considered an effective method to apply to generative design [1,2]. The current generative design, which is built on machine learning, has met the need for devising a great number of design ideas. However, in the age of big data, most automatic design methods only look at quantitative goals and constraints and ignore qualitative design information, which is hard to describe mathematically [3,4].

Mining the generative design preferences of machine learning in the field of plan layout can help determine the design inspirations of machine learning in the field of RSPL design to explain the benefits of a machine learning-assisted plan layout. Generative design preferences can guide planners in developing solutions that better meet residents' expectations. In recent years, investigation into the independent learning of generative preference design has begun regarding machine learning-assisted plan layouts. This investigation has two main paths for applying technical means and innovating research perspectives: (1) Machine learning is a technical tool used for solving research problems. For example, satellite images were used to identify land use changes [5], crime was assessed through street images [6], COVID-19 plan distribution states for urban security risk assessment were identified [7,8], and remote sensing images were used to detect forest carbon stocks to predict their carbon sink development [9]. (2) Machine learning provides innovative perspectives on extracting design elements to facilitate decisions. For instance, Silva et al. used convolutional neural networks and YOLO algorithms to identify sites and extract required elements (e.g., vegetation strips and buildings) to improve decision making for urban design development [10]. Moreover, Chinazzi et al. employed machine learning models to create a new method for generating scientific maps of knowledge, providing a scientific method for classifying urban planning and other fields [11]. Additionally, the creation of urban knowledge systems has been seen as an innovative result of the mutual representation of artificial intelligence techniques and the extraction of targets [12]. These earlier works have shown a strong link between machine learning and plan structure in recognizing, perceiving, evaluating, and predicting. Additionally, they showed that machine learning offers new ways to use technology to extract design elements to help plan layouts during autonomous learning exploration. This process allowed planners to determine the best practices for machine learning to assist with plan layout.

Each design element of an RSPL can be a generative design preference for machine learning in residential layout planning and can exhibit an application value. Residential design elements refer to each component of an RSPL, including design elements such as housing, roads, landscapes, and green spaces. These are indispensable and important components in the planning and design of residential spaces. Existing research of machine learning-assisted RSPLs only involves the study of individual design elements. For example, Xinyu Cong used CGAN to generate residential area layouts [13], Dai et al. used the Gray Wolf optimization algorithm model to improve the impact of community public space promotion from a child's perspective [14], and Elariane used a machine learning model to evaluate real estate website API data to determine the characteristics of long-term rental apartment homes [15]. Therefore, we attempt to apply machine learning to RSPLs through an experimental study of the totality of the design elements in residential planning, allowing machines to learn autonomously to determine their preferred designs of interest and their characteristic properties.

In the current artificial intelligence boom, generative adversarial networks have derived many new development-powered models such as CycleGAN [16], Pix2pix_HD [2], Pix2pix, etc. However, after combing through the literature regarding the strengths and weaknesses of each generative adversarial network model (as shown in Table 1), it was found that the Pix2pix model outperforms the others in the image transcription and classification tasks [17]. The Pix2pix model was proposed by Phillip Isola et al. in 2017 based on GANs, the earliest image recognition and generation applications. The most significant difference between the Pix2pix model and previous GAN-derived models is that Pix2pix optimizes the original input method to an imaging approach, enabling the image-from-to-image learning process [18]. Its discriminator Patch design can reduce the dimension of the input image significantly, reducing the number of parameters and increasing the operation speed. This study then generates a one-to-one site plan of the settlement to discover its design preference through labeled graphs. The Pix2pix model principle is to realize one-to-one image mapping. In addition, the external sites of residential areas have different shapes and scopes. In contrast, the Pix2pix model has no limitations regarding image scale and size, thus allowing for an increase in the scalability of the Pix2pix model. Therefore, the Pix2pix model was selected for the research in this paper. However, at the same time, the Pix2pix model has the disadvantage of generating fuzzy and conflicting images. The existing research provides the following solutions: (1) increase the details of labeled maps and (2) improve the quality of the parameters.

Table 1. Comparison of the advantages and disadvantages of generative adversarial modeling.

Model	Advantage	Disadvantage	Reference Sources
Pix2pixGAN	A generalized approach to image-to-image translation	Generates images with blurred, conflicting characteristics	Fu, B., et al. [19] Zhao, C. W., et al. [20]
CycleGAN	Solves the problem that the Pix2Pix model requires image pairing	Low quality of generated images	Zhu J Y, et al. [21]
Pix2pix_HD	Higher quality of generated images	Still needs pair data	Chen, J. S., et al. [22]
StarGAN	Realization of multi-domain style image transformation	The image's label is entered into the model so that the attributes can be modified	Shen, Y., et al. [23] Choi Y, et al. [24]
InfoGAN	The characteristics of the generated data are controlled by setting the implicit encoding of the input generator.	Training is unstable, and its performance is susceptible to the prior distribution and the number of noisy hidden variables selected.	Wan, P., et al. [25] Chen X, et al. [26]
LSGAN	Solves the problem of training instability	Lack of diversity in generated images	Mao X., et al. [27]
ProGAN	Generates high-resolution images	Very limited ability to control specific features of the generated image	Karras T., et al. [28]
SAGAN	Generated images more closely resemble the original image	Poor quality of images for generating local autocorrelation	Zhang H., et al. [29]

Current research on applying the Pix2pix model has not been extended to other residential design elements. The application of the Pix2pix model was initiated at the beginning of Chaillou's implementation of the apartment plan design process, involving 'building plan contour', 'layout within the contour', and 'addition of furnishings', using multiple Pix2pix optimization models [30]. Pix2pix models were later optimized to evaluate automated building simulation applications [31]. For example, David Newton explored the challenge of generating layouts for Corbusier-style houses with a limited sample size. He expanded the scope of analysis by introducing noise and rotation to enhance the training effectiveness of GAN models [32] Yu et al. utilized traditional Chinese architectural datasets to generate and identify building facades [33]. Additionally, Mostafavi et al. employed machine learning to predict illumination and spatial daylight autonomy based on residential building spatial layouts [34]. However, previous studies demonstrate that the Pix2pix model has not been widely used in the design of RSPLs. While Gu D. et al. used the Pix2pix model to evaluate wind damage to residential building windows for protection against wind damage [35], their study focused solely on a single residential element. Few studies have extended the application of this model to other residential design elements and explored its diverse potential within the realm of RSPL.

Previous research has confirmed that machine learning can assist in generating and optimizing RSPLs. However, given a machine in a residential site scheme, it is unclear what the preferred design of machine learning in a residential site scheme is. This makes it hard to determine where the benefits of machine learning-assisted RSPL lie. Integrating the widespread use of large-scale data-assisted plan layouts and extracting the characteristics of the design elements of residential site schemes allows researchers to look into empirical methods to understand the potential for its use in "RSPL" from a machine learning point of view. Thus, this will contribute to the urban planning discipline. Given the above, this study presents an experimental framework for exploring Chinese residential areas. This stems from the diversity of residential types in Chinese residential areas, reflecting the universality of the research results. The proposed experimental framework was applied for experimental demonstration with the Pix2pix model as the chosen generative adversarial network model. This provides a method for exploring design preferences in residential layout planning and extending the application of the Pix2pix model in RSPLs.

The rest of the paper is organized as follows: Section 2 presents the experimental framework for this study, including the residential area design element extraction process and the data processor for analysis using this framework. To verify the effectiveness of machine learning for residential layout planning, its learning results and limitations in this study, as well as insights for future work are evaluated and discussed in Section 3; Section 4 illustrates the analysis and the conclusions.

2. Materials and Methods

2.1. Study Area

The sources of the residential schemes for this study were CAD drawings of completed residential schemes collected from major Chinese design websites (shown in Appendix A Table A1). From these sources, we selected 300 design schemes for the experiment from residential areas in various provinces in China. The residential schemes we chose were all established settlements with a size ranging from 40 to 80 hectares and a predominantly rectilinear dwelling arrangement. The housing types were categorized based on their height: low-rise, multi-story, medium-high-rise, and high-rise. The plans of some of the residential schemes are illustrated in Figure 1. On average, the selected sample had 10.7 floors, a mid-level building density of 31.3%, an average floor area ratio of 1.9, and an average study area of 67.86 hectares (as shown in Table 2). We simplified the schemes and corrected them for code non-compliance and apparent errors.

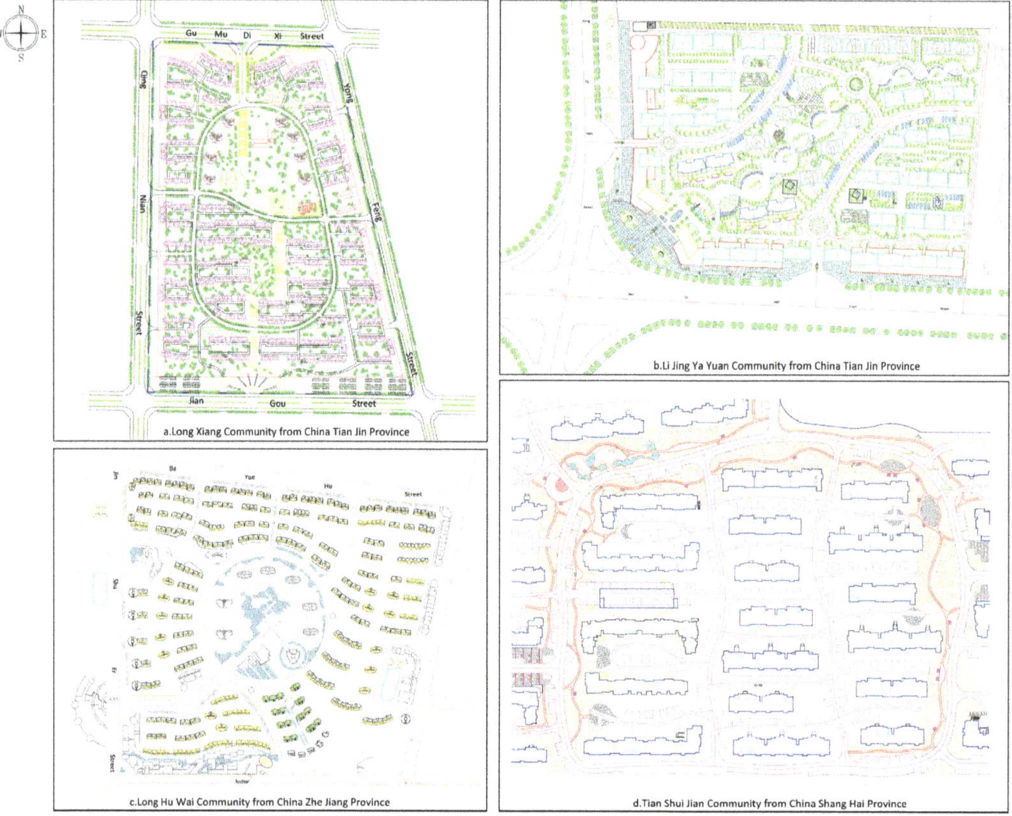

Figure 1. Scheme plans for residential areas.

Table 2. Summary of basic information in residential schemes.

Basic Information Characteristics	Classification of the Basic Information Characteristics	Count
Floors	The highest number of floors	32 F
	The lowest number of floors	1 F
	Average floors	10.7 F
Building density	Maximum building density	39.4%
	Minimum building density	20.1%
	Average building density	31.3%
Plot ratio	Maximum floor area ratio	4.4
	Minimum floor area ratio	0.7
	Average plot ratio	1.9
Floor area	Maximum floor area	87.3 ha
	Minimum floor area	41.6 ha
	The average floor area	67.86 ha

2.2. Methodological Framework

To explore the "machine learning generative design preferences in RSPL.", a framework of "extraction-translation-machine learning-evaluation" was proposed (shown in Figure 2). The experimental framework is as follows: in the first step, design elements in China's Urban Residential Planning and Design Standard GB 50180-2018 [36] (hereafter referred to as "CURPADS") were summarized into five categories: housing, green space, supporting facilities, roads, and other elements. In the second step, we translated the scheme into an image recognized by the Pix2pix model using an RGB color block assignment of the image. The Pix2pix model was used in the third step to learn the residential area scheme, aiming to obtain an optimal parameter performance and a sample augmentation solution. Subsequently, the results of the generated solutions were evaluated through standard and design dimensions in the fourth step. The evaluation process represents the preferred design determination process.

2.3. Step 1: Extraction

Based on the current classification of "CURPADS", design elements of housing, green space, and other design elements (including square, water, inlet, and outlet), supporting facilities (commercial and other supporting facilities) and road elements were extracted as the design elements of RSPL that needed to be learned via machine learning for in-depth analysis in this study (as depicted in Figure 3). "CURPADS" has modified the requirements for the residential environment and supporting facilities. It incorporates housing, green space, and public space to enhance the quality of the residential environment, and it divides supporting facilities into different levels to align with the creation of residential areas of varying scales. In the latest residential design process, there is greater emphasis on improving the quality of the environment within residential areas while meeting mandatory design standards. Simultaneously, the quality of residential planning and design is ensured through a scientifically sound, green, and ecologically balanced spatial structure. "CURPADS" serves as the standard observed in Chinese residential planning and design. The residential design elements derived from it represent the accumulated practical experience of Chinese residential planning and hold significant importance.

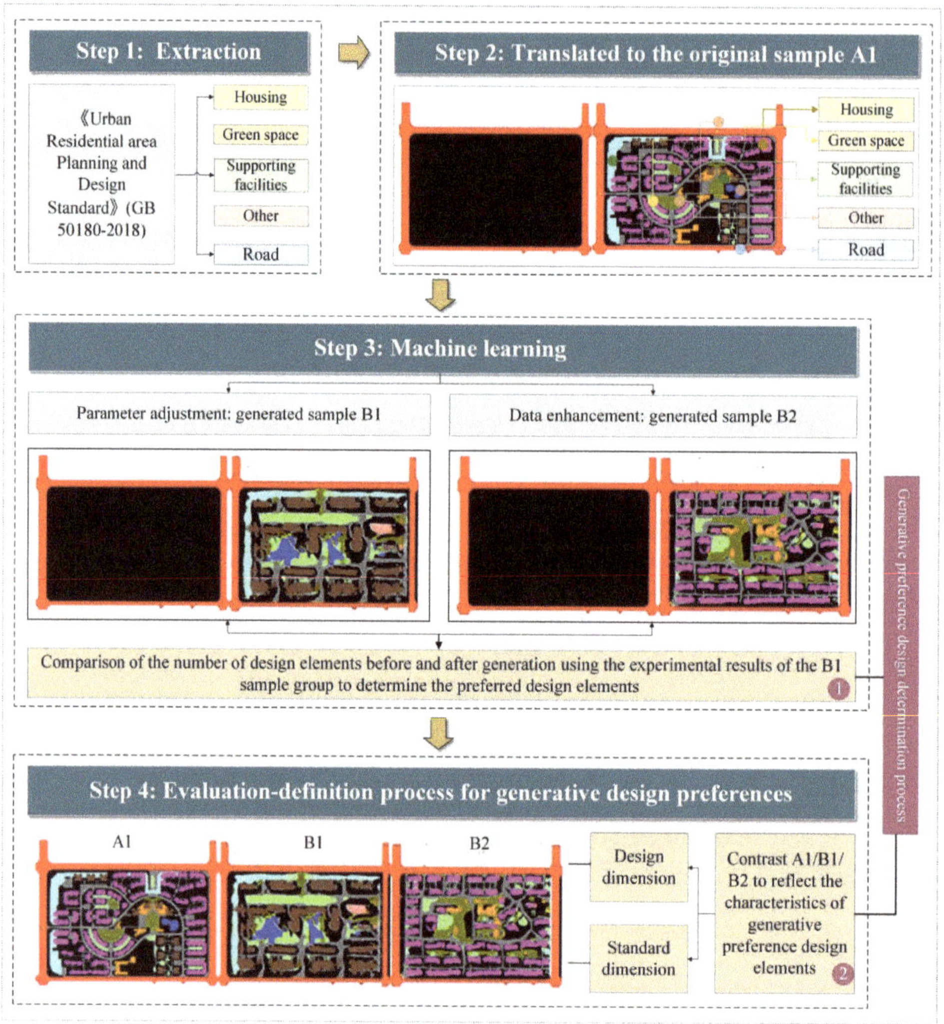

Figure 2. Experimental framework [36].

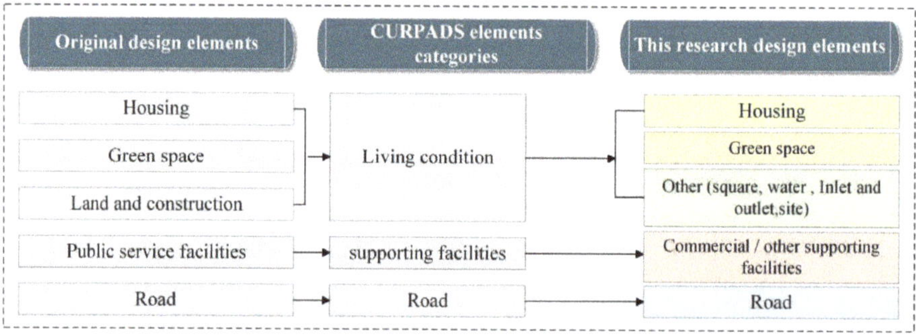

Figure 3. Classification of residential planning and design elements.

2.4. Step 2: Translation

Since the Pix2pix model performs image-to-image recognition, the residential design elements must be translated into residential images for machine learning. This is accomplished by translating the settlement design elements through assigning different RGB color values into images easily recognized and learned by the machine (as in Table 3). The final results of the labeled images are shown in a JEPG format with 512 pixels by 256 pixels and a resolution of 300 dpi.

Table 3. Different RGBs for different design elements.

Extraction Elements	Function Type of Elements	RGB Value	
Housing	Villa (1–3 F)	R:80 G:120 B:80	
	Low-rise (4–6 F)	R:255 G:0 B:255	
	Mid-rise (7–11 F)	R:150 G:100 B:75	
	Mid-rise (12–18 F)	R:180 G:0 B:255	
	High-rise (over18 F)	R:255 G:150 B:150	
Supporting facilities	Commercial supporting facilities	R:150 G:255 B:255	
	Other supporting facilities	R:255 G:150 B:0	
Road	External road	R:255 G:0 B:0	
	Internal road	R:150 G:150 B:150	
Green space	Greenery landscape	R:150 G:255 B:150	
Other	Water	R:0 G:0 B:255	
	Site	R:0 G:0 B:0	
	Square	R:150 G:150 B:0	
	Inlet and outlet	R:255 G:255 B:0	

2.5. Step 3: Machine Learning

2.5.1. Pix2pix Model

The model used for this machine learning is Pix2pix, which operates with the underlying logic of a U-NET architecture [37] and consists of 16 layers of convolutional neural networks for the generator and a PatchGAN architecture [18], as well as five layers of convolutional neural networks for the discriminator (shown in Figure 4). The generator extracts the input image information containing various elements through a convolutional neural network. It conducts it through 16 different layers of neural networks, one layer at a time, to translate the image information into computer language before passing it to the next layer. Later, after receiving the training data forward propagated by the inputter through the deconvolution layer, the generated image is transmitted to the discriminator and bridged to the input image to determine the similarity of the generated image to the input image.

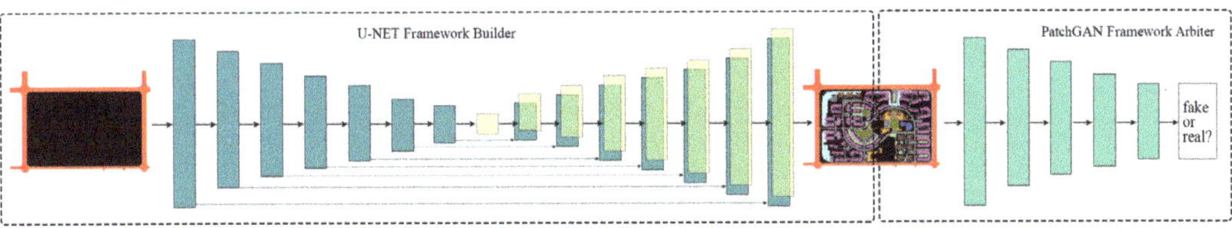

Figure 4. Model architecture diagram of Pix2pix.

2.5.2. Learning Process

To maintain the optimal learning effect of machine learning during the experiment, the Pix2pix model needs to be optimized via multiple debugging. A machine learning process under the computational mutual feedback system (shown in Figure 5) was proposed. This process comprises two parts: one is the tuning calculation, i.e., the parameter adjustment to determine its optimal parameters, and the other is the mutual data feed, i.e., the internal data augmentation to optimize its learning results.

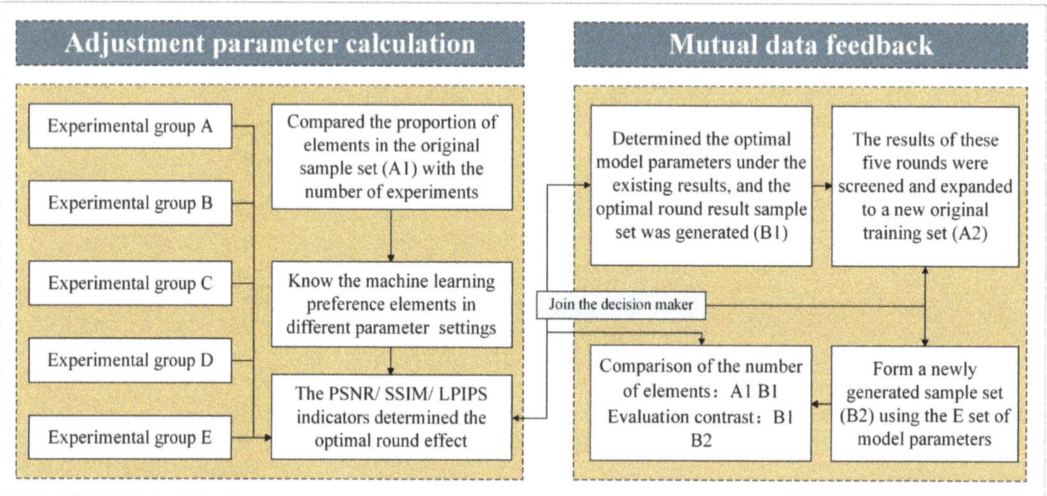

Figure 5. Flowchart of the calculated mutual feeding system.

- Parameter adjustment:

Parameters: loss function, hyperparameters, and metrics. The loss function referred to a function within the model, and the hyperparameters and metrics were used to tune the model and measure its performance, respectively. A total of five groups of tuning experiments, A, B, C, D, and E were carried out in this experiment.

Hyperparameters can influence the training and output performance of the model. Two main parameters were involved in this experiment: Epoch and Decay.

a. Epoch. The Pix2pix model learns all samples once during the learning process. A complete cycle is called one Epoch, through which the whole training process of the model is divided into several segments, and more iterations indicate a better learning effect. In this study, we selected epoch values of 100, 300, 500, and 700 for setting.

b. Decay. The decay degree represents the decay rate of the learning rate during the iterative process, and its purpose is to prevent overfitting. The optimal learning rate, which was immense initially and gradually decreased during the training process, could better approximate the optimal point. In the current work, we selected 50, 200, 250, 150, and 100 decay values for the setting.

Metrics were employed to evaluate the performance of different model algorithms. PSNR, SSIM, and LPIPS were chosen as the metrics for this experiment.

c. PSNR: Peak signal-to-noise ratio is a reference value of image quality that measures the difference between the maximum signal and background noise. It is the most common and widely used objective evaluation index for images and is usually defined by the sum mean square error (MSE) of the image. In detail, MSE is expressed as

$$\text{MSE} = \frac{1}{\text{H} \times \text{W}} \sum_{x=1}^{\text{H}} \sum_{y=1}^{\text{W}} (X(x,y) - Y(x,y))^2 \tag{1}$$

H aH and W represent, respectively, the length and width of the image, X denotes the original image, and Y indicates the generated image. $X(x,y), Y(x,y)$ represents the (x,y) pixel value of the image X, Y in coordinates. PSNR is defined as [38]

$$PSNR = 10 \cdot \log_{10}\left(\frac{MAX_L^2}{MSE}\right) \tag{2}$$

where MAX_L is the most probable maximum pixel value of the image. In the default red, green, and blue (RGB) images, this value equaled 255. MSE indicates the mean square error between the original and generated images. PSNR is measured in decibels (dB), and one of the objectives to this study is to generate image results with a high PSNR.

d. SSIM. The structural similarity index measure was used to compare the proximity of the original sample to the generated sample image with respect to brightness, contrast, and structure [39]. The SSIM algorithm was designed to consider the variation of structural information in the image in human perception [40]. The model also introduced perceptual phenomena and structural information related to perceptual variation. Structural information refers to the fact that pixels have internal dependencies on each other, especially spatially close pixel points [41]. These dependencies carry essential information about the visual perception of the target object, and therefore SSIM is more suitable than PSNR to evaluate the perceptual effects of images. Its definition is shown as

$$SSIM(x, y) = \frac{\left(2\mu_x\mu_y + \mathbb{C}_1\right)\left(2\sigma_{xy} + \mathbb{C}_2\right)}{\left(\mu_x^2 + \mu_y^2 + \mathbb{C}_1\right)\left(\sigma_x^2 + \sigma_y^2 + \mathbb{C}_2\right)} \tag{3}$$

where μ_x is the mean of x; μ_y indicates the mean of y; σ_x and σ_y are the variances of x and y; σ_{xy} is the covariance of x and y; and $\mathbb{C}_1$ and $\mathbb{C}_2$ are constants. The proposed Pix2pix model aims to make the SSIM value as close to 1 as possible.

e. LPIPS. Learning Perceptual Image Block Similarity, also known as "loss of perception", was adopted to measure the difference between two images [42]. This metric learns the reverse mapping of the generated image to Ground Truth, forcing the generator to learn to reconstruct the reverse mapping of the real image from the fake image and prioritize perceptual similarity between them. LPIPS is more consistent with human perception than traditional methods (like L2/PSNR, SSIM, and FSIM). On the other hand, LPIPS can better reflect the perception advantage [43] of the images generated by GAN. A lower value of LPIPS indicates that the two images are more similar, and vice versa, the greater the difference. For a given neural network F, Figure 6 can represent how LPIPS is computed.

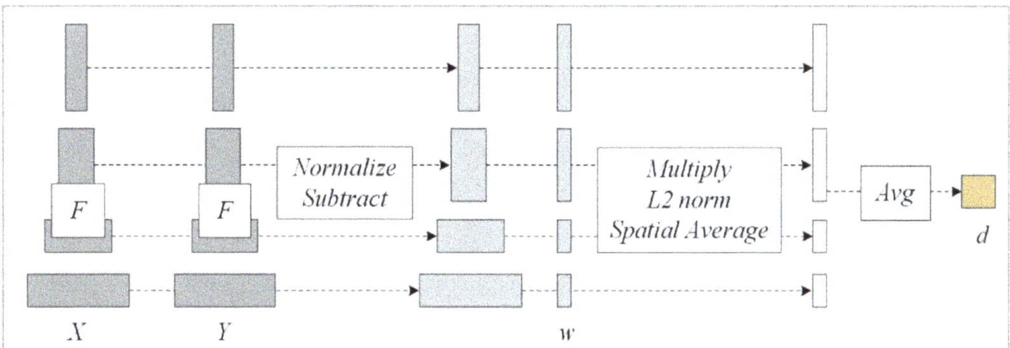

Figure 6. LPIPS operation process analysis.

The image was inputted to network F. Each convolutional layer was feature extracted and cell normalized in the channel dimension. For the L layer, the result would be written as $\hat{y}^l, \hat{y}_0^l \in R^{H_l \times W_l \times C_l}$. Meanwhile, each channel was scaled using the vector W and L2 distance was calculated. Finally, the perceptual distance result was obtained by averaging in the spatial dimension and summing in the channel dimension with the expression of

$$d(x, x_0) = \sum_{\iota} \frac{1}{H_{\iota} W_{\iota}} \sum_{\mathcal{H}, \mathcal{W}} \| \, \mathcal{W}_{\iota} \odot (\hat{y}_{\mathcal{H}\mathcal{W}}^{\iota} - \hat{y}_{0\mathcal{H}\mathcal{W}}^{\iota}) \, \|_2^2 \tag{4}$$

- Data Enhancement:

Data augmentation refers to making a limited amount of data produce more value without substantially increasing the data [44]. In this work, the solution with a better effect on the generation side of experimental groups A, B, C, D, and E was added to the original sample to achieve sample augmentation. Combining the original sample and the generated sample increases the diversity of the data set, improves the generalization ability and robustness of the training model [45], and thus enhances the value of the existing data. For the screening of sample set augmentation, the following process was mainly adopted: (1) an overall judgment was made about whether the scheme was complete and whether each design element in each scheme was easily distinguishable, etc., (2) a judgment of each design element was performed (Figure 7). If a design element could not be judged, other design elements would be combined to make a comprehensive judgment. If the generated images conformed to the judgment process, they would be mixed into the original sample set. Otherwise, the solution would be filtered and discarded. Finally, we selected 285 images from the 1500 generated results (experimental groups A, B, C, D, and E) and blended them into the original sample set for data enhancement.

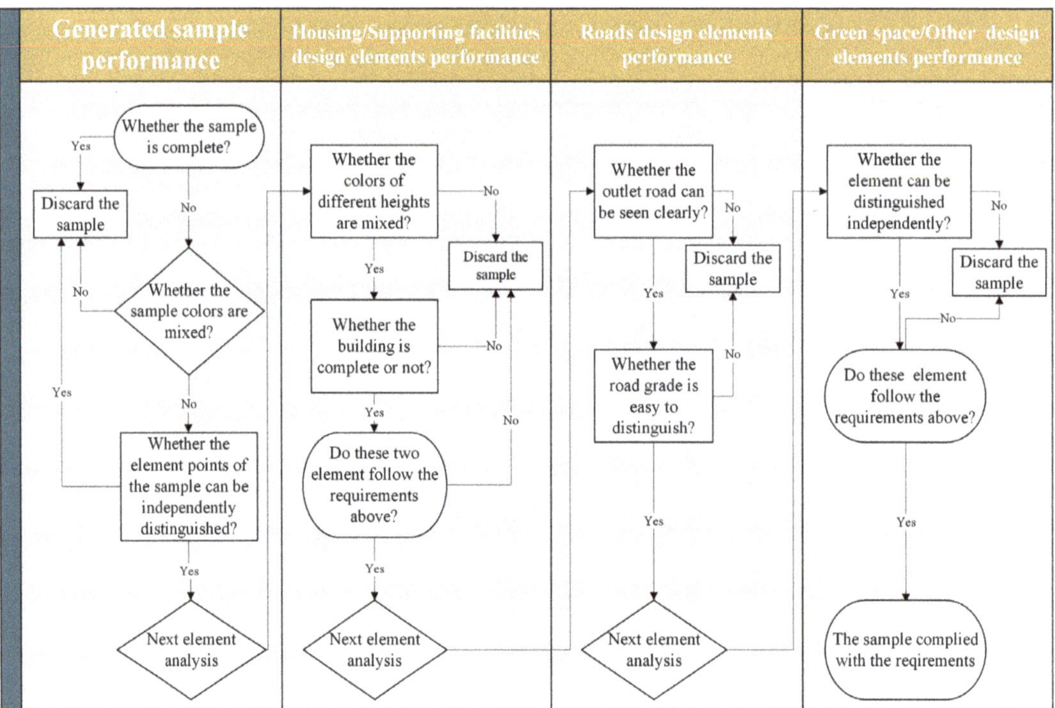

Figure 7. Sample set expansion and screening process.

2.6. Evaluation

Since machines cannot judge the quality of image design, each result in machine learning requires metrics for evaluation. According to Recio et al., it was found that in emotion, the high arousal effect performance of positive words leads to a faster visual perceptual response [46] and more easily obtained merits of the target object at the visual level. In addition, the positive evaluation words mentioned in the book "Designing Cities:

Basics-Principles-and Practice" were the first to utilize this evaluation in urban design [47] and were able to precisely identify design solutions. Hence, based on this reference, an evaluation scale for a design dimension was proposed in this study. The evaluation of this study was divided into two dimensions: the design dimension and the standard dimension. The design dimension proposes diversity, simplicity, relative property, and totality to evaluate the square paving, green landscape space, commercial facilities, and other public facilities (as shown in Table 4). For the standard dimension, three aspects of plot ratio and building density, as well as the proportion of square paving, green landscape space, commercial facilities, and other public facility activity-occupied land, were selected to evaluate the results.

Table 4. Classification of the evaluation dimensions.

Evaluative Dimension / Evaluate Elements	Square Paving	Landscape Green Space	Commercial Facilities	Other Public Facilities
Design dimension — Diversity		①Structured; ②Detailed; ③Various;		
Design dimension — Simplicity		①Well-balanced; ②Self-existed; ③Concise;		
Design dimension — Relative property		①Sequential; ②Heterogeneous;		
Design dimension — Totality		①Compact; ②Unified; ③Balanced; ④Uniform;		
Standard dimension — Plot ratio	Density of the building	The proportion of paved plazas/landscaped green areas/commercial facilities/other public facility activity sites		

3. Results and Discussion

3.1. Optimal Parameter Determination

We compared the experimental results in the five groups of parameters by selecting one of the residential area schemes (as in Figure 8). It was found that the PSNR and SSIM index scores reached the highest level in experimental group E, while the LPIPS index showed the lowest also in experimental group E. Subsequently, the generated results from the five experimental groups were compared using the mean opinion scoring method, which is a subjective image quality assessment index that rates the visual perceptual quality of the generated images on a scale from 1 (worst quality) to 5 (best quality). The final score is calculated as the arithmetic mean of the scores provided by all the raters. In this case, the highest mean score for Group E was 3.7, based on the ratings from 30 raters (see Appendix A Table A2). Based on these results, Group E is considered to have performed the best. Additionally, this group had the highest number of iterations, and the degree of decay was maintained in a gradually decreasing state. Consequently, these parameters will be used in the subsequent model training for scheme learning.

3.2. Generative Preference Design Element Determination

Once the parameters for extracting the generative design preferences of machine learning in the design elements of "RSPL" were determined, a visual method was employed to quantitatively compare the number of element changes between the original sample A1 and the generated sample B1 (as shown in Table 5). Since not all residential schemes contain all design elements, it is necessary to perform a classification count before tallying the number of element changes. The statistics are as follows: water and supporting infrastructure are classified to initially count the number of original sample sets with or without such elements. Subsequently, the number of schemes with or without these elements was counted through machine learning to discern the differences.

For example, in the original sample A1, the percentage of residential schemes with other supporting facilities was 39.7%. However, in sample B1, which was generated after machine learning, the percentage of residential schemes with commercial supporting facilities increased to 62.1%. Conversely, the proportion of residential schemes without other facilities in the original sample A1 accounted for 60.3%, while in sample B1 generated by machine learning, the proportion of residential schemes without commercial facilities decreased to 39.7%. This indicates that the number of residential schemes with other supporting facilities increased after machine learning.

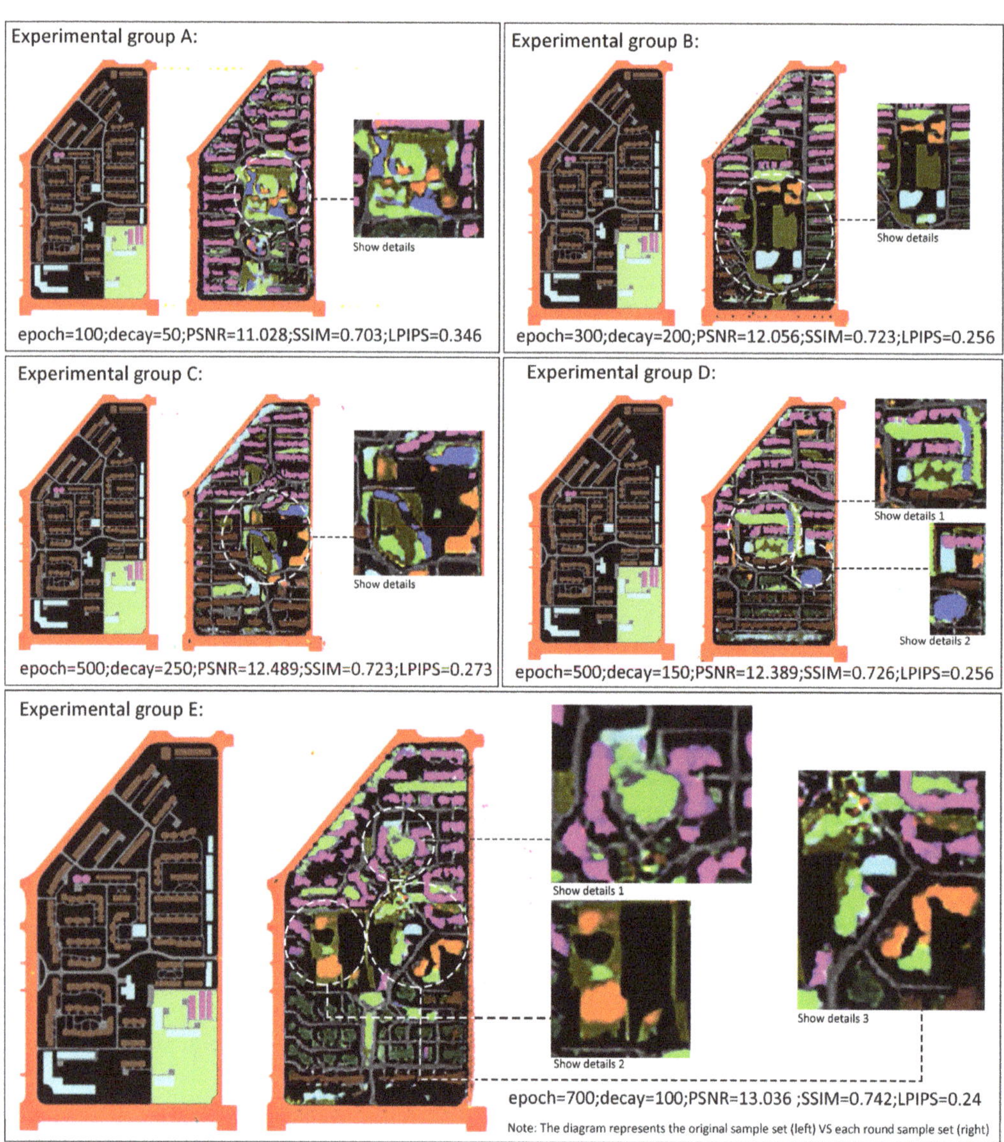

Figure 8. Partial table for generating the sample set.

Since spatial structure, landscaped green space, and road design elements are present in each sample, statistical classification is necessary for these three design elements based on their types. In this study, both elements were categorized into three groups: concentrated, dispersed, and centralized-dispersed, using a visual method. Statistics were conducted based on the difference in the number of types before and after their generation. For instance, in the original sample A1, 52.7% of residential schemes in green landscape regions were decentralized. After machine learning, 55.1% of residential schemes with decentralized landscape green areas were generated in sample B1. The proportion of residential schemes with concentrated landscape green space increased from 8.5% to 11.6%. However, the proportion of residential schemes with scattered landscape green space decreased from 38.3% to 33.3%. In the original sample A1, 29.7% of residential schemes featured axial roads, while 41.8% featured axial roads in machine-learned sample B1. Furthermore, the residential schemes in sample B1 generated by machine learning reached 41.8%. In contrast to the residential schemes with an axial-ring road (17.3% to 0.7%), the proportion of residential schemes with a ring road increased from 53.0% to 57.5%. According to the above methodologies, only public facilities and spatial structure differed above 10% in the number of changes in all categories.

Table 5. Comparison of element classification and quantitative statistical results.

Extraction Elements		Classification of Elements	The Proportion of Elements in the Original Sample A1	The Proportion of Elements in the Generated Sample B1
Water		Yes	60.6%	54.4%
		No	39.4%	45.6%
Supporting facilities	Commercial supporting facilities	Yes	60.3%	78.6%
		No	39.7%	21.4%
	Other supporting facilities	Yes	39.7%	62.1%
		No	60.3%	37.9%
Road network structure		Axis	29.7%	41.8%
		Ring	53.0%	57.5%
		Axis-ring line	17.3%	0.7%
Space structure		dispersed	32.0%	58.2%
		concentrated	28.0%	15.4%
		centralized-dispersed	40.0%	22.8%
Landscape greening structure		dispersed	52.7%	55.1%
		concentrated	8.5%	11.6%
		centralized-dispersed	38.8%	33.3%

Based on the methodological statistics mentioned above, it is evident that machine learning exhibits a preference for designing two major elements: other public facilities and spatial structure. Delving deeper into the reasons for their prominence, our concept of other public facilities in this study is defined as independent and large-area facilities such as kindergartens, elementary schools, and cultural activity centers. These facilities offer certain advantages in the image translation process when compared to other public facilities: (1) There are no additional elements surrounding them that could cause interference. In fact, our original sample set indicates that most elements of other public facilities exist in isolated corners and do not blend with other elements, minimizing interference with machine learning; (2) Due to their larger size, other public facilities are also represented by RGB pixel values in the original sample set; (3) These facilities tend to exhibit better contrast, resulting in improved machine learning results due to more pronounced shaping. This study assessed spatial structure based on the combination of square paving and green landscape elements. Furthermore, these two elements exhibited various characteristics, such as fragmented connectivity, scattered distribution, and a substantial plan area, during our labeling process. A combination of square paving and green landscape was utilized to evaluate spatial structures. Machine learning for spatial structure design elements tends to generate decentralized spatial structures. The learning effect is more favorable because these two elements exhibit diverse characteristics, including fragment connectivity, scattered distribution, and a large plan area. While Ma et al. concluded that uniformly distributed spatial facility service components were crucial in shared rental housing [48], a thorough analysis of the generated sample set by machine learning yielded a programmatic surface effect that prefers a balanced layout for each design element. This confirms that the results align with the flat characteristics of an RSPL [49].

3.3. The Design Dimension Determines Preferred Generative Design Features

Among the 12 display schemes selected (as in Figure 9), we chose two for successive comparative evaluations of the machine learning process. The original sample for Scheme 52 featured a core spatial structure comprising water and a square, with supporting facilities distributed to the right and left of the entrance. In contrast, the generated result sample 52-B1 exhibited significantly higher building density. Furthermore, it used other supporting facilities and water as the core spatial design elements within the residential area. Some commercial facilities were added to the south side to complement the design along the residential interface, but the design of the square landscape green space needed to be incorporated. On the other hand, the generated sample 52-B2 featured a core residential space composed of water, square landscaping, and other supporting facilities. It positioned commercial facilities and a larger other supporting facility on the side of the main road. The original sample 92 had a simpler design, with only water and small squares as the core space, along with some small supporting facilities distributed along the residential area. In sample 92-B1, the core residential space consisted of a large square and green area, which required more control over its scale due to its substantial size. The core residential space in sample 92-B2 was formed by other supporting facilities and a square landscape. While the inner ring road from the original sample was retained, a portion of the open square was designed by extending it along the left side of the main road towards the exterior. Upon comparing the two solutions above, it became evident that the generated sample B2 placed greater emphasis on shaping spatial structure and green landscape space compared to B1. Additionally, the overall solution was more mature, encompassing all the elements of a residential planning study. Its spatial structure and green landscape space were shaped with greater flexibility and diversity than the original sample, featuring better scale control and a more complete form.

After comparing and evaluating the generated results from parameter set E (B1) with the generated results (B2), which were obtained by mixing the training of the newly generated sample set of 285 solutions, it was evident that, from an overall perspective of diversity and contrasting learning, the design dimension was more effective in generating samples B1 and B2. This suggests that machine learning can generate innovative solutions and provide design ideas, aligning with the concept that machine learning can generate innovative building graphics and section designs through 3D models, as previously confirmed by other studies [50]. The results also verified that positive terms used for evaluating the solutions generated at the diversity level were primarily "structured" and "formally diverse". "Structured" implies that machine-learned solutions produced monocentric or polycentric spatial structures, while "diverse" signifies the diverse spatial structures formed by combining amenities, squares, and green spaces. It was noteworthy that the positive word "diverse" appeared more frequently in the sample B2 generation than in sample B1, suggesting that the performance of the data-enhanced design solutions was more inspiring to designers. At the "relative property" level, the positive words for the generated schemes were mainly "heterogeneous". Interestingly, some of the solutions that exhibited "sequential" positive words in generation sample B1 transformed into "heterogeneous" in generation sample B2 (as shown in Figure 10). This indicates that after data enhancement, machine learning for scheme results displayed more design flexibility and showcased the innovative potential of square pavement and landscaped green space to conform to the building layout. For example, Schemes 212 and 109 demonstrated the design flexibility of paving and landscaping in response to the building layout.

Figure 9. Partial display table for generating the sample set.

The results of this study primarily emphasize the autonomous exploration of RSPL in generative design preferences. In contrast, earlier studies focused on analyzing how machine learning can assist in optimizing and reconfiguring the spatial structure of planned designs [51]. This novel approach to applying machine learning in plan layouts allows for a more robust exploration of the potential of machine learning-assisted applications in RSPLs.

3.4. Standard Dimension Determines Generative Preferred Design Features

By analyzing the 12 selected display solutions and considering specification indicators such as plot ratio, building density, and active land use proportion, it became evident that the building density in the original sample set A1 and the generated sample sets B1 and B2 is identical and complies with the relevant design specifications (Figure 11a). However, the fluctuation range of the building density of the generated sample B2 is smaller than that of both the original and generated sample B1 (ranging from 25% to 39%). In contrast, the building density in the original sample A1 exhibits a fluctuation range between 20% and 39%. Although this plot ratio is lower than the original sample set, its fluctuation range is also smaller, maintained between 1.1–and 3 (as shown in Figure 11b). In contrast, the plan area ratio of the original sample A1 and the generated sample B1 has a fluctuation range exceeding 3. This phenomenon is presumably linked to the selection of more mid-rise building height solutions, indicating that the machine learning-generated solutions predominantly feature mid-rise building heights. Nevertheless, it is also confirmed that the performance of the machine-generated residential area scheme becomes more stable after data enhancement.

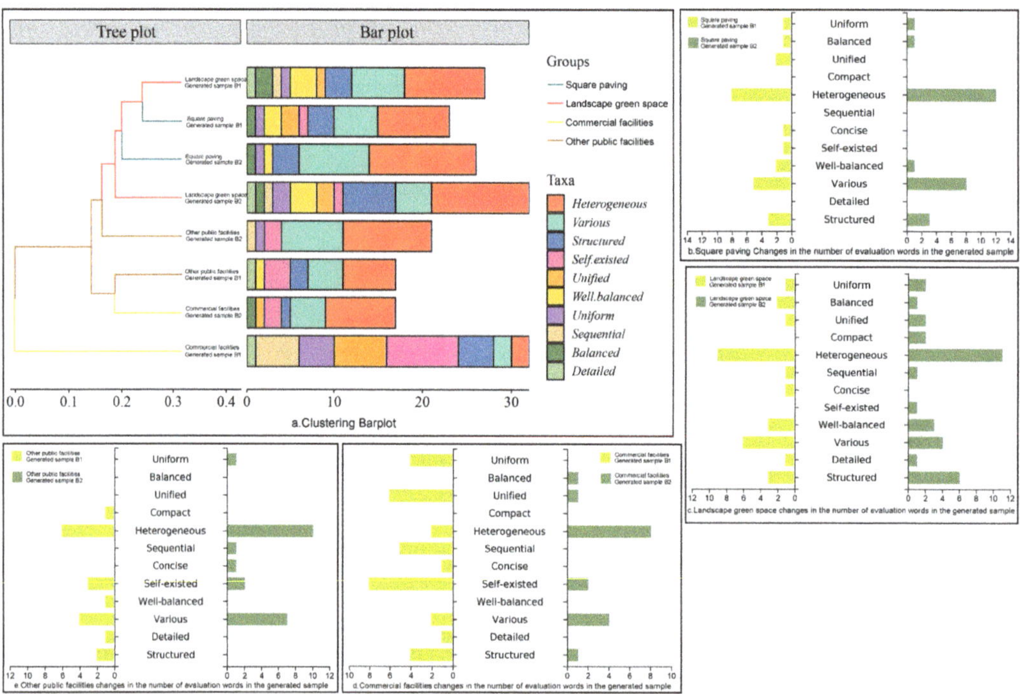

Figure 10. Evaluation results via graphs in the design dimension (Figure 9).

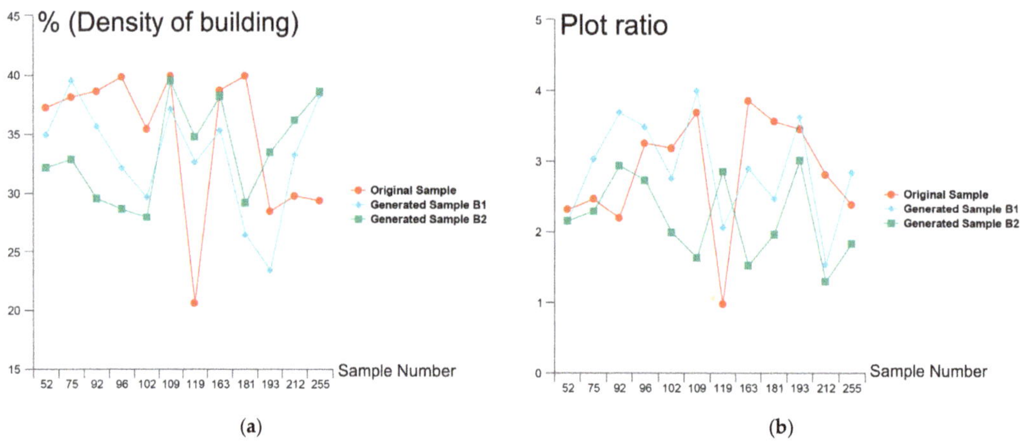

(**a**)

(**b**)

Figure 11. (**a**) Graphs of building density generated for some of the display samples in Figure 9; (**b**) Graphs of a plot ratio generated for some of the display samples in Figure 9.

Regarding the machine learning effect on the land of each activity, the generated square pavement and green landscape space became more or less concurrent compared to the original sample. In contrast, the commercial and other supporting facilities appeared smaller or converged than in the original sample (Figure 8). However, according to the statistics presented in Table 2, the commercial facilities and other supporting facilities in the generated sample B1 increased compared to the original sample. It is important to note that these statistics pertain to the proportion of occupied land, demonstrating that the scale of commercial facilities and other supporting facilities in the generated samples B1 and B2 is smaller than that in the original sample. This aligns with the previously explained machine learning-generated design scheme, which pursues a more balanced layout effect. When comparing, it becomes apparent that a significant portion of the square paving and landscaped green space in generated sample B2 converged to a greater or lesser extent when compared to generated sample B1. Commercial facilities tended to be fewer or more condensed, while other supporting facilities mostly remained unchanged (Figure 12). It appears that the machine learning effect may have been more successful in replicating the commercial facilities lined up along the street in the machine learning scheme, as most of the generated schemes did not exhibit this particular performance characteristic. This could be attributed to the fact that most of the commercial facilities in this study were commercial facilities lined up along the street. However, the results of our generation sample B2 were influenced by a mixture of the five groups A, B, C, D, and E with better experimental results, which were then re-generated. Consequently, the likelihood that the machine needed to learn about commercial facilities increased. This might explain why the generated sample B2 had fewer converging commercial facilities compared to the generated sample B1.

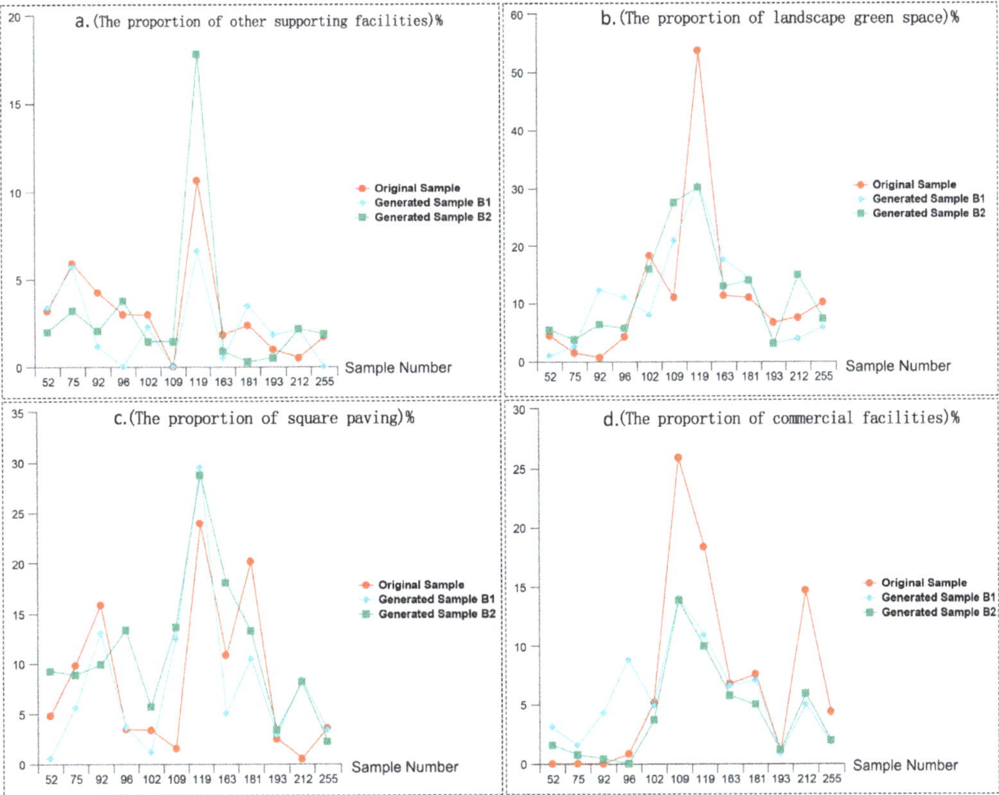

Figure 12. Each element of the active land proportion separately shown in part in Figure 9.

As mentioned earlier, following machine learning, residential design elements exhibit varying effects, with other public facilities and spatial structures proving to be the most influential and displaying diverse characteristics. Therefore, it is crucial to explore how to fully harness the potential and direction of applying design elements related to other public facilities and spatial structures. "CURPADS" focuses more on creating living circles in residential areas. It advocates the division of amenities into internal communities and settlements [52,53] and attaches more attention to public facilities to guide and define the division of residential communities. This study confirmed that the amenities in the generated scheme fulfill the role of guiding the spatial structure of the settlement that machine learning can achieve. Consequently, the machine learning preference design was considered to optimize the creation of living circles in residential areas. In studies concerning the configuration and layout of public facilities and green landscape spaces in urban planning, Khodaparasti et al. introduced an "integrated location-allocation" model to optimize the equity and efficiency of medical service facility locations [54]. Wang et al. used machine learning algorithms and POI data to select the location of elderly facilities in Wuhan [55]. The green space network of Lijiang City was constructed by Ren et al. using satellite images to design green space as a multimodal space of points, linear bars, and irregular shapes [56]. Also, the studies above can be used to analyze the design preferences and characteristics of residential areas found in this study. This allows for the studies above of public facilities, landscape green space layout, and different types of spaces to help create urban public facilities and landscape green spaces. As a result, their application potential would be expanded while contributing to the development of the urban planning discipline. In the future, we can explore additional applications of the Pix2pix model within RSPL as a specific application area, thus uncovering further value-added possibilities for the Pix2pix model.

Finally, this paper not only explores generative design preferences at the plan image design level but also considers how to enable the machine to discover generative design preferences for spatial design from the spatial planning level, which is a topic for future research experiments. Ideally, in the future, we will continue to optimize the performance of the Pix2pix model to enhance the stability of its training, leading to improved image resolution in generated designs. This will increase the generalizability of the Pix2pix model and expand its application value in the field of RSPL on a larger scale. In addition, we attempted to optimize the algorithm for LPIPS metrics to derive a machine learning visual perception evaluation metric that aligns more closely with human design thinking. The goal is to use this as a classification basis to score generated balanced layout surfaces, conduct an in-depth classification study of their balanced layout characteristics, and explore the applicability of balanced layout effects in various urban design schemes.

4. Conclusions

In this study, we conducted an experimental exploration of machine learning generative design preferences in RSPL using the Pix2pix model. The aim was to uncover machine learning's generative design preferences in RSPL and assess its feature performance, with the potential to enhance applications in residential planning and urban planning development. By analyzing design and feature performance choices, government authorities can identify the most promising urban planning areas. The following conclusions and reflections can be drawn from this experimental study on a case study of residential areas in China:

1. The experimental framework of the "extraction-translation-machine learning-evaluation" proposed in this study addressed the deficiency of simultaneously considering all design elements of residential areas within the same methodological framework. This methodological framework integrated both machine and manual computations, as well as quantitative and qualitative evaluation techniques, to jointly determine research outcomes and comprehensively characterize the scientific nature of this study.

Furthermore, this experimental framework established a methodological paradigm for machine learning-assisted plan layout explorations.

2. Machine learning favors the generation of a balanced layout and showcases the innovative design potential of various elements in harmony with housing design components. When comparing the residential area before and after machine learning, it was observed that the generated plan exhibited less fluctuation in terms of building density, floor area ratio, and active land ratio compared to the original plan. Furthermore, the comparison of two design elements, square paving and green landscape space, reveals that machine learning aligns well with the building layout and offers innovative and diverse design perspectives. This, in turn, provides inspirational ideas for residential area layout design and promotes the enhancement of environmental quality within the residential area.

3. Machine learning exhibits a more pronounced generative preference for two design elements: other public facilities and spatial structures. When comparing the generated designs before and after machine learning, there was an increase in the number of design elements. RGB pixels were assigned to form large blocks of other public facilities and spatial structures that were connected and distributed in fragments. Furthermore, the machine-learned design element of other public facilities highlights the master-centered nature of the site. In the process of learning spatial structure, both monocentric and polycentric characteristics of residential spatial structures were generated, resulting in various forms of spatial structure design. Ultimately, this can aid planners in developing schemes that better align with residents' expectations. It also contributes to the discipline of urban planning by offering design ideas for the layout of urban infrastructure, public facilities, landscaped green spaces, and diverse spatial configurations.

Author Contributions: Conceptualization, P.S. and F.Y.; methodology, P.S. and F.Y.; software, P.S. and Q.H.; validation, P.S. and Q.H.; formal analysis, P.S.; investigation, P.S.; resources, P.S.; data curation, P.S.; writing—original draft preparation, P.S.; writing—review and editing, P.S., F.Y. and H.L.; visualization, P.S. and Q.H.; supervision, F.Y.; funding acquisition, F.Y. All authors have read and agreed to the published version of the manuscript.

Funding: This research was funded by the National Natural Science Foundation of China, grant number 42341207.

Data Availability Statement: Not applicable.

Conflicts of Interest: The authors declare no conflict of interest.

Appendix A

Table A1. Residential schemes collection website source.

Residential area scheme collection website source	https://www.om.cn/	accessed on 9 April 2022
	https://www.doczhi.com/	accessed on 16 April 2022
	https://www.gstarcad.com/	accessed on 23 April 2022
	https://www.znzmo.com/	accessed on 28 April 2022

Table A2. Five sets of experiment generated results were scored.

Score Information			Group A Score	Group B Score	Group C Score	Group D Score	Group E Score
Score Identity		**Number**					
Non-urban planning major students		1	3.6	2.8	3.8	3.2	4.2
		2	1.5	2.7	3.1	3.4	3.9

Table A2. *Cont.*

Score Identity	Number	Group A Score	Group B Score	Group C Score	Group D Score	Group E Score
	3	3.5	2.9	3.2	2.3	3.4
	4	2.5	2.3	2.7	3.0	3.2
	5	3.9	3.7	4.0	4.2	4.5
	6	2.6	2.6	3.3	3.6	3.8
	7	3.7	3.4	3.8	4.1	4.8
Non-urban planning major students	8	1.9	2.3	3.4	3.7	4.4
	9	2.8	2.9	3.5	3.1	3.6
	10	3.8	4.2	4.1	4.3	4.7
	11	2.7	3.6	3.4	3.8	4.1
	12	1.6	2.4	3.0	3.3	3.7
	13	0.8	1.3	2.5	2.7	3.1
	14	2.1	2.5	2.8	3.1	3.4
	15	1.1	1.6	2.1	2.6	2.9
	16	2.0	1.8	2.3	3.5	3.9
	17	2.3	3.0	3.8	3.2	4.2
	18	1.7	2.9	2.3	3.7	3.8
	19	2.6	2.1	3.7	3.1	3.9
	20	1.9	2.8	3.6	4.3	4.7
	21	2.3	2.4	3.3	3.7	3.9
Urban planning major students	22	2.6	2.7	3.1	3.8	4.1
	23	1.2	2.1	2.7	3.3	3.7
	24	2.5	2.9	2.4	3.1	3.5
	25	2.8	3.6	3.4	4.1	4.3
	26	2.4	3.2	3.9	4.5	4.7
	27	1.4	2.2	2.7	3.0	3.3
	28	2.2	3.6	3.4	4.2	4.6
	29	1.8	2.5	3.3	2.2	3.6
	30	0.9	1.8	2.9	2.7	3.2
average value		2.25	2.3	3.35	2.95	3.7

References

1. Dan, Y.; Zhao, Y.; Li, X.; Li, S.; Hu, M.; Hu, J. Generative adversarial networks (GAN) based efficient sampling of chemical composition space for inverse design of inorganic materials. *NPJ Comput. Mater.* **2020**, *6*, 84. [CrossRef]
2. Atance, S.R.; Diez, J.V.; Engkvist, O.; Olsson, S.; Mercado, R. De novo drug design using reinforcement learning with graph-based deep generative models. *J. Chem. Inf. Model.* **2022**, *62*, 4863–4872. [CrossRef]
3. Tang, Z.; Ye, Y.; Jiang, Z.; Fu, C.; Huang, R.; Yao, D. A data-informed analytical approach to human-scale greenway planning: Integrating multi-sourced urban data with machine learning algorithms. *Urban For. Urban Green.* **2020**, *56*, 126871. [CrossRef]
4. Zhao, X.; Zhang, T.; Xiao, W. An Automated Design Method for Plane Trusses Based on User Preference Information. *Appl. Sci.* **2023**, *13*, 1543. [CrossRef]
5. Frimpong, B.F.; Koranteng, A.; Atta-Darkwa, T.; Junior, O.F.; Zawiła-Niedźwiecki, T. Land Cover Changes Utilising Landsat Satellite Imageries for the Kumasi Metropolis and Its Adjoining Municipalities in Ghana (1986–2022). *Sensors* **2023**, *23*, 2644. [CrossRef]

6. Liu, Y.; Chen, M.; Wang, M.; Huang, J.; Thomas, F.; Rahimi, K.; Mamouei, M. An interpretable machine learning framework for measuring urban perceptions from panoramic street view images. *iScience* **2023**, *26*, 106132. [CrossRef]

7. Zheng, L.; Chen, Y.; Jiang, S.; Song, J.; Zheng, J. Predicting the distribution of COVID-19 through CGAN—Taking Macau as an example. *Front. Big Data* **2023**, *6*, 1008292. [CrossRef] [PubMed]

8. Lu, Y.; Zhou, X.-H.; Xiao, H.; Li, Q. Using machine learning to predict urban canopy flows for land surface modeling. *Geophys. Res. Lett.* **2023**, *50*, e2022GL102313. [CrossRef]

9. Li, H.; Zhang, G.; Zhong, Q.; Xing, L.; Du, H. Prediction of Urban Forest Aboveground Carbon Using Machine Learning Based on Landsat 8 and Sentinel-2: A Case Study of Shanghai, China. *Remote Sens.* **2023**, *15*, 284. [CrossRef]

10. Silva, L.A.; Sales Mendes, A.; Sánchez San Blas, H.; Caetano Bastos, L.; Leopoldo Gonçalves, A.; Fabiano de Moraes, A. Active Actions in the Extraction of Urban Objects for Information Quality and Knowledge Recommendation with Machine Learning. *Sensors* **2022**, *23*, 138. [CrossRef] [PubMed]

11. Salata, S.; Ronchi, S.; Arcidiacono, A. Mapping air filtering in urban areas. A Land Use Regression model for Ecosystem Services assessment in planning. *Ecosyst. Serv.* **2017**, *28*, 341–350. [CrossRef]

12. Zhang, N.; Deng, S.; Chen, H.; Chen, X.; Chen, J.; Li, X.; Zhang, Y. Structured knowledge base as prior knowledge to improve urban data analysis. *ISPRS Int. J. Geo-Inf.* **2018**, *7*, 264. [CrossRef]

13. Cong, X. *Research on the Design Method of Generating Strong Drainage Scheme for Residential Area Based on CGAN*; Harbin Institute of Technology: Harbin, China, 2020.

14. Dai, Y.; Li, J. Impact of Participatory Community Planning on Publicity in Public Space Renewal Based on Machine Learning Algorithm Based on Child-friendly Concept. *Int. Trans. Electr. Energy Syst.* **2022**, *2022*, 5903528. [CrossRef]

15. Elariane, S.A. Location based services APIs for measuring the attractiveness of long-term rental apartment location using machine learning model. *Cities* **2022**, *122*, 103588. [CrossRef]

16. Alaçam, S.; Karadag, I.; Güzelci, O.Z. Reciprocal style and information transfer between historical Istanbul Pervititch Maps and satellite views using machine learning. *Estoa. Rev. Fac. Arquit. Urban. Univ. Cuenca* **2022**, *11*, 97–113. [CrossRef]

17. Wang, X.; Yan, H.; Huo, C.; Yu, J.; Pant, C. Enhancing Pix2Pix for remote sensing image classification. In Proceedings of the 2018 24th International Conference on Pattern Recognition (ICPR), Beijing, China, 20–24 August 2018; IEEE: Piscataway, NJ, USA, 2018; pp. 2332–2336.

18. Isola, P.; Zhu, J.Y.; Zhou, T.; Efros, A.A. Image-to-image translation with conditional adversarial networks. In Proceedings of the IEEE Conference on Computer Vision and Pattern Recognition, Honolulu, HI, USA, 21–26 July 2017; pp. 1125–1134.

19. Fu, B.; Gao, Y.; Wang, W. Dual generative adversarial networks for automated component layout design of steel frame-brace structures. *Autom. Constr.* **2023**, *146*, 104661. [CrossRef]

20. Zhao, C.W.; Yang, J.; Li, J. Generation of hospital emergency department layouts based on generative adversarial networks. *J. Build. Eng.* **2021**, *43*, 102539. [CrossRef]

21. Zhu, J.Y.; Park, T.; Isola, P.; Efros, A.A. Unpaired image-to-image translation using cycle-consistent adversarial networks. In Proceedings of the IEEE International Conference on Computer Vision, Venice, Italy, 22–29 October 2017; pp. 2223–2232.

22. Chen, J.S.; Coyner, A.S.; Chan, R.P.; Hartnett, M.E.; Moshfeghi, D.M.; Owen, L.A.; Kalpathy-Cramer, J.; Chiang, M.F.; Campbell, J.P. Deepfakes in ophthalmology: Applications and realism of synthetic retinal images from generative adversarial networks. *Ophthalmol. Sci.* **2021**, *1*, 100079. [CrossRef] [PubMed]

23. Shen, Y.; Huang, R.; Huang, W. GD-StarGAN: Multi-domain image-to-image translation in garment design. *PLoS ONE* **2020**, *15*, e0231719. [CrossRef] [PubMed]

24. Choi, Y.; Choi, M.; Kim, M.; Ha, J.W.; Kim, S.; Choo, J. Stargan: Unified generative adversarial networks for multi-domain image-to-image translation. In Proceedings of the IEEE Conference on Computer Vision and Pattern Recognition, Salt Lake City, UT, USA, 18–23 June 2018; pp. 8789–8797.

25. Wan, P.; He, H.; Guo, L.; Yang, J.; Li, J. InfoGAN-MSF: A data augmentation approach for correlative bridge monitoring factors. *Meas. Sci. Technol.* **2021**, *32*, 114008. [CrossRef]

26. Chen, X.; Duan, Y.; Houthooft, R.; Schulman, J.; Sutskever, I.; Abbeel, P. Infogan: Interpretable representation learning by information maximizing generative adversarial nets. In Proceedings of the Advances in Neural Information Processing Systems, Barcelona, Spain, 5–10 December 2016; Volume 29.

27. Mao, X.; Li, Q.; Xie, H.; Lau, R.Y.; Wang, Z.; Paul Smolley, S. Least squares generative adversarial networks. In Proceedings of the IEEE International Conference on Computer Vision, Venice, Italy, 22–29 October 2017; pp. 2794–2802.

28. Karras, T.; Aila, T.; Laine, S.; Lehtinen, J. Progressive growing of gans for improved quality, stability, and variation. *arXiv* **2017**, arXiv:1710.10196.

29. Zhang, H.; Goodfellow, I.; Metaxas, D.; Odena, A. Self-attention generative adversarial networks. In Proceedings of the International Conference on Machine Learning, PMLR, Long Beach, CA, USA, 9–15 June 2019; pp. 7354–7363.

30. Chaillou, S. *AI+Architecture: Towards a New Approach*; Harvard Graduate School of Design: Cambridge, MA, USA, 2019; pp. 35–52.

31. Yousif, S.; Bolojan, D.; Anastasia, G.; Jeroen, A.; Adam, F. Deep-Performance: Incorporating Deep Learning for Automating Building Performance Simulation in Generative Systems. *Assoc. Comput.-Aided Archit. Des. Res. Asia (CAADRIA)* **2021**, *1*, 151–160.

32. Newton, D. Deep generative learning for the generation and analysis of architectural plans with small datasets. In Proceedings of the 37th eCAADe and 23rd SIGraDi Conference, Porto, Portugal, 11–13 September 2019; Volume 2, pp. 21–28.

33. Yu, Q.; Malaeb, J.; Ma, W. Architectural facade recognition and generation through generative adversarial networks. In Proceedings of the 2020 International Conference on Big Data & Artificial Intelligence & Software Engineering (ICBASE), Bangkok, Thailand, 30 October–1 November 2020; IEEE: Piscataway, NJ, USA, 2020; pp. 310–316.
34. Mostafavi, F.; Tahsildoost, M.; Zomorodian, Z.S.; Shahrestani, S.S. An interactive assessment framework for residential space layouts using pix2pix predictive model at the early-stage building design. *Smart Sustain. Built Environ.* **2022**. [CrossRef]
35. Gu, D.; Chen, W.; Lu, X. Automated assessment of wind damage to windows of buildings at a city scale based on oblique photography, deep learning and CFD. *J. Build. Eng.* **2022**, *52*, 104355. [CrossRef]
36. *GB 50180-2018*; China's Urban Residential Planning and Design Standard. Ministry of Housing and Urban-Rural Development. China Construction Industry Press: Beijing, China, 2021.
37. Ronneberger, O.; Fischer, P.; Brox, T. U-net: Convolutional networks for biomedical image segmentation. In *Medical Image Computing and Computer-Assisted Intervention–MICCAI 2015: 18th International Conference, Munich, Germany, 5–9 October 2015, Proceedings, Part III 18*; Springer International Publishing: Cham, Switzerland, 2015; pp. 234–241.
38. Ibrahim, H.; Khattab, Z.; Khattab, T.; Abraham, R. Generative Adversarial Network Approach to Future Sermonizing of Housing Dispersal in Emerging Cities. *J. Urban Plan. Dev.* **2022**, *148*, 04021067. [CrossRef]
39. Wang, Z.; Bovik, A.C.; Sheikh, H.R.; Simoncelli, E.P. Image quality assessment: From error visibility to structural similarity. *IEEE Trans. Image Process.* **2004**, *13*, 600–612. [CrossRef]
40. Rajeswari, G.; Ithaya Rani, P. Face occlusion removal for face recognition using the related face by structural similarity index measure and principal component analysis. *J. Intell. Fuzzy Syst.* **2022**, *42*, 5335–5350. [CrossRef]
41. Singh, R.; Aggarwal, N. A distortion-agnostic video quality metric based on multi-scale spatio-temporal structural information. *Signal Process. Image Commun.* **2019**, *74*, 299–308. [CrossRef]
42. Zhang, R.; Isola, P.; Efros, A.A.; Shechtman, E.; Wang, O. The unreasonable effectiveness of deep features as a perceptual metric. In Proceedings of the IEEE Conference on Computer Vision and Pattern Recognition, Salt Lake City, UT, USA, 18–23 June 2018; pp. 586–595.
43. Lee, K.W.; Chin, R.K.Y. Diverse COVID-19 CT Image-to-Image Translation with Stacked Residual Dropout. *Bioengineering* **2022**, *9*, 698. [CrossRef] [PubMed]
44. Jungang, Y.; Xiaotao, H.; Tian, J.; Guoyi, X.; Zhimin, Z. An interpolated phase adjustment by contrast enhancement algorithm for SAR. *IEEE Geosci. Remote Sens. Lett.* **2010**, *8*, 211–215. [CrossRef]
45. Cubuk, E.D.; Zoph, B.; Mane, D.; Vasudevan, V.; Le, Q.V. Autoaugment: Learning augmentation strategies from data. In Proceedings of the IEEE/CVF Conference on Computer Vision and Pattern Recognition, Long Beach, CA, USA, 15–20 June 2019; pp. 113–123.
46. Recio, G.; Conrad, M.; Hansen, L.B.; Jacobs, A.M. Pleas. On pleasure and thrill: The interplay between arousal and valence during visual word recognition. *Brain Lang.* **2014**, *134*, 34–43. [CrossRef]
47. Schenk, L. *Designing Cities: Basics, Principles and Practice*; Engineering and Construction Industry Press, China State Construction: Beijing, China, 2021.
48. Hong, M.; An, C.; Guo, K. A study of service design-based shared housing spaces: Focusing on Chinese "poyu" shared housing. *Prod. Cult. Des. Stud.* **2022**, *69*, 163–177.
49. Yang, Y.F.; Liao, S.M.; Liu, M.B.; Wu, D.P.; Pan, W.Q.; Li, H. A new construction method for metro stations in dense urban areas in Shanghai soft ground: Open-cut shafts combined with quasi-rectangular jacking boxes. *Tunn. Undergr. Space Technol.* **2022**, *125*, 104530. [CrossRef]
50. Zhang, H. 3D model generation on architectural plan and section training through machine learning. *Technologies* **2019**, *7*, 82. [CrossRef]
51. Wang, S.; Qin, A. Design and Implementation of a Multidimensional Visualization Reconstruction System for Old Urban Spaces Based on Neural Networks. *Comput. Intell. Neurosci.* **2022**, *2022*, 4253128. [CrossRef] [PubMed]
52. Ma, W.; Wang, N.; Li, Y.; Sun, D.J. 15-min pedestrian distance life circle and sustainable community governance in Chinese metropolitan cities: A diagnosis. *Humanit. Soc. Sci. Commun.* **2023**, *10*, 364. [CrossRef]
53. Wu, H.; Wang, L.; Zhang, Z.; Gao, J. Analysis and optimization of 15-minute community life circle based on supply and demand matching: A case study of Shanghai. *PLoS ONE* **2021**, *16*, e0256904. [CrossRef] [PubMed]
54. Khodaparasti, S.; Maleki, H.R.; Bruni, M.E.; Jahedi, S.; Beraldi, P.; Conforti, D. Balancing efficiency and equity in location-allocation models with an application to strategic EMS design. *Optim. Lett.* **2016**, *10*, 1053–1070. [CrossRef]
55. Wang, Z.; Wang, X.; Dong, Z.; Li, L.; Li, W.; Li, S. More Urban Elderly Care Facilities Should Be Placed in Densely Populated Areas for an Aging Wuhan of China. *Land* **2023**, *12*, 220. [CrossRef]
56. Ren, Y.; Wang, D.; Wang, D.; Chen, F. Designing a green-space network with geospatial technology for Lijiang City. *Int. J. Sustain. Dev. World Ecol.* **2011**, *18*, 503–508. [CrossRef]

Article

Pre-Harvest Corn Grain Moisture Estimation Using Aerial Multispectral Imagery and Machine Learning Techniques

Pius Jjagwe [1,2], Abhilash K. Chandel [1,2,*] and David Langston [1]

[1] Virginia Tech Tidewater Agricultural Research and Extension Center, Suffolk, VA 23437, USA; pjjagwe@vt.edu (P.J.); dblangston@vt.edu (D.L.)
[2] Department of Biological Systems Engineering, Virginia Tech, Blacksburg, VA 24061, USA
*   Correspondence: abhilashchandel@vt.edu

Citation: Jjagwe, P.; Chandel, A.K.; Langston, D. Pre-Harvest Corn Grain Moisture Estimation Using Aerial Multispectral Imagery and Machine Learning Techniques. *Land* **2023**, *12*, 2188. https://doi.org/10.3390/land12122188

Academic Editor: Chuanrong Zhang

Received: 23 November 2023
Revised: 14 December 2023
Accepted: 15 December 2023
Published: 18 December 2023

Abstract: Corn grain moisture (CGM) is critical to estimate grain maturity status and schedule harvest. Traditional methods for determining CGM range from manual scouting, destructive laboratory analyses, and weather-based dry down estimates. Such methods are either time consuming, expensive, spatially inaccurate, or subjective, therefore they are prone to errors or limitations. Realizing that precision harvest management could be critical for extracting the maximum crop value, this study evaluates the estimation of CGM at a pre-harvest stage using high-resolution (1.3 cm/pixel) multispectral imagery and machine learning techniques. Aerial imagery data were collected in the 2022 cropping season over 116 experimental corn planted plots. A total of 24 vegetation indices (VIs) were derived from imagery data along with reflectance (REF) information in the blue, green, red, red-edge, and near-infrared imaging spectrum that was initially evaluated for inter-correlations as well as subject to principal component analysis (PCA). VIs including the Green Normalized Difference Index (GNDVI), Green Chlorophyll Index (GCI), Infrared Percentage Vegetation Index (IPVI), Simple Ratio Index (SR), Normalized Difference Red-Edge Index (NDRE), and Visible Atmospherically Resistant Index (VARI) had the highest correlations with CGM (r: 0.68–0.80). Next, two state-of-the-art statistical and four machine learning (ML) models (Stepwise Linear Regression (SLR), Partial Least Squares Regression (PLSR), Artificial Neural Network (ANN), Support Vector Machine (SVM), Random Forest (RF), and K-nearest neighbor (KNN)), and their 120 derivates (six ML models × two input groups (REFs and REFs+VIs) × 10 train–test data split ratios (starting 50:50)) were formulated and evaluated for CGM estimation. The CGM estimation accuracy was impacted by the ML model and train-test data split ratio. However, the impact was not significant for the input groups. For validation over the train and entire dataset, RF performed the best at a 95:5 split ratio, and REFs+VIs as the input variables (r_{train}: 0.97, $rRMSE_{train}$: 1.17%, r_{entire}: 0.95, $rRMSE_{entire}$: 1.37%). However, when validated for the test dataset, an increase in the train–test split ratio decreased the performances of the other ML models where SVM performed the best at a 50:50 split ratio (r = 0.70, rRMSE = 2.58%) and with REFs+VIs as the input variables. The 95:5 train–test ratio showed the best performance across all the models, which may be a suitable ratio for relatively smaller or medium-sized datasets. RF was identified to be the most stable and consistent ML model (r: 0.95, rRMSE: 1.37%). Findings in the study indicate that the integration of aerial remote sensing and ML-based data-run techniques could be useful for reliably predicting CGM at the pre-harvest stage, and developing precision corn harvest scheduling and management strategies for the growers.

Keywords: aerial multispectral sensing; corn grain moisture; machine learning; precision harvest

1. Introduction

Grain moisture is critical for determining optimum harvest schedules for crops, which has economic implications during harvest and storage. Markets and safe storages require crops to be harvested at a grain moisture content between 13 to 15.5%, depending on the crop, its variety, and storage duration [1]. Harvesting below this range leads to yield losses due to grain shrinkage, lodging, and grain dropping during harvest as well as bird interference. On the other hand, harvesting at a moisture level above this range risks fungal infection during storage, requires additional costs and infrastructure for artificial drying, and eventually discounted prices at sales points. Under both situations grain yield, quality, and net returns are at risk [2]. Corn grain moisture (CGM) decreases from about 85% during the silking stage to 30% at around maturity through dehydration [3]. This dehydration occurs in two steps in the field: (i) during maturation, and (ii) post-maturity [4]. As the grain approaches physiological maturity (i.e., maturation dehydration), the assimilates of starch and protein displace water molecules within the grain [1,5,6]. During the post-maturity stage, the grain moisture is lost through exchange with the atmosphere, and this dehydration is influenced by air temperature, relative humidity, and husk weight and thickness [6].

Conventionally, corn growers assess grain moisture indirectly by spotting the milk line and black layer around the grain to determine harvest dates. Among direct methods, cup-shaped capacitive units and portable grain analyzers are used in fields [7]. Another traditional technique is oven drying [4]. Researchers have also developed moisture detection techniques based on the electrical and dielectric characteristics of the grains [8,9]. For non-invasive estimation, generalized growing degree days-based models are used to determine grain moisture and dry down periods (GDDs) [1]. However, this approach provides minimum accountability of localized soil factors, crop varieties, crop management practices, and tillage practices that may impact CGM at spatiotemporal scales. Nonetheless, all these methods are either destructive, time consuming, spatially inaccurate, subjective, or expensive, therefore they are prone to errors or limitations [6–8,10]. Given these limitations, there is a great need for techniques that not only determine CGM non-destructively but are high-throughput in nature as well as account for spatial variability.

Remote sensing is a convenient, timely, high-throughput, and precise technique for the non-destructive assessment of crop physiology and health such as for water [11], chlorophyll or nitrogen, disease infection, and pest infestation, among others, for different crops [12]. This makes remote sensing an extremely useful tool for guiding precision agriculture operations [1]. Pertaining to corn or field crops, research has been maximally restricted to the use of remote sensing with vegetation indices (VIs) or machine learning (ML) techniques for yield predictions [13–16]. Whereas very limited explorations have been conducted for estimating CGM using remote sensing. One study so far utilized satellite-based remote sensing imagery for estimating CGM in China using vegetation indices (VIs) as inputs to the crop-physiological model and observed R^2 values of up to 0.9 [16]. However, satellite-based remote sensing is highly restricted due to fixed data acquisitions, spatial resolutions, and cloud cover issues especially in coastal ecosystems [17–19]. On the other hand, small unmanned aircraft system (SUAS) platforms are widely adopted for precision agriculture operations due to the advantages of providing on-demand data at the desired spatial resolution, and avoidance of atmospheric and cloud interferences [12,15,20,21].

The advancement of data-run techniques such as ML has revolutionized precision agriculture operations significantly in recent years by broadening the horizons for crop health estimations as well as yield forecasting [15]. Some of the most widely used ML models include the support vector machine (SVM), random forest (RF), k-nearest neighbor (KNN), and artificial neural network (ANN), among others [15]. These models deploy an approach of supervised learning, which are trained to approximate complexities between the input and output variables. This enhances robustness and generalizability of ML for estimations compared to other conventional statistical or empirical models. ML is also capable of handling overfitting, remains unaffected by collinearity, number, or non-normal distribution of the variables, and does not require scale normalizations [22].

Given the restricted research of using high-resolution remote sensing and ML mostly for yield predictions, this study addresses an important gap of estimating CGM using SUAS-based multispectral imagery and a range of statistical and ML models. This would eventually help determine precision corn harvest schedules for the growers. It is also important to note that ML techniques have been restrictively evaluated for the number of variables, and a typical range of training–testing data-split ratios. Most of the studies generally consider 70:30 or 80:20 train–test splits, which may or may not serve for the small–medium datasets [16]. Therefore, this study also evaluates the influence of those two factors on CGM estimation accuracies. Specific objectives are (1) evaluating aerial multispectral imagery-derived reflectance (REFs) and VIs for assessing CGM, (2) estimating CGM using a range of statistical as well as state-of-the-art ML models, and (3) evaluating the performance of those models when using only REFs and a combination of REFs and VIs as input variables at multiple train–test data split ratios. These evaluations will be validated over the entire dataset (100%) as well as train and test datasets independently.

2. Materials and Methods

2.1. Experimental Details

The study was conducted at an experimental farm of the Tidewater Agriculture Research and Extension Center (TAREC) of Virginia Tech (36°41′7.22″ N, 76°45′57.232″ W), located in Suffolk, VA, USA. The corn seeds were planted between 25–28 April 2022, into a total of 116 plots of 4 rows each that were 30-ft long. These plots were applied with 29 distinct rates and compositions of fungicides at a reproductive growth stage for disease control and to achieve variability in crop vigor for CGM estimation modeling. The crop was harvested on 21 September 2022 (79 DAP (days after planting)) using a plot combine harvester that recorded yield and grain moisture contents for two middle rows of each plot. The combine harvester is equipped with a capacitive-type grain moisture sensor to measure grain moisture and a load sensor to measure yield. No irrigation was applied during the course of the trial.

2.2. Aerial Image Acquisition

Aerial imagery was acquired at vegetative stage-R5 on August 25, 2022 using a DJI Phantom 4 Multispectral quadcopter drone (SZ DJI Technology Co., Shenzhen, China, Figure 1). Imagery data were acquired earlier than the harvest date (i.e., 21 September 2022) to evaluate the feasibility of CGM estimation before the actual harvest operation was deployed. In addition, this is also the stage after which the crop started senescing. The SUAS was equipped with a five-band multispectral imaging sensor with blue (450 nm ± 16 nm), green (560 nm ± 16 nm), red (650 nm ± 16 nm), red-edge (RE: 730 nm ± 16 nm), and near-infrared (NIR: 840 nm ± 26 nm) wavelength sensors of 2.08 megapixels each. DJI Ground Station Pro (DJI GS Pro, version 2.0.17, SZ DJI Technology Co., Shenzhen, China) was used as the ground station control software to set up the SUAS flight mission for an altitude of 25 m above ground level (AGL). This provided multispectral images at a spatial resolution of 1.3 cm/pixel. The multispectral imaging sensor was also configured to capture images at 80% front and 75% side overlaps for seamless orthomosaicing during stitching operations. The SUAS had a real time kinematic (RTK) sensor to receive geolocation corrections for each image as well as a skyward facing downwelling light sensor to record light irradiance during each capture. This light information is used along with the images of a calibrated reflectance panel (6×, Sentera, Inc., St. Paul, MN, USA) that were captured after each flight for radiometric calibration of imagery from the mission. This process eliminates any inconsistencies induced within images due to sunlight fluctuations during the flight mission (Figure 1). The imaging flight was conducted near solar (±2 h) noon period for high-quality crop feature retrieval. The SUAS has an SD card for the storage of acquired imagery.

2.3. Image Analysis

Pre-Processing and Feature Extraction

Initially, multispectral snapshots (1125 images: 225 per waveband) were transferred from the SUAS SD card to a photogrammetry and mapping software platform (Pix4D Mapper, Pix4D, Inc., Lausanne, Switzerland). In this platform five seamless multispectral reflectance orthomosaics pertaining to each type of sensor (blue, green, red, RE, NIR) were obtained as a result of sequential image stitching operations (Figure 2), which include keypoint feature extraction and matching, imagery optimization, georectification, point cloud generation, orthomosiacing, and radiometric calibration.

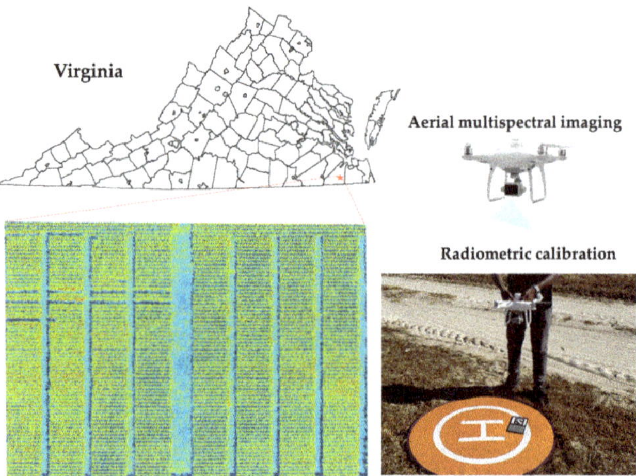

Figure 1. Corn trial plots at Tidewater Agricultural Research and Extension Center in Suffolk, VA, imaged using aerial multispectral platform.

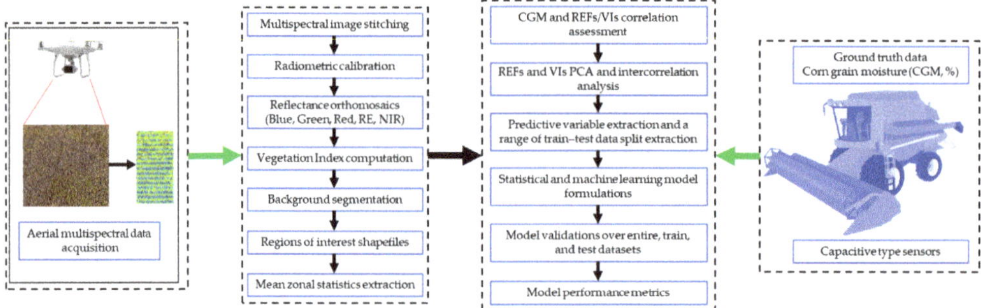

Figure 2. Flowchart showing steps of aerial multispectral image analysis and estimation of corn grain moisture using statistical and machine learning model.

The obtained REF orthomosaics were further processed in QGIS using the "Raster Calculator" toolbar (Figure 2) to obtain 24 VIs (Table 1). These VIs were selected for their significance reported in characterizing crop health under a broad range of growth and agroclimatic conditions. The soil background was segmented out from each VI raster using the histogram separation method [23,24]. Next, a shapefile polygon layer was created, where rectangular areas of interest (AOI) of equal dimensions were drawn around the two central rows of each trial plot. Using this shapefile and the "Zonal Statistics" toolbar, mean REF and VI values for each AOI (of all non-zero and not-a-number pixels) were extracted, which were then exported in the "*.xls" format for further analysis (Figure 2).

Table 1. Vegetation indices extracted from aerial multispectral imagery for corn grain moisture assessments.

Vegetation Index	Equation	Reference
Normalized Difference Vegetation Index (NDVI)	$(NIR - R)/(NIR + R)$	[25]
Infrared Percentage Vegetation Index (IPVI)	$(NIR)/(NIR + R)$	[26]
Green Normal Difference Vegetation Index (GNDVI)	$(NIR - G)/(NIR + G)$	[27]
Green Difference Vegetation Index (GDVI)	$NIR - G$	[28]
Enhanced Vegetation Index (EVI)	$2.5 \times (NIR - R)/(NIR + 6 \times R - 7.5 \times B + 1)$	[29]
Leaf Area Index (LAI)	$3.618 \times EVI - 0.118$	[30]
Modified Non-Linear Index (MNLI)	$(NIR^2 - R) \times (1 + L)/(NIR^2 + R + L)$	[31]
Soil Adjusted Vegetation Index (SAVI)	$1.5 \times (NIR - R)/(NIR + R + 0.5)$	[32]
Optimized Soil Adjusted Vegetation Index (OSAVI)	$(NIR - R)/(NIR + R + 0.16)$	[33]
Green Soil Adjusted Vegetation Index (GSAVI)	$(NIR - G)/(NIR + G + 0.5)$	[32]
Green Optimized Soil Adjusted Vegetation Index (GOSAVI)	$(NIR - G)/(NIR + G + 0.16)$	[32]
Modified Soil Adjusted Vegetation Index (MSAVI2)	$(2 \times NIR + 1 - sqrt((2 \times NIR + 1)^2 - 8 \times (NIR - R)))/2$	[34]
Normalized Difference Red-edge Index (NDRE)	$(NIR - RE)/(NIR + RE)$	[35]
Green Ratio Vegetation Index (GRVI)	NIR/G	[28]
Green Chlorophyll Index (GCI)	$(NIR/G) - 1$	[36]
Green Leaf Index (GLI)	$((G - R) + (G - B))/((2 \times G) + R + B)$	[37]
Simple Ratio (SR)	NIR/R	[38]
Modified Simple Ratio (MSR)	$((NIR/R) - 1)/(sqrt(NIR/R) + 1)$	[39]
Renormalized Difference Vegetation Index (RDVI)	$(NIR - R)/sqrt(NIR + R)$	[40]
Transformed Difference Vegetation Index (TDVI)	$1.5 \times ((NIR - R)/sqrt(NIR + R + 0.5))$	[41]
Visible Atmospherically Resistant Index (VARI)	$(G - R)/(G + R - B)$	[42]
Wide Dynamic Range Vegetation Index (WDRVI)	$(a \times NIR - R)/(a \times NIR + R)$	[43]

R, G, B, RE, and NIR are pixel values of the spectral responses in red, green, blue, red-edge, and near-infrared images.

2.4. Data Analysis and CGM Estimation

A dataset containing CGM measurements (%) along with five REF and 24 VI features was derived for 116 plots. Firstly, data normality was checked, and all the data followed a normal distribution. Then, a Pearson correlation analysis was conducted to identify the association between the CGM and all the derived REF and VI features.

Next, four ML models and two statistical models were formulated for CGM estimation. These models include stepwise linear regression (SLR), partial least-squares regression (PLSR), random forest (RF), k-nearest neighbor (KNN), support vector machine (SVM), and artificial neural network (ANN). In SLR, the variable with the maximum sum of squares of regression is selected first, and then binary regression is formed by selecting an additional variable from the remaining variables. This process repeats until all non-significant variables are eliminated that could induce cofounding effects [44,45]. PLSR combines basic multiple linear regression functions and performs correlation and PCA to eliminate collinearity between variables and maintains relationships with dependent variables, i.e., CGM [46,47]. PLSR also has the capability of avoiding non-normal data. RF is a highly used ML model for agricultural operations that assembles multiple decision trees to estimate a result. The strength of RF is its ability to handle complex datasets and mitigate overfitting for predictive modeling. In this study, the RF model was initially tested with 1000 trees for all dependent variables and optimum trees were identified in the ranges of 300–400 where the prediction accuracy was almost saturated. This hyperparameter tuning was achieved by setting "five variables selected at random" as candidates for each iteration of tuning [48]. KNN performs its function by approximating the association between the independent and dependent variables by averaging the observations in the same neighborhood. In this study, for KNN, repeated cross validation was adopted with three repeats or iterations for up to 30 neighbors. Once the least mean square error was obtained for a particular number of neighbors, that number was used for final model training [49]. SVM identifies a hyperplane in an n-dimensional space that distinctly classifies the data points. This hyperplane is developed iteratively such that the misclassification error is minimal while predicting continuous outputs [50,51]. ANN is a supervised ML model that comprises node layers, namely, an input layer, one or more hidden layers, and an output layer. The structure of ANN is inspired by the brain where each node connects to another with an associated weight and threshold. If the output of any node is above the threshold, that node gets activated and sends the data to the next layer of the network. This process repeats for user-defined iterations until the network's output error reaches the desired value [50,52]. The major advantage of ANN over other statistical or linear models is that it flexibly computes the complicated or non-linear relationships between the input and the outputs. In this study, two hidden layers were selected with ten and three nodes, respectively.

Prior to implementing these models, significant variables that would be used as inputs were identified among the derived 29 REF and VI features. This was completed to complement reduced overfitting and enhanced robustness of ML models for CGM estimation. For this, firstly a principal component analysis was conducted to identify the collinear variables. Two primary axes that explained the main variability, intercorrelations, and dominating pattern of VIs in the data matrix were used to generate the PCA biplots for dimensionality reduction. Next, a pair-wise correlation analysis was conducted between all REF and VI features to reduce the number of variables. A threshold of 0.99 was defined in this pair-wise correlation analysis and variables with correlations above this threshold were identified and variables with largest mean absolute correlation were removed.

In the next step, two groups of input variables, (1) REFs and (2) REFs+VIs, as well as ten training–testing datasets were defined. These training–testing datasets were based on ten split ratios starting from 50:50 up until 95:5 at a 5% increment of the training dataset. These sets of train–test splits were developed to identify and evaluate appropriate training data sizes for the best model performance, especially for small- to medium-sized datasets as in this study (i.e., total 116 data points). For evaluating the estimation model performances, the trained models were implemented on the entire dataset, the testing dataset, as well as the training dataset. Metrics of Pearson correlation (r) and relative root mean square error (rRMSE, %, Equation (1)) were computed to evaluate the model performance and accuracy of CGM estimation. All the ML and statistical modeling, metrics (r and rRMSE) computations, and other analyses were performed with the R statistical computing software (version 4.3.1; RStudio, Inc. Boston, MA, USA) with all statistical analyses inferred at 5% significance.

$$\text{rRMSE } (\%) = 100 \times \frac{\sqrt{\frac{\sum_{i=1}^{n}\left(\text{CGM}_\text{E}-\text{CGM}_\text{m}\right)^2}{n}}}{\text{mean}(\text{CGM}_\text{m})} \tag{1}$$

where CGM_E is the estimated CGM and CGM_m is the measured CGM.

3. Results

3.1. Crop Reflectance and Vegetation Index Feature Evaluation

Pearson's correlation (r) analysis (Figure 3 and Table 2) showed that CGM had strong and significant correlations with REFs in RE, and NIR, and the derived 24 VIs (0.68–0.80). The correlation with REFs in the red band was moderate (r = −0.52). Among the VIs, the highest correlation was observed for GNDVI (r = 0.80) and the lowest for VARI (0.68). Correlations with the REFs in blue and green wavebands were the lowest (−0.27 and 0.05).

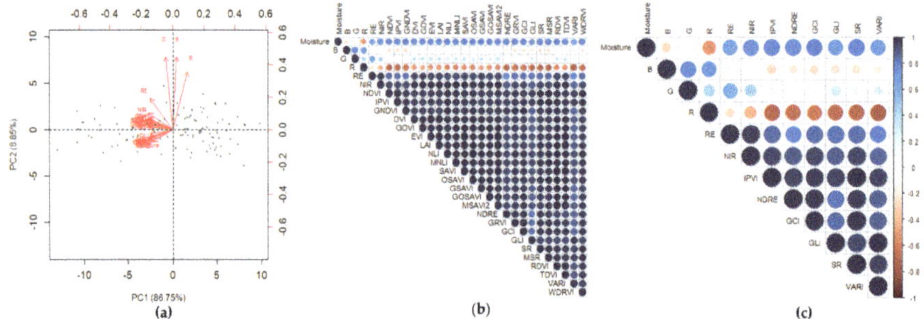

Figure 3. (**a**) Principal component analysis biplot of 24 vegetation indices and five reflectance features accounting for a total of 95.60% of the variability in the data, (**b**) intercorrelation heat map between spectral features, and (**c**) final selected input features after dimensionality reduction.

Table 2. Correlations of reflectance and vegetation indices with corn grain moisture.

Vegetation Index	Pearson Correlation (r)
Blue	−0.27
Green	0.05
Red	−0.52
Red Edge	0.66
Near Infrared	0.74
Normalized Difference Vegetation Index (NDVI)	0.77
Infrared Percentage Vegetation Index (IPVI)	0.77
Green Normal Difference Vegetation Index (GNDVI)	0.80
Difference Vegetation Index (DVI)	0.76
Green Difference Vegetation Index (GDVI)	0.76
Enhanced Vegetation Index (EVI)	0.77
Leaf Area Index (LAI)	0.77
Non-Linear Index (NLI)	0.78
Modified Non-Linear Index (MNLI)	0.76
Soil Adjusted Vegetation Index (SAVI)	0.77
Optimized Soil Adjusted Vegetation Index (OSAVI)	0.78
Green Soil Adjusted Vegetation Index (GSAVI)	0.78
Green Optimized Soil Adjusted Vegetation Index (GOSAVI)	0.79
Modified Soil Adjusted Vegetation Index (MSAVI2)	0.77
Normalized Difference Red-edge Index (NDRE)	0.76
Green Ratio Vegetation Index (GRVI)	0.79
Green Chlorophyll Index (GCI)	0.79
Green Leaf Index (GLI)	0.69
Simple Ratio (SR)	0.77
Modified Simple Ratio (MSR)	0.78
Renormalized Difference Vegetation Index (RDVI)	0.77
Transformed Difference Vegetation Index (TDVI)	0.78
Visible Atmospherically Resistant Index (VARI)	0.68
Wide Dynamic Range Vegetation Index (WDRVI)	0.78

R, G, B, RE, and NIR are reflectance in red, green, blue, red-edge, and NIR images. Correlation coefficients are significant at $p < 0.001$.

3.2. Non-Invasive CGM Estimation with ML

3.2.1. Input Feature Selection

In the PCA, two primary PCs comprising 24 VIs and five REFs accounted for the variability of 86.75% and 8.85% (Total = 95.60%, (Figure 3a)). The eigenvectors for the REF in blue, green, and red wavelengths tended towards the top of the biplot (Figure 3a), so they could be inferred to have more influence on PC-2 while the REFs in RE and NIR wavelengths, as well as all other VIs, formed a dense cluster towards the extreme left, top-left, or lower-left region, so they could be inferred to have more influence on PC-1. The PCA could also visualize numerous VIs that completely coincided or were colinear with other VIs. This observation was also supported by Figure 3b that shows complete intercorrelations (i.e., r = 1) between such VIs. Next, using the function "findCorrelation" in RStudio, we were able to identify the groups of VIs that had complete intercorrelations and among them drop VIs that had the largest mean absolute correlation. The function considers the absolute values of pair-wise correlations between variables and removes the variable with the largest mean absolute correlation. This is similar to removing variables that have lower loadings (determined through PCA) or less representation of the variability in data compared to its colinear variable(s). The process determined five REFs and six VIs that were not collinear and included B, G, R, RE, NIR, IPVI, NDRE, GCI, GLI, SR, and VARI. These were finally used for CGM estimation through statistical and ML models (Figure 3c).

3.2.2. Using Reflectance Features as Inputs

The CGM through statistical and ML models was initially estimated using only the REF features as the predictor variables (Table 3). For validation over the test dataset,

SLR performed the best at a 50:50 split (r = 0.74, rRMSE = 2.43%), followed by PLSR, SVM (50:50), RF, and KNN, and ANN was the weakest performer at the same split ratio (r = 0.61, rRMSE = 4.43%). For validation over the train dataset, RF performed best at a 95:5 split (r = 0.96, rRMSE = 1.31%), followed by ANN (70:30), PLSR (75:25), KNN (75:25), and SVM (70:30). SLR was the weakest performer at a 75:25 split (r = 0.8, rRMSE = 2.37%). For validation over the entire dataset, RF (r = 0.94, rRMSE = 1.51%) performed the best followed by ANN, PLSR, SLR, SVM, while KNN (r = 0.79, rRMSE = 2.6%) was the weakest at a 95:5 split.

Table 3. Comparison of model analysis using reflectance and a combination of reflectance and VIs at different train–test ratios.

Parameters			Dataset: Entire			Dataset: Test			Dataset: Train		
Train:Test Ratio	Input Group	Best Model	r	rRMSE (%)	Best Model	r	rRMSE (%)	Best Model	r	rRMSE (%)	
50:50	REFs	RF	0.86	2.14	SLR	0.74	2.43	RF	0.96	1.59	
	REFs+VIs		0.87	2.08	SVM	0.70	2.58		0.97	1.34	
55:45	REFs	RF	0.87	2.11	SLR	0.74	2.51	RF	0.96	1.47	
	REFs+VIs		0.87	2.08	SVM	0.69	2.68	ANN	0.97	1.26	
60:40	REFs	RF	0.88	2.05	SLR	0.70	2.27	RF	0.96	1.47	
	REFs+VIs		0.88	2.03	SVM	0.67	2.64		0.97	1.26	
65:35	REFs	RF	0.88	2.02	SLR	0.66	2.67	RF	0.96	1.41	
	REFs+VIs		0.88	2.02	SVM	0.64	2.78		0.97	1.22	
70:30	REFs	RF	0.89	1.95	SLR	0.61	2.76	RF	0.96	1.43	
	REFs+VIs		0.89	1.93	SVM	0.60	2.92		0.97	1.21	
75:25	REFs	RF	0.89	1.92	ANN	0.62	2.82	RF	0.96	1.35	
	REFs+VIs		0.90	1.86	SVM	0.60	3.08		0.97	1.17	
80:20	REFs	RF	0.91	1.86	PLSR	0.65	2.70	RF	0.96	1.34	
	REFs+VIs		0.92	1.73	SLR	0.71	2.74		0.96	1.21	
85:15	REFs	RF	0.93	1.69	PLSR	0.62	2.82	RF	0.96	1.33	
	REFs+VIs		0.92	1.65	SLR	0.67	2.69		0.97	1.20	
90:10	REFs	RF	0.94	1.55	PLSR	0.43	2.91	RF	0.96	1.32	
	REFs+VIs		0.94	1.45	SLR	0.51	2.84		0.97	1.16	
95:5	REFs	RF	0.94	1.51	KNN	0.69	3.25	RF	0.96	1.31	
	REFs+VIs		0.95	1.37	SLR	0.77	2.59		0.97	1.17	

REFs is the reflectance-only input group, REFs+VIs is the reflectance and selected vegetation indices input group.

3.2.3. Using Reflectance and Vegetation Index Features as Inputs

In the second stage of CGM estimation, five selected REFs and six VIs as a result of dimensionality reduction process were used as the inputs (Table 3). For the validation over the test dataset, SVM performed the best at a 50:50 split (r = 0.70, rRMSE = 2.58%), followed by RF, SLR, ANN, and PLSR, and KNN was the weakest performer at the same 50:50 split (r = 0.62, rRMSE = 2.69%). For the validation over the train dataset, RF performed best (r = 0.97, rRMSE = 1.17%), followed by ANN, SLR, PLSR, and SVM at a 95:5 split while PLSR was the weakest performer at that split (r = 0.82, rRMSE = 2.54%). For the validation over the entire dataset, RF at a 95:5 split (r = 0.95, rRMSE = 1.37%) performed best followed by ANN, SLR, SVM, and PLSR, while KNN (r = 0.77, rRMSE = 2.74%) was the weakest performer at a 95:5 split.

3.2.4. Impact of Training and Testing Data Split Ratios

As the training dataset size increased for training the models, the CGM estimation accuracy also increased when validated over the train and entire datasets (r_{train}: 0.61–0.97, $rRMSE_{train}$: 1.15–2.86%, r_{entire}: 0.76–0.95, $rRMSE_{entire}$: 1.37–3.31%) also increased (Figures 4b,c and 5, Table 3) and decreased when validated over the test dataset (r_{test}: −0.17–0.77, $rRMSE_{test}$: 2.27–5.59%, Figures 4a and 5, Table 3). For the train–test split ratio of 95:5, the accuracy of CGM estimation was the best and RF was the best performing model when validated over the train and entire datasets (r_{train} = 0.97, $rRMSE_{train}$ = 1.17%, r_{entire} = 0.95, $rRMSE_{entire}$ = 1.37%, Figure 4b,c and Figure 5, Table 3) with REFs+VIs as the input group. In addition, SLR performed the best at a 95:5 split ratio when validated over the test dataset with REFs+VIs as the input group. At the same split ratio, even when using REFs as the input group and validation over train and entire datasets, RF performed the best (r_{train} = 0.96, rRMSE$_{train}$ = 1.31%, r_{entire} = 0.94, rRMSE$_{entire}$ = 1.51%). SLR performed the best (r = 0.74, rRMSE = 2.43%) with REFs as the input group and SVM performed the best (r = 0.70, rRMSE = 2.58%) with REFs+VIs as the input group for the train–test split ratio of 50:50 when those models were validated over the test dataset. ANN also improved its performance at a train–test split ratio of 55:45 when using REFs+VIs as the input group and was validated over the train dataset (r_{train} = 0.97, rRMSE$_{train}$ = 1.26%, Table 3). When validated over the test dataset, ANN improved its performance for the split ratio of 75:25 and with REFs as inputs (r_{test} = 0.62, rRMSE$_{test}$ = 2.82%).

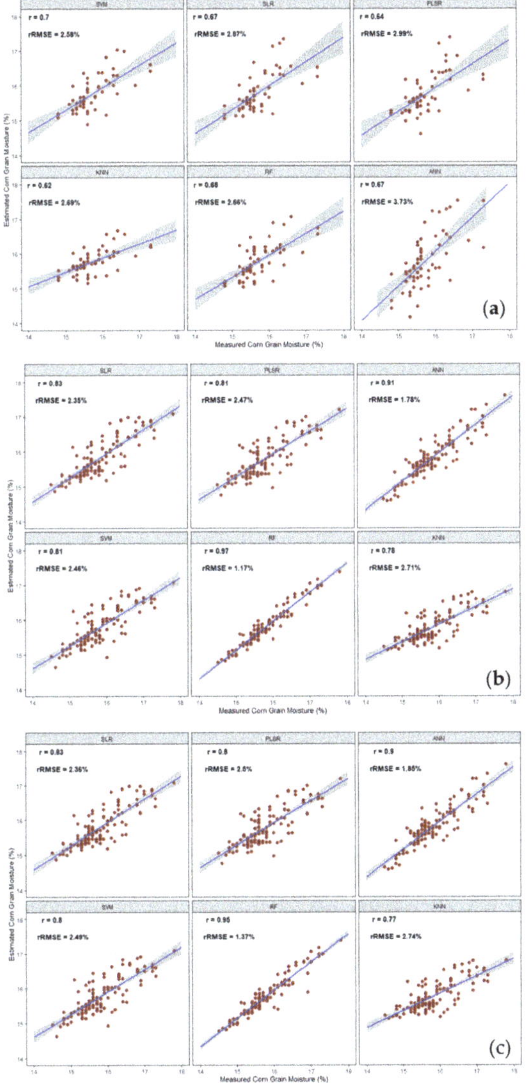

Figure 4. Plots showing measured and estimated CGM using REFs+VIs as input group for models validated over (**a**) the test dataset at 50:50, (**b**) train dataset at 95:5, and (**c**) entire dataset at 95:5 splits.

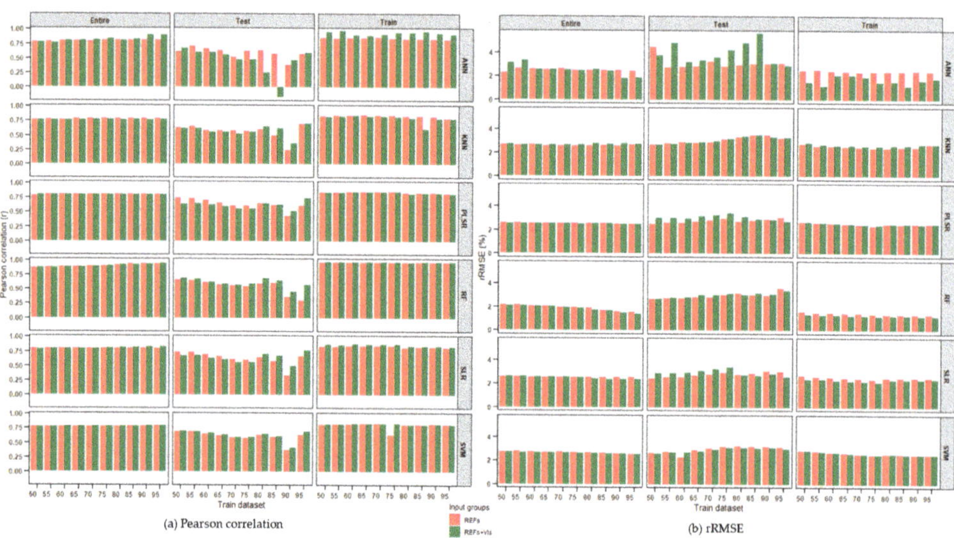

Figure 5. Plots of (**a**) Pearson correlation (r), and (**b**) Relative root mean square error (rRMSE) summarizing the performance of six corn grain moisture estimation models for ten train–test data split ratios, and for two input groups (REFs, REFs+VIs) over three validation datasets (entire, test, train).

4. Discussion

Among the REFs in five wavebands, NIR had the highest correlation with CGM (r = 0.74) depicting sensitivity to the chlorophyll light absorption feature of plants [53]. Correlations with REF in blue and green (r = −0.27, r = 0.05) wavebands were the lowest, possibly due to a low signal-to-noise ratio [44,54]. Among the total 24 derived VIs, GNDVI had the highest correlation with CGM, followed by GCI, GRVI, GOSAVI, and WDRVI, among others (r = 0.78–0.80), while VARI had the lowest correlation (r = 0.68). GNDVI is derived using NIR (840 ± 20 nm), which is more sensitive to chlorophyll content, supporting a strong correlation [44]. On the other hand, VARI had low correlation due to its nonlinear mathematical operation as well as its derivation using blue and green wavebands that had low correlations with CGM. VIs such as IPVI and GCI had stronger correlations with CGM as those take into consideration the dynamic variations in the visible–NIR region pertaining to canopy water, chlorophyll, and nitrogen contents [55]. In this study, GCI and GNDVI outperformed VIs that use reflectance in the red band such as IPVI, NDVI, TDVI, RDVI, and others, as the reflectance in the green band (560 ± 10 nm) is relatively more sensitive to chlorophyll and crop moisture contents [16,42]. This was also corroborated by observations made by Kayad et al. [56] where VIs computed using green band reflectance outperformed others in estimating corn grain yield. These VIs may also perform well for CGM estimation using simple or multiple linear regression or other statistical models (Table 2) as also supported by previous studies [16]. Nonetheless, using REF or VI feature as independent inputs may lack robustness when evaluated under other agroclimatic conditions [57,58]. Therefore, this study advanced research towards the estimation of CGM using statistical and ML models as those have the capability to robustly approximate complex and non-linear relationships between the inputs (VIs or REFs) and outputs (i.e., CGM).

From the process of conducting PCA and eliminating collinear variables, blue, green, red, RE, NIR, IPVI, NDRE, GCI, GLI, SR, and VARI were identified as not to have absolute correlations (i.e., r = 1) among each other. IPVI had collinearity with NDVI but was selected over the latter for its capability to overcome the limitations of NDVI, which can become saturated for higher biomass, and is also subjected to relatively higher noise from atmospheric and soil background conditions [19]. Studies have also reported IPVI to perform superior to NDVI for estimating crop nitrogen status and grain yield across different growth stages [21,59]. Although GNDVI had a higher correlation with CGM compared to GCI, it was not selected most probably due to it having a higher mean absolute correlation compared to GCI [60]. An only study conducted thus far for CGM estimation reported canopy chlorophyll content representing LAI as a strong input variable [16]. Most of the crop health status estimations such as chlorophyll content, water content, or yield have utilized not only the REFs as inputs to ML or statistical models but also the VIs [16,58]. Studies that have performed predictive modeling using ML have neither evaluated inter-correlationships between the input variables (i.e., VIs) nor eliminated the collinear variables before estimating the output [61,62]. This may often lead to model overfittings, compromised robustness, and require extensive computations from a user's practical standpoint [63]. Most of the ML-based prediction studies have by default utilized 70:30 or 80:20 as the train–test data split ratios for model training and validations [5,64]. However, the consideration of the entire data size as well as impact of varying training data proportions to identify the best train–test data split ratio have been minimally assessed. This may also impact model over-fitness and robustness [50].

For these reasons as well as by identifying our dataset to be of medium size, our study not only eliminated the collinear input variables but also identified the best train–test split ratio(s) for statistical or ML models for CGM estimation. It was observed that model performances improved (Figure 5) when validated over the train and entire datasets, for increasing proportions of training data [16,62]. This observation was consistent when using both input groups, REFs or REFs+VIs. Although REFs+VIs improved the model performance compared to REFs as inputs, the impact was not significant ($p = 0.374$, Table 4). Apparently, in the maximum cases irrespective of both input groups, ML models outperformed statistical models for estimating CGM (Figure 5) when validated over the training and entire datasets.

Table 4. Effect of input parameters on performance of models for corn grain moisture estimation.

Variable	p **Value (r)**	p **Value (rRMSE)**
Model	<0.001	<0.001
Train–test split	<0.001	0.619
Dataset	<0.001	<0.001
Input group	0.374	0.725
Train–test split: Dataset	<0.001	<0.001
Train–test split: Input group	0.189	0.290
Dataset: Input group	0.450	0.002
Train–test split: Dataset: Input group	0.204	0.544

Where Dataset (train, test, entire), Input groups (REF, REF+VIs), and Model (SLR, PLSR, ANN, SVM, RF, KNN).

Interestingly, SVM and SLR at a 50:50 split ratio when validated over test datasets using either of the input groups performed the best as those have been reported for smaller datasets [50]. SVM is computationally expensive to work with large data as the algorithm often fails while determining optimum boundary hyperplanes, making it more accurate and robust for small data sizes, which has also been supported by other studies [65,66]. By removing the collinearity of inputs, the cofounding effects on the estimation of CGM was also eliminated, thereby improving performances of statistical models such as SLR and PLSR [67]. This study's data size was relatively small compared to what ANNs generally require, and this was the reason why ANN was the least good performer amongst all other evaluated models in this study [50,62,68]. Overall, RF performed the best compared to all other models as it is capable of withstanding the overfitting problem unlike other statistical linear models, and it was relatively less reliant on dataset size compared to other ML models [53,69]. This is because RF is a decision-tree-based model that employs several sub-models and bagging techniques, for increased stability and resilience of the prediction outcomes [70].

This study demonstrated the feasibility of SUAS and integrated ML techniques for CGM estimation, which has not been explored thus far. The performance of the model may further be improved by collecting data over multiple cropping seasons as well as agroclimatic conditions. Identifying the earliest stage where accurate CGM could be predicted as well as their translation to satellite imaging platforms are the next goals in our efforts. Those estimates can be later converted into maps (raster or shapefiles) for the corn growers who can develop precision harvest scheduling and management strategies for enhanced crop value.

5. Conclusions

This study investigated the use of aerial multispectral imagery for assessing CGM as well as estimating it using state-of-the-art ML and statistical models. To the best of our knowledge, this was the first investigation of its kind to estimate grain moisture contents. Using Pearson correlation, the REFs and VIs derived from the SUAS imagery data were found to have a strong correlation between CGM and GNDVI, GCI, and IPVI, among others with the highest correlations (r: 0.68–0.80). PCA and pairwise correlation analysis identified REFs in blue, green, red, RE, NIR and VIs such as GCI, IPVI, NDRE, GLI, SR, and VARI as potential inputs to estimate CGM using statistical and ML models.

All four evaluated ML models and two statistical models for estimating CGM improved in performance with the increase in size of training datasets. While most ML models performed well overall, RF was observed to be the most stable (r: 0.86–0.97, rRMSE: 2.14–1.17%). It was observed that the input groups (only REFs or REFs+VIs) for CGM estimation did not impact model performances. However, the train–test split ratios did impact the model performances significantly with 50:50, 50:45, 60:40, 80:20, and 95:5 being among the split ratios that yielded strong performances. The 95:5 train–test split ratio was the best when models were validated over the train and entire datasets while the 50:50 split ratio was the best when models were validated over the test dataset. The statistical models i.e., SLR and PLSR, also yielded comparable performances to most of the ML models (r: 0.61–0.74, rRMSE: 2.76–2.43%) while ANN could not be the best-performing model of the study except at a 55:45 split ratio and for validation over train and entire datasets.

Overall, our study demonstrated that aerial multispectral imagery when integrated with ML models could suitably estimate CGM even for small–medium dataset sizes. These computations are critical for the corn growers to non-invasively as well as spatially map CGM status for scheduling and managing harvest schedules and resources. We will be further evaluating the models tested in the study over different growth stages to identify the earliest time when CGM near optimum harvest could be estimated. Moreover, these models could be translated in the form of webtools that farmers could utilize for planning and executing precision operations on the ground and for extracting the best economic value of their crop.

Author Contributions: Conceptualization, P.J. and A.K.C.; methodology, P.J., A.K.C. and D.L.; software, P.J. and A.K.C.; validation, P.J., A.K.C. and D.L.; formal analysis, P.J. and A.K.C.; investigation, P.J., A.K.C. and D.L.; resources, A.K.C. and D.L.; data curation, P.J., A.K.C. and D.L.; writing—original draft preparation, P.J. and A.K.C.; writing—review and editing, P.J., A.K.C. and D.L.; visualization, A.K.C. and D.L.; supervision, A.K.C.; project administration, A.K.C. and D.L.; funding acquisition, A.K.C. and D.L. All authors have read and agreed to the published version of the manuscript.

Funding: This research was funded in part by USDA NIFA Project # 420110, Hatch Project # VA160181, Multistate Hatch Project #VA136412, and Faculty startup.

Data Availability Statement: All collected data and pertaining analysis has been included in the manuscript.

Acknowledgments: We would like to thank the technicians of the plant pathology laboratory as well as Tidewater Agricultural Research and Extension Center, Suffolk, VA for their help in managing trials and collecting ground truth data.

Conflicts of Interest: The authors declare no conflict of interest.

References

1. Martinez-Feria, R.A.; Licht, M.A.; Ordóñez, R.A.; Hatfield, J.L.; Coulter, J.A.; Archontoulis, S.V. Evaluating Maize and Soybean Grain Dry-down in the Field with Predictive Algorithms and Genotype-by-Environment Analysis. *Sci. Rep.* **2019**, *9*, 7167. [CrossRef] [PubMed]
2. Agyei, B.; Andresen, J.; Singh, M.P. Evaluation of a Handheld Near-Infrared Spectroscopy Sensor for Rapid Corn Kernel Moisture Estimation. *Crop Forage Turfgrass Manag.* **2023**, *9*, e20235. [CrossRef]
3. Pordesimo, L.O.; Sokhansanj, S.; Edens, W.C. Moisture and Yield of Corn Stover Fractions before and after Grain Maturity. *Trans. ASAE* **2004**, *47*, 1597–1603. [CrossRef]
4. Fan, L.-F.; Chai, Z.-Q.; Zhao, P.-F.; Tian, Z.-F.; Wen, S.-Q.; Li, S.-M.; Wang, Z.-Y.; Huang, L. Nondestructive Measurement of Husk-Covered Corn Kernel Layer Dynamic Moisture Content in the Field. *Comput. Electron. Agric.* **2021**, *182*, 106034. [CrossRef]
5. Pham, B.T.; Son, L.H.; Hoang, T.A.; Nguyen, D.M.; Tien Bui, D. Prediction of Shear Strength of Soft Soil Using Machine Learning Methods. *Catena* **2018**, *166*, 181–191. [CrossRef]
6. Maiorano, A.; Fanchini, D.; Donatelli, M. MIMYCS. Moisture, a Process-Based Model of Moisture Content in Developing Maize Kernels. *Eur. J. Agron.* **2014**, *59*, 86–95. [CrossRef]
7. Sadaka, S.; Rosentrater, K.A. Tips on Examining the Accuracy of On-Farm Grain Moisture Meters. In *Agriculture and Natural Resources*; UAEX: Fayetteville, AR, USA, 2019; pp. 1–5.
8. Nelson, S.O.; Trabelsi, S. A Century of Grain and Seed Moisture Measurement by Sensing Electrical Properties. *Trans. ASABE* **2012**, *55*, 629–636. [CrossRef]
9. Soltani, M.; Alimardani, R. Prediction of Corn and Lentil Moisture Content Using Dielectric Properties. *J. Agric. Technol.* **2011**, *7*, 1223–1232.
10. Zhang, H.L.; Ma, Q.; Fan, L.F.; Zhao, P.F.; Wang, J.X.; Zhang, X.D.; Zhu, D.H.; Huang, L.; Zhao, D.J.; Wang, Z.Y. Nondestructive in Situ Measurement Method for Kernel Moisture Content in Corn Ear. *Sensors* **2016**, *16*, 2196. [CrossRef]
11. Clevers, J.G.P.W.; Kooistra, L.; Schaepman, M.E. Estimating Canopy Water Content Using Hyperspectral Remote Sensing Data. *Int. J. Appl. Earth Obs. Geoinf.* **2010**, *12*, 119–125. [CrossRef]
12. Croft, H.; Chen, J.M.; Zhang, Y.; Simic, A. Modelling Leaf Chlorophyll Content in Broadleaf and Needle Leaf Canopies from Ground, CASI, Landsat TM 5 and MERIS Reflectance Data. *Remote Sens. Environ.* **2013**, *133*, 128–140. [CrossRef]
13. Khanal, S.; Klopfenstein, A.; Kushal, K.C.; Ramarao, V.; Fulton, J.; Douridas, N.; Shearer, S.A. Assessing the Impact of Agricultural Field Traffic on Corn Grain Yield Using Remote Sensing and Machine Learning. *Soil Tillage Res.* **2021**, *208*, 104880. [CrossRef]
14. Shajahan, S.; Cho, J.; Guinness, J.; van Aardt, J.; Czymmek, K.J.; Ketterings, Q.M. Corn Grain Yield Prediction and Mapping from Unmanned Aerial System (Uas) Multispectral Imagery. *Remote Sens.* **2021**, *13*, 3948.
15. Pinto, A.A.; Zerbato, C.; de Souza Rolim, G.; Barbosa Júnior, M.R.; da Silva, L.F.V.; de Oliveira, R.P. Corn Grain Yield Forecasting by Satellite Remote Sensing and Machine-Learning Models. *Agron. J.* **2022**, *114*, 2956–2968. [CrossRef]
16. Xu, J.; Meng, J.; Quackenbush, L.J. Use of Remote Sensing to Predict the Optimal Harvest Date of Corn. *Field Crops Res.* **2019**, *236*, 1–13. [CrossRef]
17. Zhang, L.; Zhang, H.; Niu, Y.; Han, W. Mapping Maizewater Stress Based on UAV Multispectral Remote Sensing. *Remote Sens.* **2019**, *11*, 605. [CrossRef]
18. Yu, N.; Li, L.; Schmitz, N.; Tian, L.F.; Greenberg, J.A.; Diers, B.W. Development of Methods to Improve Soybean Yield Estimation and Predict Plant Maturity with an Unmanned Aerial Vehicle Based Platform. *Remote Sens. Environ.* **2016**, *187*, 91–101. [CrossRef]
19. Ranjan, R.; Chandel, A.K.; Khot, L.R.; Bahlol, H.Y.; Zhou, J.; Boydston, R.A.; Miklas, P.N. Irrigated Pinto Bean Crop Stress and Yield Assessment Using Ground Based Low Altitude Remote Sensing Technology. *Inf. Process. Agric.* **2019**, *6*, 502–514. [CrossRef]
20. Moeinizade, S.; Pham, H.; Han, Y.; Dobbels, A.; Hu, G. An Applied Deep Learning Approach for Estimating Soybean Relative Maturity from UAV Imagery to Aid Plant Breeding Decisions. *Mach. Learn. Appl.* **2022**, *7*, 100233. [CrossRef]
21. Qi, H.; Wu, Z.; Zhang, L.; Li, J.; Zhou, J.; Jun, Z.; Zhu, B. Monitoring of Peanut Leaves Chlorophyll Content Based on Drone-Based Multispectral Image Feature Extraction. *Comput. Electron. Agric.* **2021**, *187*, 106292. [CrossRef]
22. Ali, I.; Greifeneder, F.; Stamenkovic, J.; Neumann, M.; Notarnicola, C. Review of Machine Learning Approaches for Biomass and Soil Moisture Retrievals from Remote Sensing Data. *Remote Sens.* **2015**, *7*, 16398–16421. [CrossRef]
23. Cazenave, A.B.; Shah, K.; Trammell, T.; Komp, M.; Hoffman, J.; Motes, C.M.; Monteros, M.J. High-Throughput Approaches for Phenotyping Alfalfa Germplasm under Abiotic Stress in the Field. *Plant Phenome J.* **2019**, *2*, 1–13. [CrossRef]
24. Montandon, L.M.; Small, E.E. The Impact of Soil Reflectance on the Quantification of the Green Vegetation Fraction from NDVI. *Remote Sens. Environ.* **2008**, *112*, 1835–1845. [CrossRef]
25. Rouse, J.W.; Haas, R.H.; Schell, J.A.; Deering, D.W. Monitoring Vegetation Systems in the Great Plains with ERTS. *NASA Spec. Publ.* **1974**, *351*, 309–317.
26. Crippen, R.E. Calculating the Vegetation Index Faster. *Remote Sens. Environ.* **1990**, *34*, 71–73. [CrossRef]

27. Gitelson, A.A.; Merzlyak, M.N. Remote Sensing of Chlorophyll Concentration in Higher Plant Leaves. *Adv. Space Res.* **1998**, *22*, 689–692. [CrossRef]

28. Sripada, R.P.; Heiniger, R.W.; White, J.G.; Meijer, A.D. Aerial Color Infrared Photography for Determining Early In-Season Nitrogen Requirements in Corn. *Agron. J.* **2006**, *98*, 968–977. [CrossRef]

29. Huete, A.; Didan, K.; Miura, T.; Rodriguez, E.P.; Gao, X.; Ferreira, L.G. Overview of the Radiometric and Biophysical Performance of the MODIS Vegetation Indices. *Remote Sens. Environ.* **2002**, *83*, 195–213. [CrossRef]

30. Boegh, E.; Soegaard, H.; Broge, N.; Hasager, C.B.; Jensen, N.O.; Schelde, K.; Thomsen, A. Airborne Multispectral Data for Quantifying Leaf Area Index, Nitrogen Concentration, and Photosynthetic Efficiency in Agriculture. *Remote Sens. Environ.* **2002**, *81*, 179–193. [CrossRef]

31. Yang, Z.; Willis, P.; Mueller, R. Impact of band-ratio enhanced awifs image to crop classification accuracy. In Proceedings of the Pecora 17—The Future of Land Imaging. . .Going Operational, Denver, CO, USA, 18–20 November 2008.

32. Sripada, R.P. *Determining In-Season Nitrogen Requirements for Corn Using Aerial Color-Infrared Photography*; North Carolina State University: Raleigh, NC, USA, 2005.

33. Rondeaux, G.; Steven, M.; Baret, F. Optimization of Soil-Adjusted Vegetation Indices. *Remote Sens. Environ.* **1996**, *55*, 95–107. [CrossRef]

34. Qi, J.; Chehbouni, A.; Huete, A.R.; Kerr, Y.H.; Sorooshian, S. A Modified Soil Adjusted Vegetation Index. *Remote Sens. Environ.* **1994**, *48*, 119–126. [CrossRef]

35. Leprieur, C.; Kerr, Y.H.; Pichon, J.M. Critical Assessment of Vegetation Indices from Avhrr in a Semi-Arid Environment. *Int. J. Remote Sens.* **1996**, *17*, 2549–2563. [CrossRef]

36. Gitelson, A.A.; Gritz, Y.; Merzlyak, M.N. Relationships between Leaf Chlorophyll Content and Spectral Reflectance and Algorithms for Non-Destructive Chlorophyll Assessment in Higher Plant Leaves. *J. Plant Physiol.* **2003**, *160*, 271–282. [CrossRef] [PubMed]

37. Louhaichi, M.; Borman, M.M.; Johnson, D.E. Spatially Located Platform and Aerial Photography for Documentation of Grazing Impacts on Wheat. *Geocarto Int.* **2001**, *16*, 65–70. [CrossRef]

38. Birth, G.S.; McVey, G.R. Measuring the Color of Growing Turf with a Reflectance Spectrophotometer 1. *Agron. J.* **1968**, *60*, 640–643. [CrossRef]

39. Chen, J.M. Evaluation of Vegetation Indices and a Modified Simple Ratio for Boreal Applications. *Can. J. Remote Sens.* **1996**, *22*, 229–242. [CrossRef]

40. Tucker, C.J. Red and Photographic Infrared Linear Combinations for Monitoring Vegetation. *Remote Sens. Environ.* **1979**, *8*, 127–150. [CrossRef]

41. Bannari, A.; Asalhi, H.; Teillet, P.M. Transformed Difference Vegetation Index (TDVI) for Vegetation Cover Mapping. In Proceedings of the IEEE International Geoscience and Remote Sensing Symposium, Toronto, ON, Canada, 24–28 June 2002; IEEE: Piscataway, NJ, USA, 2002; Volume 5, pp. 3053–3055.

42. Gitelson, A.A.; Stark, R.; Grits, U.; Rundquist, D.; Kaufman, Y.; Derry, D. Vegetation and Soil Lines in Visible Spectral Space: A Concept and Technique for Remote Estimation of Vegetation Fraction. *Int. J. Remote Sens.* **2002**, *23*, 2537–2562. [CrossRef]

43. Gitelson, A.A. Wide Dynamic Range Vegetation Index for Remote Quantification of Biophysical Characteristics of Vegetation. *J. Plant Physiol.* **2004**, *161*, 165–173. [CrossRef]

44. Chandel, A.K.; Khot, L.R.; Yu, L.-X. Alfalfa (*Medicago sativa* L.) Crop Vigor and Yield Characterization Using High-Resolution Aerial 1 Multispectral and Thermal Infrared Imaging Technique. *Comput. Electron. Agric.* **2021**, *182*, 105999. [CrossRef]

45. Yu, X.; Liu, Q.; Wang, Y.; Liu, X.; Liu, X. Evaluation of MLSR and PLSR for Estimating Soil Element Contents Using Visible/near-Infrared Spectroscopy in Apple Orchards on the Jiaodong Peninsula. *Catena* **2016**, *137*, 340–349. [CrossRef]

46. Wold, S.; Sjostrom, M.; Eriksson, L. PLS-Regression. A Basic Tool of Chemometrics. *Chemom. Intell. Lab. Syst.* **2001**, *58*, 109–130. [CrossRef]

47. Nijat, K.; Shi, Q.; Wang, J.; Rukeya, S.; Ilyas, N.; Gulnur, I. Estimation of Spring Wheat Chlorophyll Content Based on Hyperspectral Features and PLSR Model. *Trans. Chin. Soc. Agric. Eng.* **2017**, *33*, 208–216.

48. Marques Ramos, A.P.; Prado Osco, L.; Elis Garcia Furuya, D.; Nunes Gonçalves, W.; Cordeiro Santana, D.; Pereira Ribeiro Teodoro, L.; Antonio da Silva Junior, C.; Fernando Capristo-Silva, G.; Li, J.; Henrique Rojo Baio, F.; et al. A Random Forest Ranking Approach to Predict Yield in Maize with Uav-Based Vegetation Spectral Indices. *Comput. Electron. Agric.* **2020**, *178*, 105791. [CrossRef]

49. Abdulridha, J.; Batuman, O.; Ampatzidis, Y. UAV-Based Remote Sensing Technique to Detect Citrus Canker Disease Utilizing Hyperspectral Imaging and Machine Learning. *Remote Sens.* **2019**, *11*, 1373. [CrossRef]

50. Sharma, P.; Leigh, L.; Chang, J.; Maimaitijiang, M.; Caffé, M. Above-Ground Biomass Estimation in Oats Using UAV Remote Sensing and Machine Learning. *Sensors* **2022**, *22*, 601. [CrossRef] [PubMed]

51. Mountrakis, G.; Im, J.; Ogole, C. Support Vector Machines in Remote Sensing: A Review. *ISPRS J. Photogramm. Remote Sens.* **2011**, *66*, 247–259. [CrossRef]

52. Zou, J.; Han, Y.; So, S.S. Overview of Artificial Neural Networks. In *Artificial Neural Networks: Methods and Applications*; Humana Press: Totowa, NJ, USA, 2009; pp. 14–22.

53. Ngie, A.; Ahmed, F. Estimation of Maize Grain Yield Using Multispectral Satellite Data Sets (SPOT 5) and the Random Forest Algorithm. *S. Afr. J. Geomat.* **2018**, *7*, 11. [CrossRef]

54. Jiang, Z.; Huete, A.R.; Didan, K.; Miura, T. Development of a Two-Band Enhanced Vegetation Index without a Blue Band. *Remote Sens. Environ.* **2008**, *112*, 3833–3845. [CrossRef]
55. Xue, J.; Su, B. Significant Remote Sensing Vegetation Indices: A Review of Developments and Applications. *J. Sens.* **2017**, *2017*, 1353691. [CrossRef]
56. Kayad, A.; Sozzi, M.; Gatto, S.; Marinello, F.; Pirotti, F. Monitoring Within-Field Variability of Corn Yield Using Sentinel-2 and Machine Learning Techniques. *Remote Sens.* **2019**, *11*, 2873. [CrossRef]
57. Zhang, Y.; Ta, N.; Guo, S.; Chen, Q.; Zhao, L.; Li, F.; Chang, Q. Combining Spectral and Textural Information from UAV RGB Images for Leaf Area Index Monitoring in Kiwifruit Orchard. *Remote Sens.* **2022**, *14*, 1063. [CrossRef]
58. Habibi, L.N.; Watanabe, T.; Matsui, T.; Tanaka, T.S.T. Machine Learning Techniques to Predict Soybean Plant Density Using UAV and Satellite-Based Remote Sensing. *Remote Sens.* **2021**, *13*, 2548. [CrossRef]
59. Fei, S.; Hassan, M.A.; He, Z.; Chen, Z.; Shu, M.; Wang, J.; Li, C.; Xiao, Y. Assessment of Ensemble Learning to Predict Wheat Grain Yield Based on UAV-Multispectral Reflectance. *Remote Sens.* **2021**, *13*, 2338. [CrossRef]
60. Kuhn, M. Caret Package. *J. Stat. Softw.* **2008**, *28*, 1–26.
61. Zhou, Y.; Lao, C.; Yang, Y.; Zhang, Z.; Chen, H.; Chen, Y.; Chen, J.; Ning, J.; Yang, N. Diagnosis of Winter-Wheat Water Stress Based on UAV-Borne Multispectral Image Texture and Vegetation Indices. *Agric. Water Manag.* **2021**, *256*, 107076. [CrossRef]
62. Yue, J.; Yang, G.; Tian, Q.; Feng, H.; Xu, K.; Zhou, C. Estimate of Winter-Wheat above-Ground Biomass Based on UAV Ultrahigh-Ground-Resolution Image Textures and Vegetation Indices. *ISPRS J. Photogramm. Remote Sens.* **2019**, *150*, 226–244. [CrossRef]
63. Yue, J.; Feng, H.; Yang, G.; Li, Z. A Comparison of Regression Techniques for Estimation of Above-Ground Winter Wheat Biomass Using near-Surface Spectroscopy. *Remote Sens.* **2018**, *10*, 66. [CrossRef]
64. Gill, W.R.; Asae, M. Influence of Compaction Hardening of Soil on Penetration Resistance. *Trans. ASAE* **1968**, *11*, 741–0745. [CrossRef]
65. Hota, S.; Tewari, V.K.; Chandel, A.K. Workload Assessment of Tractor Operations with Ergonomic Transducers and Machine Learning Techniques. *Sensors* **2023**, *23*, 1408. [CrossRef]
66. Adugna, T.; Xu, W.; Fan, J. Comparison of Random Forest and Support Vector Machine Classifiers for Regional Land Cover Mapping Using Coarse Resolution FY-3C Images. *Remote Sens.* **2022**, *14*, 574. [CrossRef]
67. Fu, Z.; Jiang, J.; Gao, Y.; Krienke, B.; Wang, M.; Zhong, K.; Cao, Q.; Tian, Y.; Zhu, Y.; Cao, W.; et al. Wheat Growth Monitoring and Yield Estimation Based on Multi-Rotor Unmanned Aerial Vehicle. *Remote Sens.* **2020**, *12*, 508. [CrossRef]
68. Nguyen, Q.H.; Ly, H.B.; Ho, L.S.; Al-Ansari, N.; Van Le, H.; Tran, V.Q.; Prakash, I.; Pham, B.T. Influence of Data Splitting on Performance of Machine Learning Models in Prediction of Shear Strength of Soil. *Math. Probl. Eng.* **2021**, *2021*, 4832864. [CrossRef]
69. Palmer, D.S.; O'Boyle, N.M.; Glen, R.C.; Mitchell, J.B.O. Random Forest Models to Predict Aqueous Solubility. *J. Chem. Inf. Model.* **2007**, *47*, 150–158. [CrossRef]
70. Prasad, A.M.; Iverson, L.R.; Liaw, A. Newer Classification and Regression Tree Techniques: Bagging and Random Forests for Ecological Prediction. *Ecosystems* **2006**, *9*, 181–199. [CrossRef]

MDPI

St. Alban-Anlage 66

4052 Basel

Switzerland

www.mdpi.com

Land Editorial Office

E-mail: land@mdpi.com

www.mdpi.com/journal/land